Analytical Population Dynamics

Population and Community Biology Series

Principal Editor

M. B. Usher *Chief Scientific Adviser and Director of Research and Advisory Services, Scottish Natural Heritage, UK*

Editors

D. L. DeAngelis *Department of Biology, University of Florida, USA* and **B. F. J. Manly** *Director, Centre for Applications of Statistics and Mathematics, University of Otago, New Zealand*

The study of both populations and communities is central to the science of ecology. This series of books explores many facets of population biology and the processes that determine the structure and dynamics of communities. Although individual authors are given freedom to develop their subjects in their own way, these books are scientifically rigorous and a quantitative approach to analysing population and community phenomena is often used.

Already published

1. **Population Dynamics of Infectious Diseases: Theory and Applications**
   R. M. Anderson (ed.) (1982) 368pp. Hb. Out of print

2. **Food Webs**
   S. L. Pimm (1982) 219pp. Hb/Pb. Out of print.

3. **Predation**
   R. J. Taylor (1984) 166pp. Hb/Pb. Out of print.

4. **The Statistics of Natural Selection**
   B. F. J. Manly (1985) 484pp. Hb/Pb. Hb out of print.

5. **Multivariate Analysis of Ecological Communities**
   P. Digby & R. Kempton (1987) 206pp. Hb/Pb. Hb out of print.

6. **Competition**
   P. Keddy (1989) 202pp. Hb/Pb. Hb out of print.

7. **Stage-Structured Populations: Sampling, Analysis and Simulation**
   B. F. J. Manly (1990) 200pp. Hb.

8. **Habitat Structure: The Physical Arrangement of Objects in Space**
   S. S. Bell, E. D. McCoy & H. R. Mushinsky (1991, 1994) 452pp. Hb.

9. **Dynamics of Nutrient Cycling and Food Webs**
   D. L. DeAngelis (1992) 285pp. Pb.

10. **Analytical Population Dynamics**
    T. Royama (1992) 387pp. Hb. (1996) Pb.

11. **Plant Succession: Theory and Prediction**
    D. C. Glenn-Lewin, R. K. Peet & T. T. Veblen (1992) 361pp. Hb.

12. **Risk Assessment in Conservation Biology**
    M. A. Burgman, S. Ferson & R. Akcakaya (1993) 324pp. Hb.

13. **Rarity**
    K. Gaston (1994) 192pp. Hb/Pb.

14. **Fire and Plants**
    W. J. Bond & B. W. van Wilgen (1996) ca. 288pp. Hb.

ANALYTICAL POPULATION DYNAMICS

T. Royama

Canadian Forest Service–Maritimes Region
New Brunswick Canada

To Mark,

with compliments.

Tom

CHAPMAN & HALL

London · Glasgow · Weinheim · New York · Tokyo · Melbourne · Madras

Published by Chapman & Hall, 2-6 Boundary Row, London SE1 8HN, UK

Chapman & Hall, 2-6 Boundary Row, London SE1 8HN, UK

Blackie Academic & Professional, Wester Cleddens Road, Bishopbriggs, Glasgow G64 2NZ, UK

Chapman & Hall GmbH, Pappelallee 3, 69469 Weinheim, Germany

Chapman & Hall USA., 115 Fifth Avenue, New York, NY 10003, USA

Chapman & Hall Japan, ITP-Japan, Kyowa Building, 3F, 2-2-1 Hirakawacho, Chiyoda-ku, Tokyo 102, Japan

Chapman & Hall Australia, 102 Dodds Street, South Melbourne, Victoria 3205, Australia

Chapman & Hall India, R. Seshadri, 32 Second Main Road, CIT East, Madras 600 035, India

First edition 1992
First Pubished in paperback 1996

© 1992 T.Royama

Typeset in 10/12pt Times by Interprint Ltd, Malta
Printed in Great Britain by Page Bros (Norwich) Ltd

ISBN 0 412 24320 2 (HB)   0 412 75570 X (PB)

A Catalogue record for this book is available from the British Library

Library of Congress Cataloging-in-Publication Data available

∞ Printed on permanent acid-free text paper, manufactured in accordance with ANSI/NISO Z39.48-1992 and ANSI/NISO Z39.48-1984 (Permanence of Paper)

To the memory of Dr David Lack

Contents

Acknowledgements

The whole or part of the following figures, taken (whole or in part) from various journals, books, or reports, are reproduced (some being modified or rearranged) in this book with permission from the respective publishers or copyright holders and with consent from as many original authors as I could contact. Specific citations are given in text or in figure captions. I am grateful to all for their generosity:

Figures 3.19, 9.2–9.7, 9.11, 9.13–9.17, 9.19 and 9.20 with permission from the Ecological Society of America (© 1981 and © 1984); Figures 4.5, 8.1–8.5, 8.11, 8.12, 8.14 and 8.15 with permission from Kluwer Academic Publishers (© 1990); Figures 5.4 and 6.4 with permission from the Wildlife Society (© 1972, ©1978 and © 1979); Figure 6.1 with permission from The University of Toronto Press and from The Wildlife Society (© 1984); Figure 6.14 with permission from National Research Council of Canada; Figure 6.15 with permission from Blackwell Scientific Publications Ltd (© 1988); Figure 6.16 with permission from the Ecological Society of America (© 1985); Figure 6.22 with permission from The University of Wisconsin Press (© 1963 by the Regents of the University of Wisconsin) and from Manitoba Depatment of Natural Resources; Figure 8.6 with permission from Ecological Entomology and Springer-Verlag (© 1988); Figure 8.21 with permission from Canadian Journal of Fisheries and Aquatic Sciences.

Preface

The aim of this book is to provide guidelines for the analysis of animal population processes with intent to motivate the readers to develop or improve their own method of data analysis and interpretation rather than to provide them with recipes. My primary interest is to develop the methodology of applied population dynamics of forest and agricultural entomology, fisheries and wildlife management, etc., although many topics are relevant to theoretical ecology. I assume that readers have taken an undergraduate course in population ecology and are familiar with elementary algebra, analysis, probability and statistics.

Now that the first, hard-cover edition is out of print, this new edition is printed in paperback form with reduced price so that it is more affordable by individual purchasers. Also, most typographical errors in the first edition have been corrected. There is no revision of the contents except for a few minor changes in wording.

Through book reviews and personal communications, I have learned that Chapter 1 is rather hard to digest. This might be because the fundamental concepts introduced in this chapter that replace the conventional, intuitively conceived notions, e.g., population regulation and density dependence, cannot be formulated without mathematics. Since those concepts are essential for the study of population dynamics, a reader who finds it difficult to follow some formal inferences may be advised to consult a friend with an appropriate background or read the chapter together.

One critic of the first edition thought that Part One (theories) and Part Two (applications) were only loosely connected. It might have looked superficially so because not all methods of time series analysis introduced in Part One are directly applied in the analyses of actual data in Part Two. However, the primary aim of introducing time series analysis in this book is to formulate some fundamental notions of population dynamics with which the researchers can judge what can be extracted from the data at hand, rather than to provide recipes readily applicable to data. In this respect, some of my interpretations of the classic cases in Part Two, even though formal analyses were not always applied, would not have been possible without those notions based on the theoretical investigations in Part One.

A recent trend in population ecology to apply the formalism of time series analysis often overlooks the conditions of applicability, e.g. the mean and variance of the data series should remain stationary over time and the series should be sufficiently long, inasmuch as data in any statistical treatment should be consistent and sufficiently large. Those readers who intend to apply time series analysis to their data should be well aware of these conditions and some potential pitfalls discussed in Chapter 3.

I owe much to Dr R. C. J. Carling, Senior Editor, Life Sciences, Chapman and Hall, and to Professor Michael B. Usher, the Principal Editor of the *Population and Community Biology Series*, for the publication of this new edition.

T. Royama
Fredericton, N. B.

PART ONE

Theoretical bases of population dynamics

'Why does the animal behave as it does?'

Nikolaas Tinbergen (*The Study of Instinct*, 1951)

An ultimate goal of the study of population dynamics is to answer the question: Why do animal populations fluctuate as they do? One way to comprehend the observed fluctuation is to compare its pattern with that of a theoretical model, just as we compare an observed spatial distribution of organisms with a certain theoretical model such as the well-known Poisson distribution.

The Poisson model, which formally defines random distribution, is fundamentally important because, without it, we would be unable to know what an ideally random distribution ought to be like. On the other hand, it is a simple, basic model and is inflexible; it cannot describe more complex patterns of distribution that many organisms exhibit in their natural habitats. However, we can generalize and elaborate the model to obtain a higher descriptive power. On careful comparison with an observed pattern, a good theoretical model will provide insight into the mechanism underlying the pattern. We can employ the same method in the analysis of population dynamics.

The aim of Part One, comprising the first four chapters, is to develop a theoretical system to provide logical criteria for our inferences, analyses and interpretations of observed population dynamics as stochastic processes in time. In Chapter 1, I shall introduce the concepts of persistence and regulation of animal populations. From these concepts will emerge the idea of density dependence and density independence; this, in turn, provides a basis for structuring theoretical population processes.

Chapter 2 discusses basic structures of population processes, using comparatively simple, abstract models. The population dynamics that these models generate illustrates the association of a certain pattern in population fluctuation with a certain type of generating mechanism.

Chapter 3 develops methods to analyse data so as to identify mechanisms underlying observed population dynamics. Some important concepts and results in time series analysis are introduced and adapted to the analysis of population time series.

Chapter 4 deals with certain concrete population models: the logistic model, predator–prey models and their generalizations, with a particular reference to their ecological meanings. The ultimate purpose of this chapter is to develop the models that I use for the analysis of the observed population dynamics in Part Two.

Most arguments in Part One are mathematical. It is important not to become embroiled in technical details of the arguments, so the reader may choose to glance over Part One, or even skip technical sections, at first reading.

1 Basic properties and structure of population processes

'A definition is a purely arbitrary thing. If I choose
to define a triangle as a plane figure bounded by four
sides and having four angles; and if, also, I define a
quadrilateral as a plane figure bounded by three sides
and having three angles, I shall run into no logical
conflicts; my geometry need in no wise depart from
that of Euclid; I shall need to make no changes in
existing works on geometry, beyond that of
substituting throughout the word triangle for the word
quadrilateral, and vice versa ...

'But the framing of definitions at times involves
more subtle consideration of expediency, so subtle in
fact that they may be overlooked, or misunderstood,
and a problem which is, in truth, a problem of
definition, falsely masquerades as a problem of fact.'

Alfred James Lotka (*Elements of Physical Biology*, 1925)

1.1 INTRODUCTION

A theoretical system is usually developed from a few premises. In this book,
these are the notions of persistence and regulation of populations. The
particular intention here is to deduce what conditions are required for a given
population to be regulated so as to persist. In deducing such conditions, we
shall gain insight into the basic properties and structure of animal population
processes. The practical value of our deduction depends, of course, on how we
set up the premises.

I shall first discuss and define the notion of population persistence in
section 1.2. Then I deduce in two steps the conditions that ensure persistence. The first step is to deduce general statistical conditions (section 1.3)
which will lead to a formal definition of population regulation (section 1.4).
We gain further insight about those statistical conditions if we consider an
idealized situation in which the rate of change in population density is in a
state of statistical equilibrium. Section 1.5 introduces some basic knowledge
of stochastic processes required for this idealization.

The second step, the topic of section 1.6, is to deduce ecological mechanisms which satisfy the above statistical conditions. In this section, I introduce the notion of robust, as against fragile, population regulation, and those of density-dependent and density-independent factors and processes. These notions play important roles in structuring population processes as a step towards building theoretical models needed for the analysis of observed processes.

In section 1.7, I shall elaborate on the structure of population processes to reveal the distinction between the effects of three different types of density-independent factors: the vertical, lateral, and nonlinear perturbation effects on population dynamics.

The concept of density dependence that I introduce in this chapter differs, in some important aspects, from definitions usually given in standard text books. In section 1.8, I shall discuss how and why my definitions differ from the usual ones.

1.2 PERSISTENCE OF POPULATIONS

1.2.1 Empirical perception

We know by experience that animal populations tend to persist in a suitable environment. Although their numbers may change enormously from year to year and from place to place, it seems that populations in large areas are some-how 'regulated' inasmuch as the probability of their extinction is minimal. On a local scale, population extinction does occur more frequently, although a new population may be established by immigrants from other local populations. I shall now idealize our perception of population persistence to define it formally.

1.2.2 Idealization and definition

Consider that we sample a population at every equally spaced point in time. Let population density be denoted by x ($0 \leqslant x < \infty$). In particular, let x_t be the density observed at time t. Further, let x be transformed to a logarithm so that $\log x$ takes any real number ($-\infty < \log x < \infty$).

Suppose we have an ideally long record of population fluctuation $\{x_t; t = 1, 2, \ldots, T\}$ in a given locality (Fig. 1.1a). As we accumulate our record, and so long as the population persists, we will see that the frequency distribution of $\log x_t$ (falling in a class interval of $\log x$) tends to cluster about a certain interval and to diminish as $\log x$ deviates from it in either direction (Fig. 1.1b). [Note that the frequency distribution might have more than one peak.]

An appropriate parameter with which to characterize such a central tendency in the frequency distribution is the mean square deviation of $\log x_t$ about an arbitrary finite value, say m. [Usually, in statistics, the average of total observations is taken as m. In the present argument, however, it is convenient to leave it as arbitrary.] In a persistent state, the mean square deviation

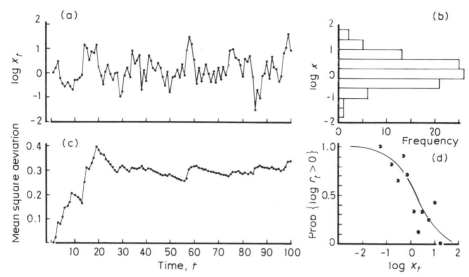

Figure 1.1   An idealized example of population fluctuation. Graph a: time plot of log population density (log x_t). Graph b: frequency distribution of log x_t in graph a. Graph c: time plot of mean square deviation of log x_t from m (chosen to be zero) defined on the left-hand side of the constraint (1.1). Graph d: probability of log $r_t > 0$, plotted against log x_t; the curve is hand-drawn through the proportions (circles) of log $r_t > 0$ realized in graph a.

should be bounded, that is, it should not exceed an arbitrary positive finite number, say A, no matter how long we observe the population, i.e.

$$\sum_{t=1}^{T} (\log x_t - m)^2 / T \leqslant A. \tag{1.1}$$

Figure 1.1c shows a time plot of the statistic on the left-hand side of (1.1) ($m = 0$ conveniently) calculated from the population series in graph a.

The above argument assumes that we have one observed series of population densities. If so, the constraint (1.1) is a natural way to define the notion of population persistence; (1.1) is also convenient for illustrating population persistence by numerical simulation as in Fig. 1.1. However, defining persistence in terms of the finite mean square in one series of population densities is not convenient for a theoretical investigation into the statistical property of a persistent population in section 1.3. We need a definition modified for that purpose.

A time series of probabilistic events is considered to be one of many (generally, infinitely many) possible realizations of a given stochastic process (section 1.5.1). Thus, the occurrence of an event at a given time, say t, is characterized by a set of moments, the first- and second-order ones being particularly important. With the population density at time t being x_t, those moments of $\log x_t$ about m are the expectations $E(\log x_t - m)$ and $E(\log x_t - m)^2$. The first moment defines the average position of $\log x_t$ relative to m, and the second moment is a measure of the dispersion of $\log x_t$ about m. Then, we may define a persistent population as a stochastic process in which $\log x_t$ has a finite expected dispersion about m for all t, i.e.

$$E(\log x_t - m)^2 \leqslant A. \tag{1.2}$$

[The average on the left-hand side of (1.1) can be asymptotically equivalent to the expectation on the same side of (1.2) if the latter has the same value for all t. This happens if the process $\{x_t\}$ is in a state of statistical equilibrium. I shall explain this fact in section 1.5. For the moment, I shall use either definition depending on the context of the argument.]

Conforming to the constraint given by either (1.1) or (1.2) does not ensure absolute persistence. This is because persistence defined in terms of a finite second-order moment does not imply that every realized density x_t is confined within limits. The constraint may permit $\log x_t$ to become extremely low. The chance for the population to recover from such a low density may depend on a local ecological condition. Thus, I interpret the notion of population persistence (or persistent state of a population) as a close resemblance of the behaviour of the population, until its accidental extinction, to the behaviour of a model process that conforms to the constraint on its second-order moment.

It is often instructive to look at a population series in terms of the rate of change in density from time t to $t + 1$ denoted by r_t, i.e.

$$r_t = x_{t+1}/x_t. \tag{1.3a}$$

Constraint (1.1) or (1.2) implies that, when x_t becomes very large, x_{t+1} will most certainly be less than x_t, and vice versa when x_t becomes small. Figure 1.1d illustrates these situations in terms of the probability of $\log r_t > 0$ $(r_t > 1)$. The probability tends to 1 when $\log x_t$ goes to negative infinity (x_t becoming infinitesimal), and to 0 when $\log x_t$ becomes infinitely large. This implies that the rate of change in a persistent population is not statistically independent of population density. This property of the persistent population will be elaborated, in a more precise manner, into the important concept of density dependence as my argument develops.

Choosing an appropriate interval between adjacent time steps is a matter of expediency. In this book, I shall consider year-to-year or generation-to-generation dynamics, for the following reason. Many species have distinct annual cycles or seasonality. In particular, they tend to have well-defined breeding seasons, although during one such season they may have several broods. Likewise, even in a confined experimental situation with uniform environmental conditions, many insect species have distinct generations, although the span of a generation may differ from species to species and in different environmental conditions. In these circumstances, one year or one generation is an obviously convenient time unit. If we chose a shorter unit, we would have to deal with complications arising from the seasonality or life-stage heterogeneity. If we chose a longer time unit, we would lose information.

The components of r_t are birth and death rates (and, if any, immigration and emigration) in the interval between t and $t + 1$. From now on, I shall call r_t the '(net) reproductive rate' of the population. Often, it is even more convenient to transform the relationship (1.3a) to logarithms:

$$R_t = X_{t+1} - X_t \qquad (1.3b)$$

where $X \equiv \log x$ and $R \equiv \log r$. Figure 1.2 illustrates the relationship (1.3b).

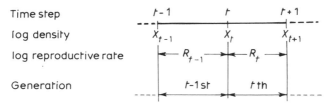

Figure 1.2 Diagram of the relationship defined in (1.3b)

1.2.3 Control of populations

In the above argument, I have not assumed any particular values for the constants m and A in (1.1)/(1.2) except that they are finite. Thus, a persistent population could fluctuate so much as to appear uncontrolled. However, one might wish to find, as in the study of biological control, a mechanism which maintains the population at a desired level and suppresses the amplitude of fluctuation. In such circumstances, we would specify appropriate values for m and A. If the population conforms to the specifications, we say that it is under control.

At this stage of argument, the reader should not be tempted to test a population series for persistence by calculating the statistic on the left-hand side of (1.1) and time-plotting it as in Fig. 1.1c. An observed series is often too short for the mere appearance of the time plot to be reliable. Even in a persistent series, it often needs a large number of time steps for the statistic to converge to a certain value. Conversely, the statistic in a non-persistent series may temporarily appear to have converged. As I discuss in section 1.8.3, limited data sets pose some practical and theoretical problems in testing population regulation.

In the following section, I shall investigate what statistical attributes the net reproductive rate must possess in order to maintain a persistent state. For this purpose, I assume that the series are long enough to pose no theoretical problems.

1.3 STATISTICAL REQUIREMENTS FOR POPULATION PERSISTENCE

Broadly speaking, the pattern of population fluctuation can be decomposed into two major components: trend and deviation from it (or, if there is no trend, from a given value). If either of the two components is unchecked, the population cannot persist, that is, constraint (1.1)/(1.2) is violated. In other words, for the population to remain persistent, it must check the trend and regulate its deviations from the trend.

1.3.1 First requirement: checking the trend

For mathematical convenience, let us set $t \equiv t - 1$ in (1.3b) and rearrange the relationship as:

$$X_t = R_{t-1} + X_{t-1}. \tag{1.4}$$

The right-hand side is, in turn, equal to $R_{t-1} + (R_{t-2} + X_{t-2})$. Recursive applications of the same operation yield

$$X_t = \sum_{i=0}^{t-1} R_i + X_0. \tag{1.5}$$

Now, let the average of log reproductive rates be denoted by $\bar{R}$, i.e.

$$\sum_{i=0}^{t-1} R_i / t = \bar{R}. \tag{1.6}$$

If $\bar{R}$ does not vanish no matter how large t is, then the substitution of $t\bar{R}$ for ΣR_i in (1.5) gives

$$X_t = t\bar{R} + X_0, \tag{1.7}$$

so that, inevitably, $|X_t - X_0| \to \infty$ as $t \to \infty$. This, of course, violates constraint (1.1)/(1.2). The average $\bar{R}$ not converging to zero for $t \to \infty$ means, in practice, that the population exhibits an unchecked trend. Thus, for the population to be persistent, it is at least necessary that log reproductive rates average out to zero in the limit, i.e.

$$\lim_{t \to \infty} \bar{R} = 0. \tag{1.8a}$$

In other words, as is often emphasized (Slobodkin, 1961; Smith, 1980) the gain and loss of population members must cancel out each other in the long run.

A more precise statement of (1.8a), needed for a further argument in section 1.3.3, is the following. For the population process to conform to constraint (1.1)/(1.2), it is necessary that the term $t\bar{R}$ on the right-hand side of (1.7) be bounded. In other words, for an arbitrary positive finite number A', it is necessary that $t|\bar{R}| \leqslant A'$, and, hence,

$$|\bar{R}| \leqslant A'/t. \tag{1.8b}$$

This means that the average $\bar{R}$ must be of the order t^{-1} at most and, hence, will vanish in the limit.

The requirement (1.8), stipulating that the population should have no unchecked trend, is a first and necessary requirement for a population to be in a state of persistence. However, this requirement is not adequate to conform to constraint (1.1)/(1.2) because it stipulates nothing about deviations from the trend (or if there is no trend, from a given value), a point often overlooked. The following argument illustrates the point.

1.3.2 Random walk in population fluctuation

Suppose we generate a series of independent, identically distributed random numbers $\{u_t\}$ with the mean zero and variance σ_u^2. If we take the ith number u_i to be a realization of the log reproductive rate R_i, and substitute it in (1.5), we have

$$X_t - X_0 = \sum_{i=0}^{t-1} u_i. \tag{1.9}$$

Because I have chosen the mean of the random variable u to be zero, the average $\Sigma u_i/t$ will asymptotically vanish, satisfying the requirement (1.8). However, the mean square deviation of X_t from the initial value X_0 will be unbounded as shown below.

From (1.9), we find that

$$(X_t - X_0)^2 = \left(\sum_{i=0}^{t-1} u_i \right)^2$$

$$= \sum_{i=0}^{t-1} u_i^2 + 2 \sum_{i=1}^{t-1} \sum_{j=0}^{i-1} u_i u_j. \tag{1.10}$$

The sum of the product terms $(u_i u_j)$, $i > j$, on the right-hand side is expected to vanish asymptotically because u are independent random numbers with zero mean. Thus, the right-hand side increases indefinitely as the sum of square terms – which is expected to be $t\sigma_u^2$ – increases without bound. This violates constraint (1.1)/(1.2).

Figure 1.3 illustrates the above situation. The series $\{u_t\}$ as $\{R_t\}$ (graph a) is a series of independent, identically distributed random numbers with the mean $= 0$, satisfying the requirement (1.8). Graph b shows the fluctuation in the first 100 points of the series $\{X_t\}$ generated by (1.9), which drifts away from the initial value X_0. Correspondingly, the mean square deviation of the X about X_0, calculated as on the left-hand side of (1.1) in which $m = X_0$, increases with time (graph c). [Compare this with Fig. 1.1c which is generated by a persistent model process.] In Fig. 1.3, I further extend the series $\{X_t\}$ in graph b to $t = 1000$ in graph d, and the corresponding mean square deviation in graph c to graph e. We see no sign of the generated series $\{X_t\}$ stabilizing, but drifting increasingly away from the origin with the elapse of time.

The process in Fig. 1.3 is known as a random walk or, more aptly, the drunkard's walk. A drunkard, though trying to navigate him along a straight path, has no memory about the pathway he has actually taken. He tries to go straight only in reference to his latest position, no matter how this position has deviated from the intended straight path. Consequently, he drifts back and forth along the straight line in an uncontrolled manner and deviates from the line with an ever increasing amplitude. As a matter of fact, the series in Fig. 1.3b/d will eventually return to the initial value X_0 (the t axis), but it may do so within a reasonably short period of time or only after an infinitely long period of time. I have generated several more series up to $t = 100$ in Fig. 1.4 to illustrate a variety of random-walk paths.

However disorganized it may seem, a random walk obeys a law as to the fraction of time the path stays on one side of the horizontal axis at the initial value X_0 (Feller, 1952). In particular, the average return time of many such paths to their initial value increases with time unboundedly. This means that, as time elapses, a given path tends to stay on one or the other side of the horizontal axis at X_0 for most of the time rather than spend, on average, half its time on one side and half on the other. In fact, as illustrated in Fig. 1.4, a majority of random-walk paths drift away from the initial value X_0; most of them spend a disproportionately large fraction of time on one side of the axis at X_0, and, as time advances, many of them will, in practice, never return to the axis. [The initial value is a convenient reference point of drift. Any given point in the path would serve as a reference point.] Thus, a random walk is a typical model of unregulated population processes that do not conform to constraint (1.1)/(1.2), even though the first requirement (1.8) for persistence is met.

In general, if the mean of the u in (1.9) is not zero, so that a generated series $\{X_t\}$ has an unchecked trend, the series performs a random walk about the trend so long as the u are independent random numbers. So, in order to maintain a persistent state of the population, the series of log reproductive rates $\{R_t\}$ requires a second stipulation that prohibits an unbounded drift. In fact, any tendency for a population to deviate

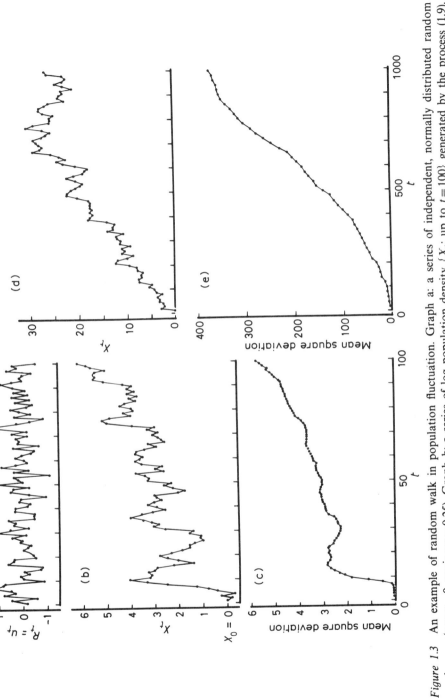

Figure 1.3 An example of random walk in population fluctuation. Graph a: a series of independent, normally distributed random numbers u_t (mean = 0, variance = 0.25). Graph b: a series of log population density $\{X_t;$ up to $t = 100\}$ generated by the process (1.9). Graph c: mean square deviation of X_t from the origin X_0. Graph d: an extension of the series in graph b up to $t = 1000$, but every 10th point is plotted. Graph e: an extension of graph c, corresponding to graph d. Notice differences in scale between graphs b and d and between c and e.

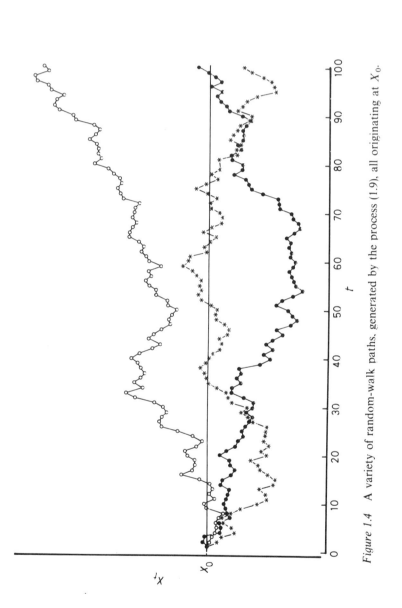

Figure 1.4 A variety of random-walk paths, generated by the process (1.9), all originating at X_0.

unboundedly from a trend (or, if there is no trend, from a given reference point) needs to be prevented.

1.3.3 Second requirement: preventing unbounded deviations

From the relationship (1.5), and letting $\bar{R}$ be the average of log reproductive rates as in (1.6), we can decompose the left-hand side of the persistence constraint (1.2). After a little algebra, and by setting $m = X_0$ (for convenience without loss of generality), we have:

$$E(X_t - X_0)^2 = E\left\{\sum_{i=0}^{t-1}[R_i - E(\bar{R})]^2 \right.$$

$$\left. +2\sum_{i=1}^{t-1}\sum_{j=0}^{i-1}[R_i - E(\bar{R})][R_j - E(\bar{R})]+[tE(\bar{R})]^2\right\}. \qquad (1.11)$$

We now see that two conditions are required for the right-hand side of (1.11) to be bounded to conform to the constraint (1.2). First, the last term $[tE(\bar{R})]^2$ must be bounded to satisfy the first (trend-checking) requirement (1.8b). Second, the total of the remaining terms on the right of (1.11) must be bounded as well. Thus, for an arbitrary positive finite number A'', it is required that:

$$\sum_{i=0}^{t-1}E[R_i - E(\bar{R})]^2 + 2\sum_{i=1}^{t-1}\sum_{j=0}^{i-1}E[R_i - E(\bar{R})][R_j - E(\bar{R})] \leqslant A''. \qquad (1.12)$$

Because the sum of squares $\Sigma[R_i - E(\bar{R})]^2$ increases without bound as t increases, the sum of product terms $2\Sigma\Sigma[R_i - E(\bar{R})][R_j - E(\bar{R})]$ must decrease so as to compensate for the unbounded increase. This implies that R_i and R_j $(i \neq j)$ in the series must be **accordingly** correlated with each other (autocorrelated). This is the second requirement for persistence. [I shall discuss 'accordingly autocorrelated' in section 1.5.] Failure to conform to the requirement (1.12) implies unbounded deviations from a trend or, if there is no trend, from an arbitrary reference point. In fact, a trend may be considered to be such a reference point consistently shifting with time. In the random-walk example (1.10), u_i is uncorrelated with u_j $(j \neq i)$ so that (1.12) is not satisfied.

To summarize, a population is said to be in a persistent state if it exhibits neither an unchecked trend as stipulated by the first requirement, nor an unbounded tendency to deviate from the trend (or, if no trend, from a given reference point) as stipulated by the second requirement.

1.4 REGULATION OF POPULATIONS

If the first requirement is violated, i.e. the expectation $E(\bar{R})$ in (1.12) is not zero, the population exhibits an unchecked trend. Nonetheless, the requirement (1.12) can be satisfied. This implies that the population would fluctuate about the trend but would not deviate or drift unboundedly away from it (I shall give an example of this in section 2.4.5). Then, it would be appropriate to say that the population fluctuation is **regulated** about the trend, even if the population is not in a **persistent** state because its trend is unchecked. Thus, we can use the second requirement (1.12) for persistence as a formal definition of population regulation.

Usually, the term 'regulation' is used as meaning the regulation of a population about an equilibrium density (Varley *et al.*, 1973; Smith, 1980). If the equilibrium density (to be formally defined in section 1.6.5) is fixed, 'regulation' becomes synonymous with 'persistence' as is often tacitly understood. However, an equilibrium density is subject to change in response to environmental changes, resulting in a trend in temporal fluctuation in population sizes (sections 1.7 and 2.4.5). The requirement (1.12) only stipulates regulation about a given reference point (e.g. an equilibrium density) regardless of the point being fixed or changing. In other words, even if a population exhibits a declining trend (as the environment deteriorates) and can no longer maintain its persistent state (it may become extinct if the trend continues), the fluctuation can still be regulated about the declining equilibrium point. Thus, in my definitions, 'persistence' implies 'regulation' but not necessarily vice versa.

A population trend due to environmental changes must be distinguished from a trend exhibited by a population which is heading towards a fixed equilibrium density as in the classical logistic growth of populations. Analysis of trends under different circumstances will be discussed in section 3.2.6.

1.5 STATISTICAL EQUILIBRIUM STATE

In deriving the requirement (1.12) for population regulation, I have made no particular assumptions as to the statistical properties of the log reproductive rates, R, other than that R_i and R_j ($j \neq i$) be 'accordingly autocorrelated' in order to regulate the population fluctuation. Therefore, (1.12) is the most general statement of the requirement. But, because of its generality, we cannot make the stipulation 'accordingly autocorrelated' any more particular and quantitative.

However, if we assume that the time series $\{R_t\}$ is in **statistical equilibrium**, we can readily quantify the stipulation. Also, the concept of a statistical equilibrium state continues to play a major role in the theoretical investigations in Chapters 2 and 3. Therefore, I shall explain it in some detail now. First, I shall introduce the notion of time and ensemble averages.

1.5.1 Time and ensemble averages

Suppose that a series of random numbers $\{n_t; t = 1, 2, \ldots, T\}$ is a realization of (or generated by) a given stochastic process. [This generating process is a conceptual entity, existing only in an idealized world or model.] Suppose further that there are S such series realized independently of one another. Thus, denoting the value of n in the sth series at time t by n_{st} ($s = 1, 2, \ldots, S$; $t = 1, 2, \ldots, T$), we have an $S \times T$ array (Table 1.1). A given row of the array, e.g. the sth row, represents a realized time series which has the average

$$\bar{n}_{s.} = \sum_{t=1}^{T} n_{st}/T. \tag{1.13}$$

This row average is called a time average. Similarly, each column has an average, e.g.

$$\bar{n}_{.t} = \sum_{s=1}^{S} n_{st}/S. \tag{1.14}$$

Now, consider an idealized situation in which these S series constitute the ensemble of all possible realizations (of length T) of the stochastic process

Table 1.1 S independent realizations of a random process $\{n_t; t = 1, 2, \ldots, T\}$.

		Time point				Av.	Var.
	1	2	.... t	.... T			
1	n_{11}	$n_{12}\ldots$	$n_{1t}\ldots$	n_{1T}		$\bar{n}_{1.}$	$v_{1.}$
2	n_{21}	$n_{22}\ldots$	$n_{2t}\ldots$	n_{2T}		$\bar{n}_{2.}$	$v_{2.}$
.		$\ldots\ldots\ldots\ldots\ldots\ldots\ldots\ldots$				$\ldots$	$\ldots$
s	n_{s1}	$n_{s2}\ldots\ldots$	$n_{st}\ldots$	n_{sT}		$\bar{n}_{s.}$	$v_{s.}$
.		$\ldots\ldots\ldots\ldots\ldots\ldots\ldots\ldots$				$\ldots$	$\ldots$
S	n_{S1}	$n_{S2}\ldots\ldots$	$n_{St}\ldots$	$\bar{n}_{ST}$		$n_{S.}$	$v_{S.}$
Av.	$\bar{n}_{.1}$	$\bar{n}_{.2}\ldots\ldots$	$\bar{n}_{.t}\ldots$	$\bar{n}_{.T}$			
Var.	$v_{.1}$	$v_{.2}\ldots\ldots$	$v_{.t}\ldots$	$v_{.t}$			

under the same physical condition. This means that S is usually infinitely large (we cannot see the whole ensemble). Then, a given column, say the tth one, embraces every realizable value of n at time t of the stochastic process. Thus, the column average (1.14) is called an ensemble average (expectation) of the stochastic process at time t. Similarly, a sample second-order moment as a time average is distinguished from an expected second-order moment as an ensemble average.

[Note that ensemble averages are conceptual entities and should not be confused with averages taken over space in the field. It is improbable that several sets of data taken over space are independent realizations of the same stochastic process. Such a data set constitutes a 'space series' (like 'time series') of events as a realization of a stochastic process; that is, **space** merely replaces **time**. Therefore, its average (call it a **spatial average**) is conceptually equivalent to a time average, not to an ensemble average. See section 4.6 for the equivalence between time- and space-series data.]

In the following argument, I shall talk about the expectation of the value n_t, n_t^2, or $n_t n_{t+j}$ ($j \neq 0$) that is realizable at time t of a time series, denoted by $E(n_t)$, $E(n_t^2)$, or $E(n_t n_{t+j})$, respectively. By such expectations, I mean ensemble averages of the stochastic process as distinguished from time averages of a single realization.

Thus, the constraint (1.1) for persistence (section 1.2.2) is defined in terms of time averages, while the constraint (1.2) is stated in terms of ensemble averages. We also see, in stating the trend checking requirement (1.8), that 'birth and death must in the long run cancel each other', we are primarily speaking of the time average $\bar{R}$ to be zero in the limit. On the other hand, the stipulation (1.12) for regulation is stated in terms of the second-order moments as ensemble averages.

If the ensemble averages depend on time t, then, there could be as many distinct variances as the length of the process and accordingly many autocovariances with a given lag. For example, the covariance between R_i and R_{i+j} may differ from that between R_h and R_{h+j} for $h \neq i$, although both are covariances with lag j. If so, it is practically impossible to make the stipulation (1.12) particular because all variances and covariances may differ from each other. Moreover, in the real world, we can only observe a realized series. Thus, we can only deal with time averages, and we would have no idea about the ensemble averages if they depend on time t and differ from each other.

However, if we assume that the stochastic process is 'stationary' and 'ergodic' (as will be explained below), we can circumvent the difficulties with some loss of generality. Insight we gain from such particular assumptions is far more important than maintaining generality.

1.5.2 Stationarity and ergodicity

Like usual random variables, such as a measurement of body-weight, a time series of certain events, e.g. a series of population densities or reproductive rates, here generically denoted by $\{n_t; t = 1, 2, \ldots, T\}$, is characterized by the moments of certain orders of the stochastic process that generates the series. Particularly important moments are mean and covariances. A covariance in time series (as an ensemble average) is one between two values that are apart from each other by a given number of time steps j ($= 0, 1, 2, \ldots$), i.e. $\mathrm{Cov}(n_t, n_{t+j})$, and is called an autocovariance with lag j; the variance, $\mathrm{Var}(n_t)$, is a special case in which $j = 0$. A covariance depends on the unit with which n is measured, e.g. a number per square metre or per square kilometre. However, dividing a covariance by the variance, e.g., $\text{Cov}(n_t, n_{t+j})/\text{Var}(n_t)$, normalizes the moment so that it does not depend on the unit of measurement. The above ratio is called an autocorrelation with lag j.

Consider now that the stochastic process has, after a certain time step (say $t > t'$), become 'stabilized' so that its mean and variance no longer change with t; and so that the autocovariance with lag j depends only on j but not on t. Then the following values are constants independent of $t > t'$:

Mean: $\qquad\qquad E(n_t) = \mu_n$ $\qquad\qquad\qquad\qquad\qquad$ (1.15a)

Autocovariance: $\;\; \text{Cov}(n_t, n_{t+j}) = E(n_t - \mu_n)(n_{t+j} - \mu_n) = \gamma_{nn}(j)$ $\qquad$ (1.15b)

Variance: $\qquad\;\; \text{Var}(n_t) = E(n_t - \mu_n)^2 = \sigma_n^2 \text{ or } \gamma_{nn}(0)$ $\qquad$ (1.15c)

Autocorrelation: $\;\; \text{Cov}(n_t, n_{t+j})/\text{Var}(n_t) = \gamma_{nn}(j)/\gamma_{nn}(0) = \rho_{nn}(j)$ $\qquad$ (1.15d)

where the subscripts nn in the parameters γ and ρ indicate that they are autocovariance and autocorrelation between two values of the process $\{n_t\}$. The process $\{n_t\}$ is then said to have reached an equilibrium state. In statisticians' terminology, a stochastic process in such a state is called a stationary process (up to second-order moments). [In this book, higher-order moments will not be considered except on a few occasions.]

Now, suppose we have observed one realized series $\{n_t; t = 1, 2, \ldots, T\}$ and compute the following time averages:

$$\sum_{t=1}^{T} n_t/T = \bar{n}_T, \qquad\qquad (1.16)$$

$$\sum_{t=1}^{T-j} (n_t - \bar{n}_T)(n_{t+j} - \bar{n}_T)/(T-j); \; j = 0, 1, 2, \ldots \qquad (1.17)$$

Then, the Birkhoff–Khintchin theorem (called an ergodic theorem) states the following (after Kendall and Stuart, 1968):

(1) If the stochastic process $\{n_t\}$ is stationary, such that the ensemble averages $E(n_t) = \mu_n$ and $\text{Cov}(n_t, n_{t+j}) = \gamma_{nn}(j)$ as in (1.15), then, the time averages (1.16) and (1.17) of a series as a realization of the process will almost certainly converge to certain constants, say m_n and $c_{nn}(j)$, respectively. And (2) if and only if the average of autocorrelations of the process $\{n_t\}$ for $j = 1, 2, \ldots, k$ vanishes in the limit, i.e. if

$$\lim_{k \to \infty} \sum_{j=1}^{k} \rho_{nn}(j)/k = 0, \qquad\qquad (1.18)$$

then, the time averages m_n and $c_{nn}(j)$ are equivalent to the ensemble averages μ_n and $\gamma_{nn}(j)$, respectively.

An example of a stationary process is a series of independent, identically distributed random numbers, say $\{u_t\}$. If we consider that array $[n_{st}]$ in Table 1.1 is equated to array $[u_{st}]$, then, the distributions in the rows and columns are identical. In other words, in this case, a time average is asymptotically equivalent to the corresponding ensemble average. In general, if time averages of a single, realized series converge to the respective ensemble averages, the process is said to be ergodic, which applies to most stationary time series we encounter in practice.

Broadly speaking, if a reasonably long observed series exhibits no systematic changes in the mean and in the variance; and if a sample correlogram (a set of sample autocorrelation coefficients for lags 1, 2, ..., j, which will be discussed in section 3.2) damps down sufficiently fast as the lag j increases; then, practically, we can consider that the series is a realization of a stationary process.

Consider, now, that array $[n_{st}]$ is $[X_{st}]$ generated by $[u_{st}]$ via the relationship (1.9); that is, each row of $[X_{st}]$ represents a random walk. As already explained in section 1.3.2, every random-walk path is an unbounded drift (Fig. 1.4). Therefore, the average (1.16) does not converge to any fixed value. Also, the average (1.17) for $j=0$ increases with t without bound. Thus, a random walk is a nonstationary, non-ergodic process.

An application of the Birkhoff–Khintchin ergodic theorem to the concept of population persistence reveals the following. If the population series $\{x_t\}$ is a realization of a stationary process, the left-hand side of (1.1) most certainly converges to the left-hand side of (1.2); that is, the constraint (1.1), which is asymptotically equivalent to (1.2), is satisfied. Thus, a stationary state of the population is a well-defined class of persistent state, if not necessarily vice versa. However, even in a stationary state, the population continues to fluctuate. In this sense, the stationary process is considered to be in a state of statistical equilibrium.

Generally, the practical significance of the stationarity concept is the following. We usually have only one observed population series that is considered to be a realization of a given population process whose statistical properties are unknown – unless the series was generated by a known theoretical model. The ergodic theorem ensures that the statistical parameters (the mean, variance and covariances) of a stationary process can be consistently estimated by the corresponding sample parameters (time averages) calculated from a single realization of the time series of a sufficient length. [Note that the time series must be sufficiently long to ensure the convergence of time averages to a reasonable degree. I shall discuss this topic in Chapter 3.]

In practice, we may wish to know if two population series are similar to each other. For example, as shown in later chapters, we might compare observed population dynamics with those deduced from a theoretical model to see if they match each other, or if they did not, to see how they differ from each other. To do this, however, we naturally compare, among other things, the averages, variances and correlograms calculated from these series. In such cases, we are tacitly setting up a hypothesis that the two series are independent samples drawn from the same ensemble.

The ergodic theorem ensures that the comparison of time averages between several stationary time series is meaningful. If they were not stationary, making a direct comparison would be meaningless because there would be no appropriate criterion for the comparison. I shall discuss how to deal with a nonstationary process in section 3.2.6.

1.5.3 Requirements for population persistence under stationary reproductive rates

We are now ready to quantify the requirement for population regulation, assuming that the series of (log) reproductive rates $\{R_t\}$ is stationary. Substituting the notations in (1.15) – in which n is replaced by R – for corresponding terms in (1.11), we find:

$$E(X_t - X_0)^2 = E\left[\sum_{i=0}^{t-1}(R_i - \mu_R)^2 + 2\sum_{i=1}^{t-1}\sum_{j=0}^{i-1}(R_i - \mu_R)(R_j - \mu_R)\right] + (t\mu_R)^2$$

$$= t\gamma_{RR}(0) + 2\sum_{j=1}^{t}(t-j)\gamma_{RR}(j) + (t\mu_R)^2 \leqslant A. \tag{1.19}$$

Dividing every term by $t\gamma_{RR}(0)$, and noting that the left-hand side is a non-negative quantity, we find:

$$0 \leqslant 1 + \frac{2\Sigma(t-j)\gamma_{RR}(j)}{t\gamma_{RR}(0)} + \frac{t\mu_R^2}{\gamma_{RR}(0)} \leqslant \frac{A}{t\gamma_{RR}(0)}. \tag{1.20}$$

For $t \to \infty$, the right-hand side will vanish. It follows that the left-hand side, because it is non-negative, must also vanish. Then, recalling (1.15d) that $\gamma_{RR}(j)/\gamma_{RR}(0) = \rho_{RR}(j)$, we find the constraint (1.2) for persistence under stationary $\{R_t\}$ to be:

$$\lim_{t \to \infty}\left[1 + 2\sum_{j=1}^{t}(1 - j/t)\rho_{RR}(j) + t\mu_R^2/\gamma_{RR}(0)\right] = 0. \tag{1.21}$$

Then, by the first requirement (1.8b), μ_R is infinitesimal of order t^{-1} at most, so that the last quotient on the left-hand side will vanish. Thus, the second requirement (1.12) is reduced to:

$$\lim_{t \to \infty}\left[1 + 2\sum_{j=1}^{t}(1 - j/t)\rho_{RR}(j)\right] = 0. \tag{1.22}$$

So far, I have considered only the present-to-future section of the time series $\{R_t\}$, that is, that the time series starts at $t=0$ and extends only to the future. Under the assumption of a stationary $\{R_t\}$, it may be more convenient to assume, instead, that the time series started a long time ago, i.e. at $t=-\infty$. Then, noting that $\rho_{RR}(0)=\gamma_{RR}(0)/\gamma_{RR}(0)=1$ and that, in a stationary state, $\rho_{RR}(j)=\rho_{RR}(-j)$, we can reduce (1.22) to:

$$\lim_{t \to \infty} \sum_{j=-t}^{t} (1-|j|/t)\rho_{RR}|j|=0. \qquad (1.23a)$$

The above weighted sum of the sequence $\{\rho_{RR}(j)\}$ is called the Caesaro sum (Priestly, 1981; p. 320).

Thus, under the stationary series of reproductive rates, the second requirement for population persistence (the requirement for population regulation) stipulates that the Caesaro sum of the autocorrelations $\rho_{RR}(j)$ vanishes in the limit.

If, on the other hand, we consider only the present-to-future half of the series as in (1.22), the requirement can be restated as:

$$\lim_{t \to \infty} \sum_{j=1}^{t} (1-j/t)\rho_{RR}(j)=-\tfrac{1}{2}. \qquad (1.23b)$$

Clearly, the random-walk model (1.9), in which $R=u$ and, hence, the autocorrelations $\rho_{RR}(j)$, $j\neq0$, are identically zero, does not meet the second requirement (1.23), even if the first requirement, $\mu_R=0$, is met.

1.6 ECOLOGICAL MECHANISM UNDERLYING POPULATION REGULATION

In deducing the statistical requirements for population persistence in the preceding sections, I have made no particular assumption as to the underlying mechanism. In other words, the requirement for regulation is a statistical stipulation with which every regulated population conforms, no matter what ecological mechanism underlies it. In this section, I shall discuss the ecological mechanism which ensures that the population process conforms with the statistical stipulation. But, first, I shall introduce the notions of feasibility and robustness of population regulation. Also, to keep my argument as simple as possible without much loss of generality in conclusion, I shall assume a stationary series of reproductive rates.

1.6.1 Feasibility and robustness of population regulation

If, at a given moment under the influence of a given set of ecological factors, a population process can somehow satisfy the statistical requirement (1.23), we may say that the regulation of the population is feasible under the circumstances. Consider further that at a subsequent moment, the process was slightly (that is, not in a destructive manner) perturbed by an additional set of factors. If the process under the perturbation could still satisfy the requirement (1.23), let us say that the regulation of the population is robust. If, on the other hand, the requirement is no longer satisfied after the perturbation, we say that the regulation is fragile.

Our particular interest here is to find the type of ecological process that ensures robust regulation.

1.6.2 Density-dependent and density-independent factors

Net reproductive rate, defined in (1.3), section 1.2.2, is a consequence of an ecological process determining natality, mortality and migration (or simply survival and reproduction). Two broad categories of ecological factors, namely, density-dependent and density-independent, are involved throughout the process. We must appropriately define these categories in order to build a mathematical model of the process that ensures robust regulation.

There are two different ways of defining the categories: (1) in terms of the **effect** of a given factor on survival and reproduction; (2) in terms of the **state of existence** (as measurable by an instrument) of the factor. The first one is conventional. For the reason given below, however, I propose to employ the second.

Since Allee *et al.* (1949), standard text books (MacFadyen, 1957; Solomon, 1969; Varley *et al.*, 1973; Southwood, 1978; Price, 1984) categorize a given factor in terms of its **effect** (as the rate of birth, death, or migration) being correlated (dependent on) or uncorrelated with (independent of) population density. For instance, if a certain type of mortality is **correlated** with population density, the factor causing it is called a **density-dependent factor**. Conversely, a **density-independent factor** is one whose **effect** is uncorrelated with density.

Identifying an ecological factor in terms of its effect might seem natural because the effect is ultimately important in population dynamics. However, if we do so, difficulties often arise when building statistical models (section 1.8). We can avoid such difficulties if we classify a factor in terms of the **measure** of its state of existence (or in terms of the parameter that characterizes the state) and, then, describe its **effect** as a function of the measure (or parameter). Therefore, I propose the following definition:

If the state of existence of an ecological factor, identified in terms of its measure or parameter, is, in turn, influenced by the population density of the animal, the factor is said to be density-dependent. Otherwise, it is said to be density-independent.

Macro-climatic factors, e.g. temperature and precipitation, are examples of density-independent factors in that their measures registered in the recording instruments in a weather station are, usually, uninfluenced by the density of animals. Many physical factors belong to this category.

A typical example in the density-dependent category is mutual interference among members of a population. This social factor can be characterized by a parameter, e.g. the frequency of mutual encounters, which is governed by population density. Another example is environmental contamination with metabolic wastes. Their measure – the amount produced per unit size of the environment – is, again, governed by population density.

Resources of the habitat are either density-dependent or density-independent factors. For example, in a predator–prey interaction, the number of prey individuals at a given point of time – the measure of food supply for the predator at that point of time – is influenced by the predator density at the previous point of time. Thus, the prey population is a density-dependent factor for the predator with a time lag (section 2.3.2). On the other hand, annual production of seeds by a tree as a source of food may not be influenced by the abundance of the animals feeding on them. Then, the annual supply of seeds is a density-independent factor for the animals.

However, these resources take effect only through competition whose intensity depends on the population density. Thus, the state of competition can only be characterized by such a parameter as per-capita share of the resources. Therefore, we may call competition a conditional density-dependent factor (for details see section 1.7), that is, conditional on the resource availability.

1.6.3 Effect of ecological factors and process structure

We can now write down the structure of an ecological process determining net reproductive rate in the following general form:

$$R_t = f(\mathbf{X}, \mathbf{Z}) \tag{1.24}$$

where R_t is, as before, the log reproductive rate over the tth generation (cf. Fig. 1.2); $\mathbf{X}$ is, in general, a set of log population densities $(X_t, X_{t-1}, X_{t-2}, \ldots)$, the basic elements of the measures (parameters) of the density-dependent factors; and $\mathbf{Z}$ represents the set of measures of density-independent factors. The notation f, the usual mathematical notation for a 'function', translates these sets of the measures of the factors involved into reproductive rate and, thus, its ecological connotation is the 'effects' of the set of factors, manifesting themselves in fecundity, mortality or dispersal rate.

Important to keep in mind is that the effect of a density-dependent factor (as defined above) is not necessarily correlated, nor is that of a density-independent factor necessarily uncorrelated, with the population density of the animal concerned. This contradicts the conventional perception of ecological factors identified by their effects. But we must change the perception for the reason that will become apparent as my argument develops.

It is sometimes appropriate to call the right-hand side of (1.24) a process. If the process involves a density-dependent factor, let us call it a density-dependent process. In a theoretical investigation, we may wish to study the effect of the density set X by keeping the set Z constant. We may represent such a process by $f(X|Z)$ and call it a pure (or conditional) density-dependent process. If, on the other hand, the density set (X) has no effect at all on reproductive rate (a theoretical possibility), we may call it a pure density-independent process.

In the following, I shall show that regulation by a pure density-independent process is always fragile and that the involvement of a **certain class** of density-dependent factors is necessary to ensure robust regulation. For this purpose, I shall consider a simple, particular process in which the effects of the two types of factors can be separated from one another, such that the relationship (1.24) is written as:

$$R_t = f(X) + g(Z). \tag{1.25}$$

Here, g translates the measures Z of the density-independent factors involved into their effects on R_t. For simplicity, however, $g(Z)$ may be denoted by z.

1.6.4 Fragile regulation by a pure density-independent process

A pure density-independent process is an idealized, special case of the process (1.25) in which $f(X)$ is identically zero, i.e.

$$R_t = g(Z) = z_t. \tag{1.26}$$

Assume, for simplicity and without loss of generality in conclusion, that $\{z_t\}$ is a stationary stochastic process with the mean μ_z and autocovariances $\gamma_{zz}(j)$. Then, replacing R by z in (1.23b) and recalling the relationship (1.15d), we see that the regulation of the population requires

$$\lim_{t \to \infty} \sum_{j=1}^{t} (1 - j/t)\gamma_{zz}(j) = -(1/2)\gamma_{zz}(0). \tag{1.27}$$

It is theoretically possible that the effect z_t happens to have a statistical property satisfying the requirement (1.27) (see the example in section 1.6.6). However, such a theoretical mechanism of population regulation is **fragile** and, hence, is unlikely to exist in the real world, as explained below.

Consider that process (1.26) is subjected to perturbation by an additional set of density-independent factors measured in, say, U, whose effect u_t on R_t

is an uncorrelated random variable with the mean μ_u and variance $\gamma_{uu}(0)$. Thus, after the perturbation, the log reproductive rate R_t is determined by the sum of z_t and u_t, i.e.

$$R_t = z_t + u_t. \tag{1.28}$$

Because the u are uncorrelated random numbers, both $\text{Cov}(u_t, u_{t+j}), j \neq 0$, and $\text{Cov}(z_t, u_{t+j})$, $-\infty < j < \infty$, are identically zero. Then,

$$\gamma_{RR}(0) = \text{Var}(z_t + u_t) = \gamma_{zz}(0) + \gamma_{uu}(0)$$
$$\gamma_{RR}(j) = \text{Cov}(z_t + u_t, z_{t+j} + u_{t+j}) = \gamma_{zz}(j); \; j \neq 0.$$

Hence,

$$\rho_{RR}(j) = \gamma_{RR}(j)/\gamma_{RR}(0) = \gamma_{zz}(j)/[\gamma_{zz}(0) + \gamma_{uu}(0)].$$

Thus, for the new process (1.28) to be regulated, it is required by (1.23b) that:

$$\lim_{t \to \infty} \sum_{j=1}^{t} (1 - j/t)\gamma_{zz}(j) = -[\gamma_{zz}(0) + \gamma_{uu}(0)]/2. \tag{1.29}$$

We see that the requirement (1.29) after perturbation is incompatible with the requirement (1.27) prior to the perturbation, because they differ on the right-hand side while remaining the same on the left-hand side. In other words, if one requirement is satisfied, the other cannot be. Hence, the system cannot cope with random environmental perturbations. In fact, process (1.28) would perform a random walk, as will be illustrated by simulation in section 1.6.6.

1.6.5 Robust regulation by a density-dependent process

I shall now investigate what property the effect $f(\mathbf{X})$ must have on R_t to regulate the population. To keep the argument simple, consider a pure density-dependent process, i.e. the model (1.25) in which the effect $g(\mathbf{Z})$ is fixed at a certain arbitrary value, say z. Let us also assume that the set $\mathbf{X}$ consists only of X_t. [I shall consider in Chapter 2 a more general situation in which densities in the past generations, $X_{t-1}, X_{t-2}, \ldots$, are also effective.] Thus, the situation we consider here is:

$$R_t = f(X_t) + z; \; z = \text{arbitrary constant}. \tag{1.30}$$

Now, recall the relationship in Fig. 1.1d, section 1.2.2, that, in a persistent state, the probability of X_{t+1} being larger than X_t (i.e. $R_t > 0$) approaches 1 as X_t decreases, whereas the probability approaches 0 as X_t increases. This, if applied to (1.30), implies that:

$$f(X_t) + z < 0 \text{ or } > 0 \text{ as } X_t \rightarrow +\infty \text{ or } -\infty, \tag{1.31}$$

respectively. The condition (1.31) implies that the reproduction curve of (1.30), the plot of R_t against X_t, must cross the X_t axis on which $R_t = 0$ with a negative slope at least once somewhere within a finite interval at $X = X^*$, say. Figure 1.5 illustrates a simple case. The point X^*, at which $R_t = f(X^*) = 0$, is a log equilibrium density, given z. Moreover, because no individual animal can produce an infinite number of offspring, the reproduction curve must be bounded from above with the maximum being, say, R_m.

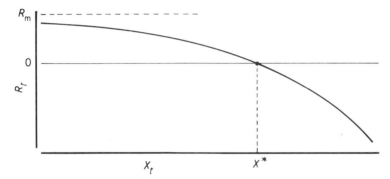

Figure 1.5   An idealized example of a simple reproduction curve of (1.30), the plot of log reproductive rate R_t against the log population density X_t. X^* is a log equilibrium density, and R_m is an upper bound of the log reproductive rate of the population, given z.

Because the maximum (R_m) of the reproduction curve (1.30) depends on z as shown in Fig. 1.6, the condition (1.31) holds only above a certain value of z, e.g. z_1 and z_2. Below a certain value, e.g. z_3, no equilibrium density can exist ($X^{***} \rightarrow -\infty$) and, hence, the condition (1.31) does not hold. In the meantime, there is an upper limit to z because no factor can enhance reproductive rate indefinitely. Usually, above an optimal value of Z (such as an optimal temperature), the effect $z = g(Z)$ would begin to decrease, lowering reproductive rate. After a further increase in Z, the effect z would become so low that no equilibrium density can exist. Thus, the condition (1.31) can hold only within a certain interval of Z. I shall elaborate on this fact in section 1.7.1.

In general, if the effect $f(X_t)$ is a decreasing function of X_t, an equilibrium density – around which a population maintains an equilibrium state – can exist within a certain interval of the measurement Z. Thus, in contrast to inherently fragile regulation by a density-independent process, regulation by a density-dependent process is robust at least within that interval of Z.

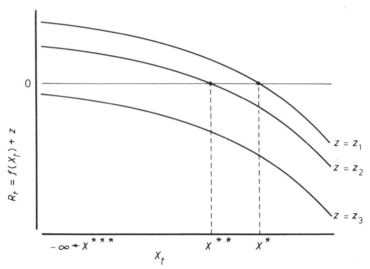

Figure 1.6 Vertical perturbation of the reproduction curve by (1.30). As z decreases from z_1 to z_3, the log equilibrium density decreases from X^* to X^{***}.

[Theoretically, with a certain interval of X, the effect $f(X_t)$ may be an increasing, rather than decreasing, function, or, in between these opposing tendencies, may even stay unchanged. In such cases, a population cannot maintain an equilibrium state about that interval of X (Chapter 2).]

It follows that the involvement of a density-dependent factor that ensures the existence of a stable equilibrium density is an ecologically necessary requirement for robust regulation of a population under perturbation by a density-independent factor.

1.6.6 Examples of fragile and robust regulation

Conceptually, the distinction between fragile and robust regulation is clear cut. However, in practice, the distinction can be subtle as the following two theoretical experiments illustrate.

First, recall the pure density-independent process (1.26),

$$R_t = g(Z) = z_t. \tag{1.32}$$

The following mechanism can regulate this population process.

Suppose that we build an environmental chamber which simulates the effect z_t. We generate a series of independent, identically distributed random numbers $\{u_t\}$ with the mean μ_u and variance σ_u^2; then store the series in the memory bank of the chamber's control box, and place a population in the chamber. Now, adjust the control so that it generates a new series

$\{u_t - u_{t-1}\}$ as $\{z_t\}$. Thus, the reproductive rate of the population is regulated as:

$$R_t = z_t$$
$$= u_t - u_{t-1}. \tag{1.33}$$

It is easy to show that a population series $\{X_t\}$ generated by this process satisfies the constraint (1.1)/(1.2) for persistence and, hence, is regulated; remember that **persistence** implies **regulation**, if not vice versa.

From the definition (1.3b) it follows that

$$X_t = \sum_{i=1}^{t-1} R_i + X_1. \tag{1.34}$$

[cf. (1.5).] Substituting the difference $(u_i - u_{i-1})$ in (1.33) for R_i in (1.34), we find:

$$X_t = u_{t-1} - u_0 + X_1 \tag{1.35}$$

which, setting $u_0 = X_1 = 0$ conveniently, is reduced to

$$X_t = u_{t-1}. \tag{1.36}$$

We see that the series $\{X_t\}$ is a series of independent, identically distributed random numbers $\{u_{t-1}\}$ (Fig. 1.7). The series $\{X_t\}$ conforms to the constraint (1.1)/(1.2) for persistence because the variance of X_t, being equal to σ_u^2, is constant. Thus, we have a theoretical model of regulation by a density-independent process.

Now, consider a second experiment with the following model:

$$R_t = -X_t + u_t. \tag{1.37}$$

This is the simplest linear model of a density-dependent process. It conforms to the condition (1.31) and, therefore, is regulated.

It is interesting that the substitution of $R_t = X_{t+1} - X_t$ in (1.37) gives

$$X_{t+1} - X_t = -X_t + u_t,$$

and, hence,

$$X_t = u_{t-1}$$

which is identical to (1.36). In other words, the density-dependent process model (1.37) generates a population series which is identical in statistical

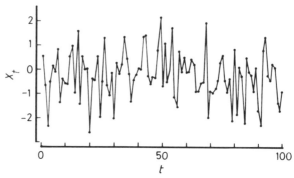

Figure 1.7 A series of log population density $\{X_t\} = \{u_{t-1}\}$ generated by the model (1.33).

properties to one generated by the density-independent process model (1.33) as in Fig. 1.7.

The point is, however, that the difference will show up when the two processes are subjected to random perturbation. As I discussed in sections 1.6.3 and 1.6.4, the structure (1.33) is fragile against such perturbation, whereas (1.37) is robust, as we see below.

Because we cannot by any practical means build a perfect control box for the environmental chamber, some error is inevitable. Let us assume that the error is uncorrelated. If the control is in error by as much as v_t (with the mean 0, variance σ_v^2, and identically zero autocovariances), the reproductive rates in the two situations are, instead of (1.33) and (1.37),

$$R_t = u_t - u_{t-1} + v_t \qquad (1.38)$$

and

$$R_t = -X_t + u_t + v_t, \qquad (1.39)$$

respectively.

The processes (1.38) and (1.39) are no longer equivalent to each other in their statistical characteristics. As shown in Fig. 1.8a, the series of log densities $\{X_t\}$ generated by (1.39) appears much the same as the pre-perturbation series in Fig. 1.7 except for a greater amplitude in fluctuation. In fact, substituting $R_t = X_{t+1} - X_t$ in (1.39), we see that the log density X_t is equal to the sum $(u_{t-1} + v_{t-1})$ which is an independent, identically distributed random number. But the sum has the variance $(\sigma_u^2 + \sigma_v^2)$, causing a greater amplitude of fluctuation in X_t. In other words, the process (1.39) differs from (1.37) only in the variance and still maintains its persistent state.

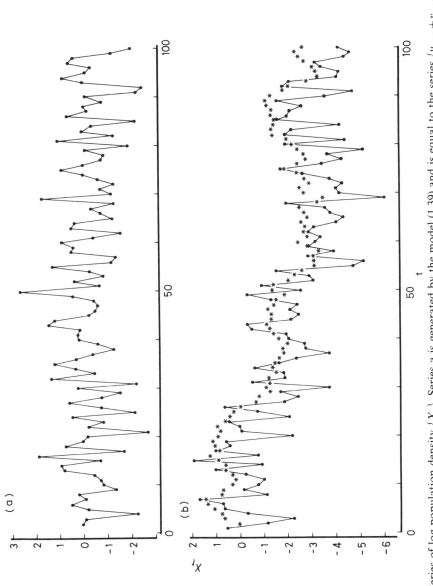

Figure 1.8 Series of log population density $\{X_t\}$. Series a is generated by the model (1.39) and is equal to the series $\{u_{t-1} + v_{t-1}\}$; Series b is generated by the model (1.38)/(1.40) and is equal to $\{u_{t-1} + \Sigma v_i\}$. The series of asterisks is the random-walk path Σv_i.

In contrast, the process (1.38) is no longer regulated because, as is easily computed,

$$\gamma_{RR}(j) = -\sigma_u^2, \ j = 1$$
$$= 0, \ j > 1,$$

and, hence, the process does not satisfy the requirement (1.27) for regulation. In fact, by substituting the relationship $R_i = X_{i+1} - X_i$ in (1.38), and summing over $i = 1$ to $t - 1$, assuming $u_0 = X_1 = 0$, we find that

$$X_t = u_{t-1} + \sum_{i=1}^{t-1} v_i \tag{1.40}$$

in which the sum Σv_i is a random walk (cf. model (1.9)). As Fig. 1.8b illustrates, a realization of the process (1.38)/(1.40) shows a clear sign of unregulated drift: its trend closely follows a random-walk path of the sum Σv_i (marked with asterisks).

The above theoretical experiments demonstrate that, under special circumstances, two population processes, which have ecologically totally different structures, can nonetheless generate population series with mutually identical distributions, so that there is no statistical way of distinguishing between the two. Although a difference does show up when the two processes are subjected to random perturbation, the degree of difference depends on the variance σ_v^2 of the perturbation factor v and on the length t of the generated series. If the variance is small and/or the length of the series is comparatively short, we may not be able to see the difference. I shall elaborate on this point a little further in the next section.

1.6.7 Negative correlation between reproductive rate and density in a persistent population

In Fig. 1.9, I regressed the log reproductive rate R_t of the series in Fig. 1.7 on the log density X_t. We see a clear negative correlation. If an observer did not know how the series $\{X_t\}$ was generated, the appearance of the regression might have prompted him/her to fit the simple linear regression model

$$R_t = aX_t + b \tag{1.41}$$

to find the least-square estimate $\hat{a} = -1.07$. Thus, the observer might have concluded that the population series was generated by a density-dependent process such as (1.37).

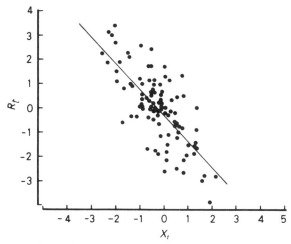

Figure 1.9   Regression of log reproductive rate R_t on log density X_t of the series in Fig. 1.6; the diagonal line is a least-square regression line: $R_t = -0.29 - 1.07X_t$ with the -0.73 correlation coefficient.

However, the above conclusion ignores the other possibility that the series was generated by the density-independent process (1.33), in which, from (1.36),

$$\text{Cov}(R_t, X_t) = \text{Cov}(u_t - u_{t-1}, u_{t-1}) = -\sigma_u^2$$

and

$$\text{Var}(X_t) = \sigma_u^2,$$

and, hence, the regression coefficient is also

$$\text{Cov}(R_t, X_t)/\text{Var}(X_t) = -1. \tag{1.42}$$

In general, whenever a population is in a persistent state, i.e. whenever the constraint (1.1)/(1.2) holds, there will, almost necessarily, be a negative correlation between reproductive rate and population density **no matter what type of factor is regulating the population**. [I justify this assertion in Appendix A at the end of this section.]

It is important to keep in mind, however, that the converse is not necessarily true: a negative correlation does not necessarily imply persistence and, hence, not regulation either. This is because such a correlation does not necessarily imply that R_t is influenced by (and, hence, is a function of) X_t. The following example illustrates the point.

As already noted, regulation by a density-independent process is fragile so that we would, in practice, never see it realized. What we would be likely to see is an unregulated series, such as one generated by (1.38)/(1.40), as illustrated in Fig. 1.8b. Figure 1.10a shows the regression of R_t on X_t of this unregulated series. A least-square estimate yields a regression coefficient equal to -0.4. Although it is much weaker than the -0.73 correlation (in Fig. 1.9) of the perfectly regulated series (Fig. 1.7), R_t of the unregulated series (Fig. 1.8b) is, nonetheless, still significantly negatively correlated with X_t. As pointed out above, however, this negative correlation does not imply the involvement of a density-dependent factor in determining the reproductive rate of the process. Therefore, the model structure (1.41) is in no way appropriate. In the following, I shall show the true mechanism of the above negative correlation.

First, I explain the principle involved in the negative correlation with the aid of a theoretical regression coefficient, i.e. the ratio between the ensemble averages $\mathrm{Cov}(R_t, X_t)$ and $\mathrm{Var}(X_t)$.

From the relationships (1.38) and (1.40), we find that

$$\mathrm{Cov}(R_t, X_t) = \mathrm{Cov}\left(u_t - u_{t-1} + v_t, u_{t-1} + \sum_{i=1}^{t-1} v_i\right)$$

$$= -\mathrm{Var}(u_{t-1}) = -\sigma_u^2 \tag{1.43}$$

$$\mathrm{Var}(X_t) = \mathrm{Var}(u_{t-1}) + \sum_{i=1}^{t-1} \mathrm{Var}(v_i) = \sigma_u^2 + (t-1)\sigma_v^2 \tag{1.44}$$

and, hence, that regression coefficient is given by

$$\mathrm{Cov}(R_t, X_t)/\mathrm{Var}(X_t) = -\sigma_u^2/[\sigma_u^2 + (t-1)\sigma_v^2] \tag{1.45}$$

which is negative, although vanishing asymptotically. We see that the negative covariance between R_t and X_t in the density-independent process (1.38)/(1.40) is due to the negative autocovariance in the series $\{R_t\}$ for lag 1, i.e. $\gamma_{RR}(1) = -\sigma_u^2$. It is not due to a negative feedback from population density as in the density-dependent process (1.39).

Dependence of the regression coefficient (1.45) on length t is retained by a sample coefficient calculated from a single series by least squares. To illustrate this, I divided the generated series in Fig. 1.8b into four sections of equal length and plotted R_t against X_t in each section (Fig. 1.10b, c, d and e). We see that the correlation is much higher (in absolute value) in each of the shorter sections than it is for the whole series in graph a. The highly correlated groups of data points in these shorter sections drift from graphs b to e as the log densities drift. As a result, when combined in graph a, the correlation deteriorates and would eventually become insignificantly different from zero should the length of the series be increased further. How quickly it becomes virtually zero depends, as seen in (1.45), on the variances σ_v^2. [Notice that the slope of each regression line in Fig. 1.10 is much steeper than the theoretical coefficient in (1.45) for a given length t. I shall explain this discrepancy in Appendix B of this section.]

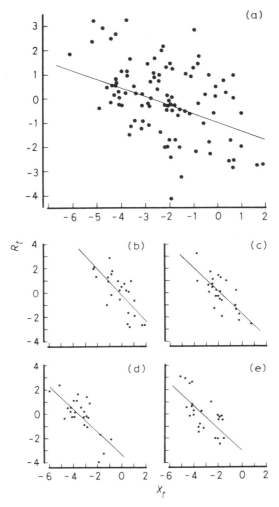

Figure 1.10 Correlations between the log reproductive rate R_t and log density X_t of the unregulated population series in Fig. 1.8b. Graph a: correlation based on the whole series. Graphs b to e: correlations in four shorter sections. The sets (correlation coefficient, slope and intercept of regression line at $X_t = 0$) are: $(-0.45, -0.40, -0.91)$, $(-0.74, -1.1, -0.19)$, $(-0.73, -0.95, -1.80)$, $(-0.73, -0.95, -3.36)$ and $(-0.68, -0.92, -3.03)$ in graphs a, b to e, respectively.

The above argument shows that a negative correlation between reproductive rate and population density would neither necessarily indicate the involvement of a density-dependent factor nor would it necessarily imply regulation of the process. To distinguish between regulated and unregulated populations, we need a much more detailed analysis than simple regression. I shall discuss this topic towards the end of this chapter.

Appendix A

A negative correlation between reproductive rate and population density when the population is in a persistent state can be readily shown with a stationary population process. Recalling that

$$X_t = \sum_{i=1}^{t-1} R_i + X_1$$

and assuming $X_1 = 0$ without loss of generality, we find that

$$\text{Cov}(R_t, X_t) = \text{Cov}\left(R_t, \sum_{i=1}^{t-1} R_i\right) = \sum_{j=1}^{t-1} \gamma_{RR}(j). \tag{A1.1}$$

Recall also that, in a persistent state, the stipulation (1.23b) for regulation is met, so that, recalling further that $\rho(j) = \gamma(j)/\gamma(0)$, and for a sufficiently large t,

$$-1/2 \sim \sum_{j=1}^{t} (1 - j/t)\rho_{RR}(j)$$

$$= \sum_{j=1}^{t} \gamma_{RR}(j)/\gamma_{RR}(0) - \sum_{j=1}^{t} j\rho_{RR}(j)/t$$

$$\sim \text{Cov}(R_t, X_t)/\gamma_{RR}(0) - \sum_{j=1}^{t} j\rho_{RR}(j)/t. \tag{A1.2}$$

In a process in which $\rho_{RR}(j)$ diminishes comparatively quickly as j increases (as is true with most stationary series), the absolute value of the weighted average of autocorrelations – the second term on the right-hand side of (A1.2) – would be small compared to the absolute value of the first term for sufficiently large t. Hence, $\text{Cov}(R_t, X_t)$ would be negative, as would the regression coefficient $\text{Cov}(R_t, X_t)/\text{Var}(X_t)$.

Appendix B

A major reason for the slopes in sample regressions in Fig. 1.10 being steeper than expected from the theoretical coefficient (1.45) is the following. The variance in (1.44) as an ensemble average is the second moment of X_t about the ensemble mean $E(X_t) = 0$. The mean is zero because the expectations of u and v in (1.40) are identically zero. In contrast, the sample (time) average $(\bar{X})$ of a given section tends to drift about as the series $\{X_t\}$ drifts, rather than centering about zero. As a result, for a comparatively short series, the deviation of X_i ($i = 1, 2, \ldots, t$) from the time average $\bar{X}$ tends to be much smaller, in absolute value, than the deviation of X_t from the ensemble mean $E(X_t)=0$. Consequently, the sample variance

$$\sum_{i=1}^{t} (X_i - \bar{X})^2/(t-1)$$

as a time average tends to be smaller than the theoretical variance in (1.44) as an ensemble average. Consequently, the slope of regression lines in Fig. 1.10 tends to be steeper than indicated by the theoretical coefficient in (1.45).

1.7 EFFECTS OF DENSITY-INDEPENDENT FACTORS

In this section, I shall elaborate on the effects of density-independent factors by classifying them into three categories, namely, vertical, lateral and nonlinear perturbation effects. [I shall elaborate on the effects of densities in Chapters 2 and 4.] The discussion of the last two categories leads to the parameterization of competition as a conditional density-dependent factor.

For simplicity without loss of generality of the argument here, I assume that there is no time delay in the effect of any ecological factor involved. Also, the set of density-independent factors Z is treated as a single factor. Thus, the set X consists only of X_t, and Z can be written as Z_t in (1.24), i.e.

$$R_t = f(X_t, Z_t). \tag{1.46}$$

1.7.1 Vertical perturbation effect

Consider that the effect of density-independent factor, say z_t, is a function of the measurement Z_t only, i.e. there is no interaction between z_t and the density X_t, such that we can write:

$$z_t = g(Z_t). \tag{1.47}$$

So, the general model (1.46) takes the linear form:

$$R_t = f(X_t) + z_t. \tag{1.48}$$

This model was illustrated in Fig. 1.6 in the context of the robust regulation in section 1.6.5. It showed that the reproduction curve, plot of R_t against X_t, is determined by the function $f(X_t)$, but the curve's relative position in the (R_t, X_t) plane is shifted up or down as z_t changes; hence, the term 'vertical perturbation effect'.

Because no animal can produce an infinite number of offspring, there must be an upper bound in R_t in (1.48). This implies, of course, that the vertical perturbation effect, z, must have a maximum somewhere across the entire spectrum of the measurement Z. An example is the effect of temperature as a density-independent factor. The factor can be measured by $Z°C$, and its effect z on reproductive rate is the function $g(Z)$ as in (1.47) (dropping the time subscript t). Now, look at Fig. 1.6 again. Model (1.48) shows that the effect of an extremely low temperature brings down the reproduction curve so low that the log reproductive rate (R) cannot be positive, i.e. no equilibrium density exists (see the curve in which $z = z_3$ and the equilibrium point $X^{***} \rightarrow -\infty$). As the temperature rises, the reproduction curve is shifted up so high that the curve can now have an equilibrium point (e.g. X^{**} of the curve in which $z = z_2$). As the temperature continues to rise, the reproduction curve will be raised to a maximum level. A further rise in temperature will have an adverse effect and the reproduction curve will be shifted down. Eventually, under an extremely high temperature, the curve is brought down so low again that no equilibrium density can exist.

In Fig. 1.11, I plotted the effect $z = g(Z)$ against the measurement $Z(°C)$. It shows that a population can persist only within a certain interval of temperatures, (Z_n, Z_m), outside which the effect z causes the log reproductive rate (R) to always become negative so that a population has no chance to persist. As Z rises above the critical temperature (Z_n), the effect z raises the reproduction curve so that persistence becomes probable. For a further rise in temperature, its effect reaches a maximum (the probability of persistence is the greatest) and, then, declines. It so declines that, above the other critical temperature (Z_m), the probability of persistence will become nil again.

Thus, under the environmental conditions that exert vertical perturbation effects, a population can persist (because robust regulation is possible) only within the interval (Z_n, Z_m). The marginal environmental conditions,

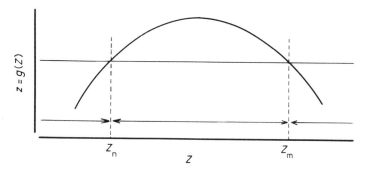

Figure 1.11   The effect $z = g(Z)$ plotted against Z. The interval (Z_n, Z_m) defines the range in which the population has a chance to persist.

Z_n and Z_m, mark the boundaries (a tolerance range) of a potential distribution of the animals concerned.

Vertical perturbation is the simplest type of perturbation by density-independent factors in that we can evaluate its effect completely independent of population density. In the following, I shall consider two other classes of density-independent factors – 'lateral' and 'nonlinear' perturbations – whose effects on reproductive rate cannot be evaluated independently of population density.

1.7.2 Lateral perturbation effect

Some hole-nesting birds, such as certain titmice of the genus *Parus*, do not excavate holes by themselves. Hence, their average breeding success depends on the supply of suitable holes in the woods. The number of holes, if much in excess of the number of breeding pairs, would practically have no effect on their nesting success. As the bird population increases, however, some degree of competition occurs sooner or later. Less competitive individuals might have to settle in holes which are comparatively poor in quality so that they might breed less successfully. Some individuals may not even be able to obtain a hole. Thus, given the supply of holes, the intensity of competition increases as population density increases and lowers the average breeding success of the population. The following simple model explains the principles in the process involving competition for the density-independent resource.

Suppose that Z' holes are distributed at random over x breeding territories. Suppose further that a pair can breed successfully if its territory contains at least one hole; otherwise, no breeding. Then, assuming a Poisson distribution with the parameter Z'/x (the mean number of holes per pair), only $1 - \exp(-Z'/x)$ proportion of the population can breed. In other words, the realized reproductive rate r_t is equal to the potential rate r_m reduced to the above proportion at time t, i.e.

$$r_t = r_m[1 - \exp(-Z_t'/x_t)]. \qquad (1.49)$$

We see that the effect of nest holes as a density-independent factor can only be evaluated jointly with the effect of bird density.

A logarithmic transformation of (1.49) can be written in the general form:

$$R_t = f(X_t - z_t') \qquad (1.50)$$

in which $R \equiv \log r$, $X \equiv \log x$ and $z' = \log Z'$. It shows that the parameter $-(X_t - z_t')$, the log per-capita share of the resource, characterizes competition as a conditional density-dependent factor. Graphically, competition determines the conditional curve $R_t = f(X_t | z_t')$, and z_t' translates the coordinates (R_t, X_t) of the curve laterally along the X_t axis (Fig. 1.12). A large supply of holes would shift the equilibrium density to the right, and a small supply to the left.

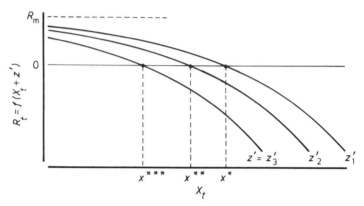

Figure 1.12 Lateral perturbation of the reproduction curve by (1.50). As z' varies from z'_1 to z'_3, the log equilibrium density shifts from X^* to X^{***}, although the maximum (potential) log reproductive rate R_m remains unchanged. All curves will coincide with each other if laterally translated along the X_t axis.

Both vertical and lateral perturbations can be called linear perturbations inasmuch as changes in z and z' do not influence the relative shape of the conditional reproduction curve, $R_t = f(X_t|z, z')$. Vertical and/or lateral translations of the coordinates (R_t, X_t) can bring all curves with different values (z, z') into one. In the following nonlinear perturbation, this cannot be done.

1.7.3 Nonlinear perturbation effect

Let Z'' be the measure of a density-independent factor. Suppose that changes in Z'' change the relative shape of the conditional reproduction curve, $R_t = f(X_t|Z'')$, so that curves with different values of Z'' cannot be brought into one by linear translations of the (R_t, X_t) coordinates. Then, this density-independent factor acts as a nonlinear perturbation.

As I mentioned in section 1.6.2, a food supply may be uninfluenced by density of the animals which feed on it. For instance, in some localities in Britain and Europe, the beech (*Fagus sylvaticus*) yields a good crop of mast in some years and a poor crop in other years. The great tit, *Parus major*, utilizes the beechmast in winter. When other types of food are scarce, the bird tends to survive better in a year of good crop than in a poor year (Perrins, 1979; Balen, 1980). The annual beech crop, though, is uninfluenced by the abundance of the animals which feed on the mast in the winter time. Therefore, the annual beech crop is a density-independent factor. However, like the nest hole example, its effect on the reproductive rate of the birds can only be evaluated jointly with the bird density. Simply, competition for the crop depends on the number of birds, given the crop.

Here, again, competition is characterized by the per-capita share of the resource, i.e. the quantity of mast each bird can consume through the winter. However, unlike the availability of nest holes, consumption per bird cannot be measured by a simple crop-to-bird ratio at the onset of the winter, because the food will be gradually depleted through the winter. Thus, the consumption per bird must be a more complicated function of both the annual crop, Z'', and the log bird density, X. I shall not attempt to find a probable function here.

However, without resorting to a particular model of consumption, we can readily see that changes in Z'' would influence the relative shape of the conditional reproduction curve $R_t = f(X_t | Z'')$. For a given X, competition will be more intense for a small Z'' than for a large Z''. As a result, the negative slope of a conditional reproduction curve for a small Z'' would become steeper more quickly as X_t increases, than it would for a large Z''. Hence, the beech crop acts as a nonlinear perturbation (Fig. 1.13).

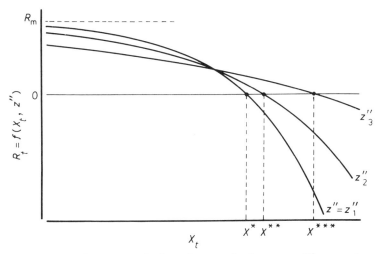

Figure 1.13 A nonlinear perturbation of a reproduction curve. The curvature of the reproduction curve changes as z'' changes from z''_1 to z''_3, causing the log equilibrium density to shift from X^* to X^{***}.

Several more examples are conceivable. (1) A multivoltine parasitoid, which requires alternate hosts, may act as a nonlinear perturbation factor on a given host, if the parasitoid density is largely governed by the densities of other alternate hosts. (2) A predator which feeds on a given prey in summer may migrate to a winter quarter. The predator density, if governed by food supply at the winter quarter, would act as a nonlinear perturbation factor on the summer prey. I shall discuss the impact of these types of predation and parasitism on the outbreak processes of the spruce budworm (*Choristoneura fumiferana*) in Chapter 9. (3) The susceptibility of animals to a density-independent environmental factor, such as a drought, may also depend on their density. The drought may reduce the production of foods for the animals. If so, the animals would suffer more severely when their density is high than when it is low. Such a weather factor would act as nonlinear perturbation.

A population process could be subject to all of these types of perturbations by density-independent factors simultaneously. Thus, letting z, z' and z'' be, respectively, vertical, lateral and nonlinear perturbation effects, we can write:

$$R_t = f[(X_t - z_t'), z_t''] + z_t. \tag{1.51}$$

Keep in mind, however, that a quantitative analysis of a process involving lateral or nonlinear perturbation is not easy. Thus, provided that their effects are comparatively small, we may – as in Chapter 2 – resort to an approximation by the simplest vertical perturbation scheme. Nonetheless, the conceptual distinction between the three types of perturbation would still be useful for qualitative understanding of population dynamics, such as the mechanism causing an insect outbreak (Chapter 9), even if a quantitative analysis is difficult.

In this section, the production of resources was assumed to be independent of the density of the animal which utilizes them. The production of some resource may be governed by the density of the animal, as in a predator–prey interaction. This topic will be discussed in Chapter 4.

1.8 CONCEPT OF DENSITY DEPENDENCE

In section 1.6.2, I remarked that my concept of density-dependent and density-independent factors differs from common usage. I now discuss the pragmatic signficance of the difference.

1.8.1 Connotations of 'density dependence'

The concept of **dependence** carries wide and narrow connotations. In the wide sense, it refers to a statistical correlation and, in the narrow sense, to a causal connection.

For instance, in describing a relationship between certain parts of the body, one might say that arm length **depends** on leg length. Here, obviously, one is speaking of a correlation between the two parts, rather than implying a functional influence of one part on the anatomy of the other. On the other hand, when we say that crop yield **depends** on precipitation, we are primarily speaking of the causal dependence of the crop on precipitation. It would sound rather silly if one said that precipitation depends on the crop. Nonetheless, the word dependence may still be used in its wide sense inasmuch as the information on the crop at hand can be used as an indicator of the amount of rainfall received during the season. The situation is so obvious that there is no danger of confusion.

In the context of population dynamics, the popular terminology 'density dependence' may likewise be used in the wide or narrow sense. However, the distinction is often so subtle that it must be made clear or confusion might result.

As already mentioned in section 1.6.2, standard text books use the term in the wide sense. Thus, any attribute of an ecological factor – be it the measure of its physical state (e.g. degree of temperature or population density) or its effect on a given population parameter (e.g. birth, death or dispersal rate) – is said to be density-dependent if it is **correlated** with the population density of the animal concerned. Conversely, it is said to be density-independent if uncorrelated.

Now a non-zero correlation between population density and the effect of a given factor on the density may result from several different situations: (1) when the physical state of the factor determining the net reproductive rate $r_t(=x_{t+1}/x_t)$ is, by itself, influenced by the density x_t, i.e. the correlation is due to a feedback between the population and the factor determining it; (2) when the state of the factor is autocorrelated, even if it is uninfluenced by population density, i.e., no feedback from population density, as demonstrated in section 1.6.7; (3) when the effect of the factor can be evaluated only jointly with the effect of population density, even though the factor itself is uninfluenced by the density, e.g. when the factor acts as lateral or nonlinear perturbation in section 1.7; (4) when the observed population series is short, even if the state of the factor is neither influenced by density nor autocorrelated (to be explained in section 3.3.3).

Evidently, the term 'density-dependent factor' used in the wide sense cannot distinguish between these situations as they all give rise to a non-zero correlation. Using it in the wide sense, we would be unable to describe the effects of those factors that cause lateral or nonlinear perturbation. We would be unable to distinguish those factors that regulate populations in the fragile manner from those that regulate in the robust manner. If we identify a factor by **effect**, we might place structurally similar mechanisms in different categories, and put different mechanisms in the same category, thereby making it difficult to identify the process structure. In other words, although a proper evaluation of the effects of the factors involved is ultimately a key to the analysis of population dynamics, it would defeat our purpose if the factors were identified in the wide sense.

Thus, as I suggested in section 1.6.2, we should use the term 'density-dependent or density-independent factor' in the narrow sense: excluding all but the first of the above four different situations from which a non-zero correlation results. Only then would we be able to describe the structure of a population process mathematically by representing the effect of a density-dependent factor as a function of density and that of a density-independent factor as vertical, lateral or nonlinear perturbation as in (1.51).

I do not imply that the expression 'density-dependent or density-independent' should always be used in the narrow sense. It is sometimes useful and convenient for describing some ecological relationships in the wide sense, if so understood. But, as a specific term to label a factor determining a population process, it must be used in the narrow sense.

To summarize my categorization of ecological factors developed in this section for the purpose of analysing population processes:

A factor influencing net reproductive rate of an animal population is said to be density-dependent if the measure of its state of existence (or the parameter that characterizes the state) – but not necessarily its effect on the reproductive rate (section 1.6.5) – is, in turn, influenced by (and, in that sense, dependent on) the population density. Conversely, a factor is said to be density-independent if its measure (parameter) is uninfluenced by (and, hence, independent of) the population density, even if its effect on reproductive rate is correlated with density.

1.8.2 Density-dependent processes and density-dependent regulation

If, as discussed in section 1.6.3, a density-dependent factor (in the narrow sense) is involved in a process which determines the net reproductive rate, let us call it a 'density-dependent process' and, otherwise, a 'density-independent process'. Then, as discussed in section 1.6.5, robust population regulation can be achieved only by the **particular class** of density-dependent processes that ensures an equilibrium state of a population within a suitable range of density-independent environmental conditions.

Accordingly, we should use the term 'density-dependent regulation' as meaning robust regulation (section 1.6.5) by a density-dependent process belonging to the above class, as opposed to fragile (hence, only theoretically possible) density-independent regulation by a density-independent process (section 1.6.4).

Note also that the distinction between **density-dependent** and **density-independent regulations** is meaningful only in the narrow sense. In the wide sense, regulation by any type of process – including the density-independent process (1.33) – would always be density-dependent. As demonstrated in section 1.6.7, a correlation between net reproductive rate and population density (after the removal of population trend, if any) is necessarily negative in every regulated population no matter what regulates it. Therefore, in the wide sense, density-independent regulation cannot exist by definition, and the term density-dependent regulation is tautological and meaningless.

1.8.3 On the idea of detecting density dependence and population regulation

To conclude the present chapter, I attempt to answer the oft-asked question: Can we detect the involvement of density-dependent factors in a population process and test population regulation?

Given the conventional notion of a density-dependent factor, the question of how to detect its involvement would simply be reduced to one of statistics: namely, how to determine a **correlation** between the **effect** of the factor and population density against the null hypothesis of no correlation (Morris, 1959; Varley and Gradwell, 1960). The method suggested by these authors is a simple regression of mortality (caused by a given factor) – or natality, dispersal rate – on the population density at the beginning of the time interval in which the factor acts. The method is widely adopted, despite a scepticism over the fact that a least-square regression coefficient tends to be negatively biased and could, thereby, distort the judgement. Nonetheless, nobody has questioned in principle the idea of appealing to a statistical method to detect density dependence in a mortality factor, because everyone has tacitly accepted the notion of density dependence in the wide sense. This is true even in more recent literature (Pollard *et al.*, 1987; Stiling, 1988; Hassell *et al.*, 1989; Reddingius and Den Boer, 1989; Den Boer and Reddingius, 1989).

[It is true that a negative bias results when: (1) the data series is short (Mealzer, 1970; St. Amant, 1970; Kuno, 1973), though the estimate is asymptotically unbiased; or (2) the series contains sampling error (Kuno, 1971). For an overview, see Southwood (1978) and Tanner (1978). Bulmer (1975) suggested a remedy to the first case. Quenouille's (1956) bias reduction method could be also effective (to be explained in section 3.5.1). The estimate in the second case would be biased even asymptotically and no formal remedy exists unless the magnitude of sampling error is known by some other means (Kennedy, 1984).]

Defined in its narrow sense, however, density dependence can no longer be sought with statistics. We must determine the nature of the factor as an ecological entity as discussed in section 1.6.2. In many cases, like those examples discussed in that section, the determination is more or less self-evident. The identification of a particular class, e.g. vertical, lateral or nonlinear perturbation effect, may require careful observation.

Once the factor has been identified, the evaluation of its effect on reproductive rate then becomes a subject of statistics. However, the evaluation requires much more than a simplistic regression scheme as illustrated in Southwood (1978, section 10.4.3). We must tackle the much broader problem of determining the structure of the population process. This is a central issue of population dynamics and will be discussed in detail in the next three chapters.

The test of regulation, on the other hand, remains essentially a statistical issue, since the concept of regulation – defined by the stipulation (1.12) in section 1.3.3 or, for a practical application, by (1.23) in section 1.5.3 – is a statistical notion. However, conventional simple regression analysis is inadequate for investigating whether or not the population conforms to the stipulation. In the conventional method, there is no clear distinction between population persistence and regulation. If the population has a trend, even if it is regulated about its trend, the correlation would be obscured. More importantly, even if the population has no trend so that the first stipulation (1.8) in section 1.3.1 is practically satisfied, a negative correlation does not imply regulation (section 1.6.7). In other words, although the stipulation (1.23) implies a negative correlation (as demonstrated in Appendix A of section 1.6), the converse is not necessarily true. Furthermore, the simple regression is useless if the effect of the factor concerned depends on densities in the past few generations (to be discussed in Chapters 2 and 3).

Why, then, should we not directly test whether or not the observed population series satisfies the stipulation (1.23)? It seems to be a natural and logical alternative. Unfortunately, it is not without difficulties, some very serious. To explain in simple terms, I assume that the series of log net reproductive rate $\{R_t\}$ is stationary. [I shall discuss in Chapter 3 a situation in which this assumption does not hold.]

Let S_t be a parameter defined by the Caesaro sum on the left-hand side of (1.23b), i.e.

$$S_t = \sum_{j=1}^{t} (1-j/t)\rho_{RR}(j) \tag{1.52}$$

and replacing $\rho_{RR}(j)$ with the sample autocorrelation coefficient $r_{RR}(j)$, we have the sample statistic:

$$\hat{S}_t = \sum_{j=1}^{t} (1-j/t)r_{RR}(j). \tag{1.53}$$

The test would be to compare $\hat{S}_t$ with S_t. But three major problems arise.

(1) While the parameter S_t of (1.52) converged to -0.5 in all regulated processes, it can, in an unregulated process, take any value in the interval $(-0.5, 0]$ – excluding the left end value as indicated by the parenthesis. For instance, the reader can readily verify that, in the unregulated process (1.38), the $\rho_{RR}(j)$ for $j > 1$ are identically zero and, therefore, that

$$S_t = \rho_{RR}(1) = -\sigma_u^2/(2\sigma_u^2 + \sigma_v^2)$$

which can take any value in the interval $(-0.5, 0]$, depending on σ_u^2 and σ_v^2. If σ_v^2 is so small that S_t is close to -0.5, it could be, in practice, difficult to distinguish from -0.5, the expected value in a regulated process.

(2) A more serious difficulty arises when the observed series is short so that the parameter S_t of (1.52) has not converged to any specific value: and, often, it converges very slowly. In that case, the value of S_t, given t, cannot be determined because it is entirely dependent on the unknown autocorrelation coefficients $\rho_{RR}(j)$ of the population process. Thus, we have no specific value of S_t to compare with the sample statistic $\hat{S}_t$ for a given t.

(3) Even if the observed series is adequately long, we encounter another difficulty. The sample statistic $\hat{S}_t$ is not necessarily a consistent estimator of S_t; it depends on the process. For instance, if the process is an unregulated random walk, as in Fig. 1.3, the expectation $E(\hat{S}_t)$ converges to -0.5, not to the true value of zero, and the variance $\text{Var}(\hat{S}_t)$ does not vanish even in the limit. On the other hand, if the process is perfectly regulated to generate a purely random population series as in Fig. 1.7, the expected value $E(\hat{S}_t)$ converges consistently to -0.5, i.e. $\text{Var}(\hat{S}_t)$ vanishes in the limit. However, the speed of convergence is often very slow. [Interested readers may confirm the above nature of the sample statistic $\hat{S}_t$ by computing its expectations. The discussion of the statistical properties of sample autocorrelations in section 3.6 provides necessary information for the computations.]

Thus, the sample statistic does not provide useful information. We would always be left with uncertainty in examining any type of process. This methodological impasse is, in fact, inherent in our attempt to summarize population regulation by means of a single parameter. The regulation of a population depends on the autocorrelation structure, and that can never be thoroughly described by a single parameter. By weighting and summing up all autocorrelation coefficients into the single parameter S_t, we are effectively killing potentially useful information about the structure of the process. The only alternative – and it is not perfect – is to look into the detail of the structure in each population process. Section 3.5 provides a rough guide for diagnosing the status of an observed population series.

But why should we want to know whether or not natural populations are regulated? An indiscriminate test of regulation just to satisfy one's curiosity, using any data series available in the literature, is uninspiring. The test would be redundant if we already know that these populations have been around for quite a while. That fact speaks for itself.

One strong motivation is, perhaps, to demonstrate the theory of density-dependent regulation of natural populations against the antithesis of the classical climatic control theory advocated by Bodenheimer (1938, 1958) and elaborated by Andrewartha and Birch (1954); see Krebs (1972) or Itô (1978) for a review of population theories.

The climatic control theory argues essentially that the fluctuations in many populations show a strong sign of climatic influences; therefore, the climatic factors involved must have been controlling the populations. However, some authors (Varley et al., 1973) contend that this reasoning was logically flawed. My argument in section 1.3.2, using model (1.39), succinctly demonstrates the point of contention. The perturbation effect $\{u_t + v_t\}$ of weather influences would show up in the population series $\{X_t\}$ only when the population is **tightly regulated** about an equilibrium level by a density-dependent process. It is so closely regulated by the density-dependent process that only the effect of density-independent weather influences show up in the population fluctuation. On the other hand, if the series of (log) reproductive rates $\{R_t\}$ is controlled entirely by the weather influences, then the population would perform a random walk as in the model (1.38). The advocates of the climatic control theory failed to understand the above cause–effect relationships. [For further discussions on this subject, see sections 3.3.1 and 3.4.1.]

It would be futile to attempt by examples to demonstrate that natural populations are regulated by density-dependent processes. A list of favourable examples, no matter how many may be in the list, does not provide a proof. Even less inspiring is an attempt to test (null-test) a logically invalid theory against observations. My logical deduction in sections 1.6.4 and 1.6.5 – that density-independent regulation is fragile and only theoretically possible – should be enough to end belief in density-independent regulation of natural populations.

A test of population regulation might have a practical value for specific purposes, e.g. diagnosing and monitoring the stability of the population of an endangered species. To do it, however, we must look into the detail of the autocorrelation structure and, eventually, into the underlying ecological mechanisms. Then, in effect, the detection of regulation simply dissolves into the study of much broader and fundamental aspects of population dynamics.

A central issue of the study of animal population dynamics is to answer the question I posed at the beginning: why do the animal populations behave as they do? In particular, we wish to know: why some populations persist at a comparatively steady level for a long time, while others fluctuate between extreme levels; why some exhibit a remarkably regular pattern, while others fluctuate irregularly; and why some populations, particularly introduced ones, fail to maintain themselves, while others establish firmly and become even a permanent nuisance. Evidently, the problem of fundamental importance is how to identify the structure of natural population processes.

The concept of population regulation discussed in this chapter has played an important theoretical role in bringing out the distinction between the two types of ecological factors and, thus, it has enabled us to appropriately structure population processes of various classes. In other words, the concept provides a basis for an investigation into the detail of population processes. But it is not a theory meant to be subject to an empirical demonstration.

Keeping all this in mind, our next step is to develop mental images of various classes of the structure of population processes and the pattern of dynamics that is associated with each class.

2 Structures and patterns of population processes

2.1 INTRODUCTION

In this chapter, I set up some basic theoretical models and illustrate the patterns that they generate. By such exercises, we gain knowledge essential for guessing the unknown generating mechanism of an observed pattern (to be discussed in Chapter 3).

A population process is, in general, governed by both density-dependent and density-independent factors. However, I shall start with idealized pure density-dependent processes in section 2.2 to illustrate the effect of population density on net reproductive rate, setting aside the effects of density-independent factors. I begin this section by introducing the concept of order of density dependence. The reproductive rate of a population depends generally on its own densities realized at a few time steps in the past (time-lags); the more time-lags that are needed to describe the process, the higher the order.

Then, I illustrate patterns of population dynamics generated by some models, ranging from the simplest first-order linear process to a more complex second-order nonlinear one. Our final goal is to understand the nonlinear dynamics that actual populations would exhibit. However, nonlinear processes are difficult to analyse. So, I use linear models to illustrate the principles. I discuss what determines the order of density dependence in section 2.3.

Section 2.4 deals with population dynamics as a stochastic process in which the effects of density-independent factors act as random perturbations of the pure density-dependent dynamics of section 2.2. Finally, in Section 2.5, I discuss a special role of density-independent factors in synchronizing the phase of several, independently fluctuating, regional populations.

2.2 PURE DENSITY-DEPENDENT PROCESSES

2.2.1 Order of density dependence

Recall the relationship (1.3b) in section 1.2.2, i.e.

$$R_t = X_{t+1} - X_t \tag{2.1}$$

where R_t is the log reproductive rate, and X_t, the log population density at time t (cf. Fig. 1.2). Recall also (1.24), i.e. $R_t = f(\mathbf{X}, \mathbf{Z})$, in which $\mathbf{X}$ is a set of log population densities, and **Z**, a set of measures of density-independent factors. In this section, I consider idealized pure density-dependent processes with the effect of **Z** being fixed.

Suppose that densities in a past few time steps are effective in determining the log reproductive rate R_t, so that we write the process explicitly with respect to X:

$$R_t = f(X_t, X_{t-1}, \ldots, X_{t-h+1} | \mathbf{Z}), \qquad (2.2)$$

and call it an hth-order (conditional or pure) density-dependent process. In particular, in a first-order process ($h=1$), R_t is uniquely determined by X_t, given **Z**; in a second-order process ($h=2$), R_t is determined by X_t as well as by X_{t-1}; and so on.

2.2.2 Linear first-order processes

The simplest of pure density-dependent processes is one in which R_t is a linear function of X_t only:

$$R_t = a_0 + a_1 X_t \qquad (2.3a)$$

in which a_0 and a_1 are constants; a_0 may be considered as the fixed effect of **Z** in (2.2). Using the relationship (2.1) and setting $t \equiv t-1$, we can write it alternatively as:

$$X_t = (1 + a_1)X_{t-1} + a_0. \qquad (2.3b)$$

Let us examine the patterns of series $\{X_t\}$ that model (2.3b) generates, given the initial value X_0. We can do this by solving (2.3b), i.e. evaluating X_t in terms of the initial value X_0 and the constants a_0 and a_1. Successive substitutions of 1, 2, ... for t in (2.3b) yield:

$$
\begin{aligned}
X_1 &= (1+a_1)X_0 + a_0, \\
X_2 &= (1+a_1)X_1 + a_0 \\
&= (1+a_1)[(1+a_1)X_0 + a_0] + a_0 \\
&= (1+a_1)^2 X_0 + [1 + (1+a_1)]a_0,
\end{aligned}
$$

$$\ldots\ldots\ldots\ldots\ldots\ldots\ldots\ldots\ldots\ldots\ldots\ldots\ldots$$
$$\ldots\ldots\ldots\ldots\ldots\ldots\ldots\ldots\ldots\ldots\ldots\ldots$$

$$X_t = (1+a_1)^t X_0 + [1 + (1+a_1) + (1+a_1)^2 + \cdots + (1+a_1)^{t-1}]a_0.$$

As is known in algebra, the geometric series in the square brackets on the right-hand side of the last equation is equal to the sum

$$[(1 + a_1)^t - 1]/a_1.$$

So, we have the solution

$$X_t = (1+a_1)^t(X_0 + a_0/a_1) - a_0/a_1. \tag{2.4}$$

Because X_0 and a_0/a_1 are given constants, the dynamics of the generated series $\{X_t\}$ depends solely on the factor $(1+a_1)^t$. We see that parameter a_0 has no influence on the process dynamics and, therefore, can be set equal to 0 without loss of generality.

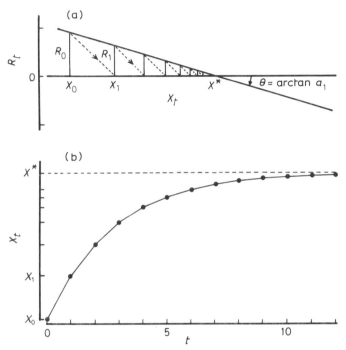

Figure 2.1   A graphic solution of the linear first-order pure density-dependent process (2.3a). Graph a: a reproduction line, the plot of R_t against X_t. Graph b: a series $\{X_t\}$ generated in graph a.

We can also solve the process (2.3a) graphically (Fig. 2.1a). The slanted line is a reproduction line, the plot of R_t against X_t. On the horizontal X_t axis, $R_t = 0$. The reproduction line intersects the X_t axis at the point marked X^*. This is an equilibrium point where $X_t = X_{t-1} = X^*$ because $R_t = 0$ there. Now, let X_0 be given on the horizontal axis. The length of the perpendicular line extending from X_0 to the intersection with the reproduction line is R_0. A 45° projection of this perpendicular line onto the horizontal axis gives X_1 because $X_1 = R_0 + X_0$ as defined in (2.1). Likewise, we can determine $X_1, X_2, \ldots$ successively. The series $\{X_t\}$ thus generated is plotted against t in Fig. 2.1b.

Parameter a_1 is the slope (tan θ) of the reproduction line. The angle θ is negative if measured clockwise from the horizontal axis, but is positive otherwise. In the Fig. 2.1a example, $-45° < \theta < 0°$ which is equivalent to $-1 < a_1 < 0$. We see that the series $\{X_t\}$ asymptotically converges to the equilibrium point X^*. It is easy to compute that $X^* = -a_0/a_1$ by substituting $R_t = 0$ and $X_t = X^*$ in (2.3a).

Figure 2.2 shows every possible pattern of series $\{X_t\}$ that model (2.3) can generate as parameter $a_1 (= \tan \theta)$ varies from positive to less than -2 in four intervals with two singular values between. In the interval $-2 < a_1 < 0$ (graphs c and d), generated series $\{X_t\}$ (graphs c' and d') converge to the equilibrium point X^*, no matter how far the initial point X_0 is away from it. In these cases, the equilibrium point X^* is said to be (globally) stable. When a_1 is either positive (graph a) or less than -2 (graph f), X^* is said to be unstable because a small deviation from it makes a series $\{X_t\}$ (graph a' or f') diverge. At the two singular values $a_1 = 0$ (graph b) and -2 (graph e), generated series $\{X_t\}$ (graphs b' and e') neither converge nor diverge, and the state of $\{X_t\}$ is solely dependent on the initial state X_0. In such cases, the equilibrium state is said to be neutral. Under random perturbation by density-independent factors, a population at a neutral equilibrium state performs a random walk, i.e. is unregulated (section 2.4.3).

In the linear scheme, a reproduction line on the left in Fig. 2.2 is symmetric about the equilibrium point X^*; accordingly, a generated series on the right is symmetric about X^*.

2.2.3 Nonlinear first-order processes

The linear model discussed above is mathematically the simplest and easiest to analyse. Ecologically, however, the model has an unrealistic property: its net reproductive rate is unbounded. As already pointed out in section 1.6.5, the reproductive rate of any animal must have an upper bound simply because no animal can produce an infinite number of offspring. Therefore, the log reproductive rate R_t must have a biologically realizable maximum, say R_m. Hence, R_t in an actual population must be a nonlinear function of log density X_t.

The following simple model has the above nonlinear property:

$$R_t = R_m - \exp(-a_0 - a_1 X_t). \tag{2.5a}$$

This is a nonlinear analogue of the linear process (2.3a). [I used model (2.5a) to illustrate Fig. 1.5. Its ecological meaning will be discussed in Chapter 4.]

A mathematical analysis of the dynamics of a nonlinear process is generally difficult. A graphical analysis of model (2.5a) provides some idea about first-order nonlinear dynamics. For this purpose, we may reduce (2.5a) to a canonical form by setting $R_m = 1$, $a_0 = 0$ and $a_1 = a_1^*$:

$$R_t = 1 - \exp(-a_1^* X_t). \tag{2.5b}$$

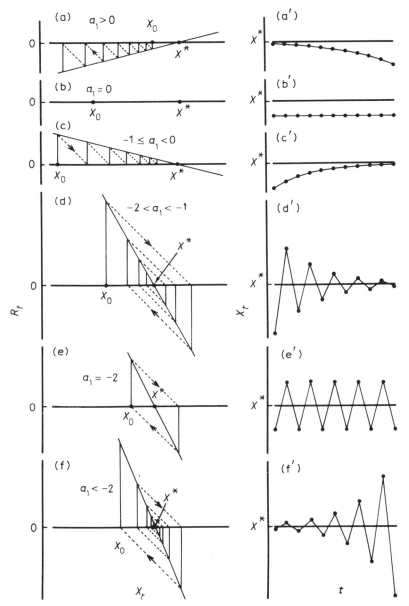

Figure 2.2 Same as Fig. 2.1 but for four intervals and two singular values in parameter a_1, showing the series of log densities $\{X_t\}$ (right graphs) generated by the corresponding reproduction lines (left graphs).

Parameter a_1^* is the slope of a reproduction curve at the equilibrium point X^*. Thus, the dynamics of process (2.5b) in the vicinity of X^* is directly compared with its linear analogue (2.3) already shown in Fig. 2.2. To make a comparison, set a_1^* of (2.5b) equal to a_1 of (2.3).

For the non-negative domain of a_1^*, the dynamics of the process (2.5b) can be easily deduced. For $a_1^* > 0$, the process diverges asymptotically at a constant rate as R_t increases to the maximum rate R_m ($= 1$): it does not diverge at an exponential rate as in the linear scheme with $a_1 > 0$ (Fig. 2.2a). For $a_1^* = 0$, the equilibrium point X^* is indeterminate, i.e. neutral; the process stays at a neutral equilibrium state as in the linear process in which $a_1 = 0$ (Fig. 2.2b).

Figure 2.3 illustrates the dynamics of process (2.5b) in the negative domain of a_1^*, corresponding to Fig. 2.2c to f of the linear scheme. In the interval $-2 < a_1^* < 0$ (Fig. 2.3a and b), the dynamics are much the same as those of the linear counterparts in $-2 < a_1 < 0$ (Fig. 2.2c and d), all uniformly converging to X^*. The nonlinear dynamics, though, are no longer symmetric about X^* like their linear analogues.

As a_1^* decreases below -2 in the nonlinear process (2.5b) (Fig. 2.3c and d), some important differences from its linear analogue (Fig. 2.2f) begin to show. The nonlinear process, though divergent near the unstable equilibrium point X^*, never diverges unboundedly like the linear scheme. Sooner or later, the amplitude of an oscillation will reach a certain limit, and the series $\{X_t\}$ remains more or less there, oscillating perpetually. If, in particular, the pattern of oscillation repeats itself after an initial time period as in Fig. 2.3c', it is said to have reached a limit cycle.

As the slope of a reproduction curve becomes steeper as in Fig. 2.3d, a degree of irregularity might show (Fig. 2.3d') though an oscillation remains within a certain domain away from, but about, the unstable equilibrium point X^*. If the initial point of a series $\{X_t\}$ is outside the domain of such a bounded, undamped oscillation (Fig. 2.3c and d), a generated series (graphs c' and d') will, sooner or later, converge into the domain. So, unlike the neutral oscillation of the linear scheme of Fig. 2.2e', these undamped, bounded oscillations are robust against random perturbations.

Thus, we can readily understand that an actual population, in which the reproductive rate has an upper bound, would seldom diverge without bound as long as its reproduction curve has a negative slope a_1^* at the equilibrium point X^*; and so long as the slope does not become too steep within the finite range of log density above the equilibrium point X^*. If the slope becomes very steep, the population could crash in such a way that it would be unlikely to recover in an actual situation.

An undamped, stable oscillation in the present nonlinear scheme tends to alternate between high and low densities in two successive time steps if a_1^* is not much lower than -2 (Fig. 2.3c and c'). The pattern of oscillation changes if a_1^* decreases further. Density tends to increase for a few successive generations followed by a crash decline – a relaxation oscillation (Fig. 2.3d'). At the same time, it tends to become increasingly more irregular, the so-called deterministically chaotic behaviour of a nonlinear dynamic system (May, 1976; Jensen, 1987). Note that the slope of the reproduction curve of (2.5b) at an arbitrary value of X_t is $a_1^* \exp(-a_1^* X_t)$: since $a_1^* < 0$ here, the

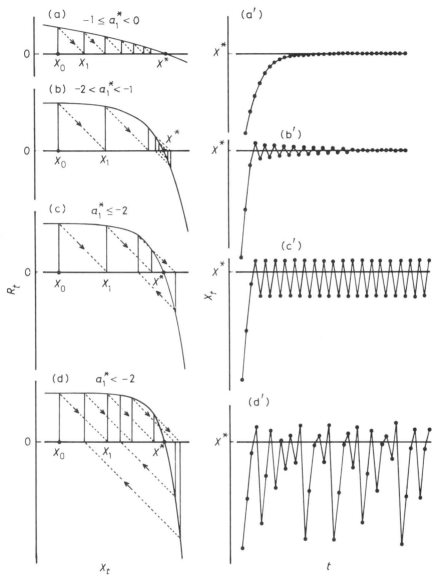

Figure 2.3. Graphic representations, similar to Fig. 2.2, of the nonlinear first-order pure density-dependent process (2.5b). Parameter a_1^* is the tangent of a reproduction curve at equilibrium point X^*.

slope tends to minus infinity (perpendicular) as X_t tends to infinity. This particular feature of model (2.5) tends readily to chaos as in Fig. 2.3d'. I shall show in section 4.3.4 an alternative model, in which the slope does not tend to perpendicular as X_t tends to infinity.

As already mentioned, the nonlinear process (2.5b) diverges if a_1^* is positive, analogous to Fig. 2.2a. However, it is inconceivable that an actual population has a reproduction curve whose slope is non-negative for the entire spectrum of population density. An adverse effect is bound to make the reproductive rate decrease at some point as the population increases. This property leads to the idea of multiple equilibrium processes (Ricker, 1954; Morris, 1963b).

Figure 2.4 illustrates simple multiple equilibrium models: each one has three equilibrium points X^*, X^{**} and X^{***}, at which the slopes of the reproduction curve are $a^* < 0$, $a^{**} > 0$ and $a^{***} < 0$, respectively. Figure 2.4a illustrates a case in which the slopes a^* and a^{***} are not less than -2, so that the generated series $\{X_t\}$ converges either to X^* or to X^{***},

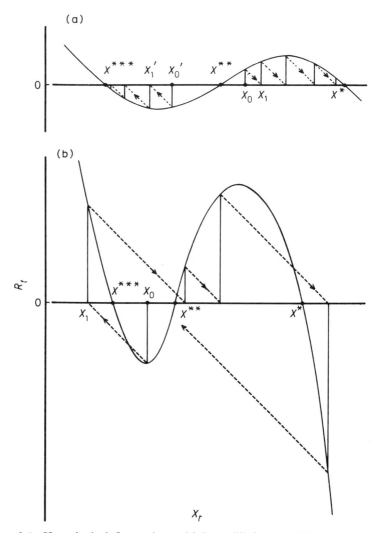

Figure 2.4   Hypothetical, first-order multiple equilibrium models.

depending on the relative position of the initial point X_0 or X'_0. Only under random perturbation could the population equilibrium shift between X^* and X^{***}. For much steeper slopes a^* and a^{***} as in Fig. 2.4b, the series $\{X_t\}$ cannot converge to any one point and may exhibit bounded, but rather complex, undamped oscillations.

Ricker (1954) proposed this scheme to explain the population dynamics of the pink salmon (*Onchorhyncus gorbuscha*), and Morris (1963b) applied it as a model of outbreak processes of the spruce budworm, (*Choristoneura fumiferana*). However, because there are substantial differences between these two species, they cannot be explained by one type of model. My alternative explanations for the salmon (Chapter 8) and for the budworm (Chapter 9) will reveal that the Ricker scheme applies to neither species. Therefore, I do not further elaborate on this model.

Be it a linear or nonlinear scheme, a first-order pure density-dependent process is characterized by the fact that, given X_t, X_{t+1} is uniquely determined. However, there are many natural populations that exhibit a consistent upward trend for a few generations followed by a consistent decline for another few generations; that is, given X_t, there appear to be two solutions for X_{t+1}, one being larger, and the other smaller, than X_t.

If the density-dependent component of a population process is first order, the source of the above pattern has to be sought in an influence from an extrinsic, density-independent factor that exhibits a similar pattern. However, as will be discussed at length in Chapter 5, there is little evidence that suggests that such an extrinsic influence is prevalent among natural populations. On the other hand, as I discuss in section 2.3, there are several reasons to believe that, in many natural populations, R_t is second-order in density dependence, i.e. dependent not only on X_t but also on X_{t-1}. This type of density dependence can generate the pattern of gradual oscillations without resorting to unsubstantiated density-independent influences.

2.2.4 Linear second-order processes

Adding an extra lag term, X_{t-1}, to the first-order model (2.3) gives a second-order model,

$$R_t = a_0 + a_1 X_t + a_2 X_{t-1} \tag{2.6a}$$

or equivalently

$$X_t = a_0 + (1 + a_1)X_{t-1} + a_2 X_{t-2} \tag{2.6b}$$

in which a_0, a_1 and a_2 are constant parameters. Given these constants and an arbitrary pair of initial points (X_0, X_1), we can successively compute, as we did in the first-order scheme, the subsequent points $X_2, X_3, \ldots$ by (2.6b). However, unlike the first-order scheme, such computations quickly lead to unmanageably cumbersome formulae. Neither is a graphic analysis as easy and useful as in the first-order scheme. The log reproductive rate R_t in the second-order process (2.6a) is represented by a plane in a three-dimensional (R_t, X_t, X_{t-1}) coordinate space. Such a graph is awkward to analyse.

Thus, for the analysis of the process, we have to resort to a special mathematical method of solving equation (2.6b); i.e. to evaluate X_t in terms of the constants a_0, a_1 and a_2, given a pair of initial states (X_0, X_1).

Because the dynamics of model (2.6) about the equilibrium point X^* is not affected by parameter a_0, we can assume it to be zero without loss of generality. Thus, we concentrate on solving

$$X_t = (1 + a_1)X_{t-1} + a_2 X_{t-2}. \tag{2.7}$$

An equation of the above form is known as a homogeneous second-order difference equation and can be solved by the following steps.

First, suppose there exists a factor λ such that $X_t = \lambda X_{t-1}$ holds true. Then, successive substitutions of λX_{i-1} for X_i, $i = 2, 3, \ldots, t$, yield:

$$X_t = \lambda^t X_0. \tag{2.8}$$

Further, substituting $\lambda^i X_0$ for X_i ($i = t$, $t-1$, and $t-2$) in (2.7), we have, for $X_0 \neq 0$:

$$\lambda^t - (1 + a_1)\lambda^{t-1} - a_2 \lambda^{t-2} = 0,$$

and, for a nontrivial solution ($\lambda \neq 0$):

$$\lambda^2 - (1 + a_1)\lambda - a_2 = 0. \tag{2.9}$$

This quadratic equation, called the auxiliary equation of (2.7), has two roots, say λ_1 and λ_2. These are given by the well-known formulae

$$\lambda_1 = [(1 + a_1) + \sqrt{(1 + a_1)^2 + 4a_2}]/2 \tag{2.10a}$$

$$\lambda_2 = [(1 + a_1) - \sqrt{(1 + a_1)^2 + 4a_2}]/2 \tag{2.10b}$$

The general solution of X_t is, then, given by the sum of constant multiples of the particular solutions, λ_1^t and λ_2^t:

$$X_t = k_1 \lambda_1^t + k_2 \lambda_2^t. \tag{2.11}$$

Setting $t = 0$ and 1 in (2.11), we obtain, respectively,

$$X_0 = k_1 + k_2$$

$$X_1 = k_1 \lambda_1 + k_2 \lambda_2. \tag{2.12}$$

If $\lambda_1 \neq \lambda_2$, we can eliminate k_1 and k_2 from (2.11) and (2.12) to obtain

$$X_t = [(\lambda_2 X_0 - X_1)\lambda_1^t - (\lambda_1 X_0 - X_1)\lambda_2^t]/(\lambda_2 - \lambda_1) \qquad (2.13)$$

in which λ_1 and λ_2 have been given in (2.10). Thus, we have solved (2.7), i.e. evaluated X_t in terms of a_1, a_2, X_0 and X_1. The solution enables us to examine how a generated series $\{X_t\}$ behaves according to the values assigned to a_1 and a_2, given X_0 and X_1. [If $\lambda_1 = \lambda_2$, take the limit $\lambda_1 \to \lambda_2$ in (2.13) by l'Hôpital's rule. The result is not shown as it is not essential in the present argument.]

Figure 2.5 illustrates various regions on the $(1 + a_1, a_2)$ parameter plane, divided in accordance with the distinct patterns that a series $\{X_t\}$ generated by (2.7) exhibits (Fig. 2.6).

Consider, in Fig. 2.5, the regions (I, I', II and II') above the parabola P where

$$(1 + a_1)^2 + 4a_2 > 0. \qquad (2.14)$$

The quantity on the left in (2.14) is the number under the square-root operator in (2.10). Because it is positive, λ_1 and λ_2 are distinct real roots in these regions.

Consider, in particular, regions I and I', i.e. right to the vertical line L_4. Because $1 + a_1 > 0$ in these regions, it is easy to see in (2.10) that $\lambda_1 > |\lambda_2|$. We see then, in (2.13), that X_t is determined by λ_1^t for large t, since λ_1^t numerically dominates over λ_2^t. In region I, $a_2 < -a_1$ and $|a_1| < 1$. Then, from (2.10a), $0 < \lambda_1 < 1$. Thus, λ_1^t asymptotically converges to zero in the region and, hence, so does X_t (graph I, Fig. 2.6). In region I', where $\lambda_1 > 1$, X_t exponentially diverges (graph I', Fig. 2.6). On the side (L_1) of the triangle (Fig. 2.5), X_t neither converges nor diverges, i.e. it stays at a neutral equilibrium state. This applies everywhere on the line $a_1 + a_2 = 0$ (an extension of L_1 beyond the triangle) because the equilibrium point X^* is indeterminate there. On the line, therefore, the process $\{X_t\}$ would perform a random walk under perturbation by density-independent factors (section 2.4.3).

Similarly, we find that, in regions II and II' – left to the vertical line L_4 above the parabola P (Fig. 2.5), X_t is determined by λ_2^t for large t since λ_2 is dominant over λ_1 in these regions. However, because $\lambda_2 < 0$ there, λ_2^t alternates between negative (when t is odd) and positive (when t is even) values. Accordingly, a generated series $\{X_t\}$ oscillates alternately between high and low values. The series converges in region II where $|\lambda_2| < 1$ but diverges in region II' because $|\lambda_2| > 1$ there (graphs II and II', Fig. 2.6). On the side (L_2) of the triangle, X_t is at a neutral equilibrium state.

Below parabola P in Fig. 2.5, the quantity under the square-root operator in (2.10) is negative, that is, the two roots λ_1 and λ_2 are complex conjugates which, written in polar form, are given by

$$\lambda_1 = r(\cos w + i \sin w)$$

$$\lambda_2 = r(\cos w - i \sin w) \qquad (2.15)$$

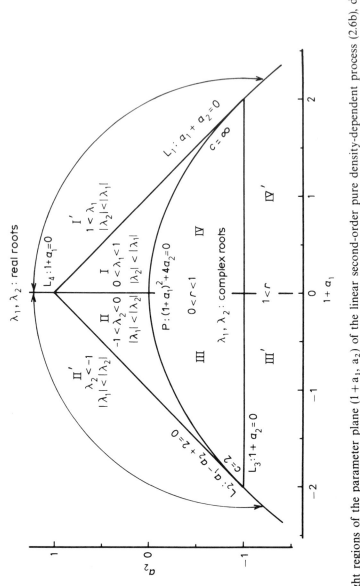

Figure 2.5 Eight regions of the parameter plane $(1 + a_1, a_2)$ of the linear second-order pure density-dependent process (2.6b), divided according to the values of the two roots, λ_1 and λ_2, of the auxilliary equation (2.9), given by the formulae (2.10).

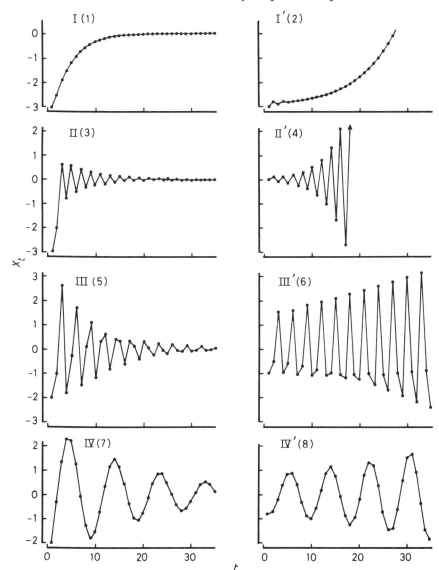

Figure 2.6   Typical patterns exhibited by the series of log densities $\{X_t\}$ generated by the linear second-order pure density-dependent process (2.6) from the eight regions, i.e. I to IV inside, and I' to IV' outside the triangle in the parameter plane $(1 + a_1, a_2)$ of Fig. 2.5. The Arabic numeral in parentheses following each Roman numeral indicates the point (parameter coordinates) marked in Fig. 2.9.

where r is the modulus of the complex roots and i is the imaginary number $\sqrt{-1}$. Eliminating λ_1 and λ_2 from (2.15) and (2.13), using de Moivre's theorem, i.e.

$$(\cos w + i \sin w)^t = \cos tw + i \sin tw,$$

we find, after a little algebra, that

$$X_t = r^{t-1}[X_1 \sin tw - rX_0 \sin(t-1)w]/\sin w. \tag{2.16}$$

[Note that the imaginary terms cancel each other in the final solution, so that X_t is a real number as it should be.] Clearly, a series $\{X_t\}$ converges if $r<1$ but diverges if $r>1$. However, from (2.15) and (2.10), we find that $r^2 = \lambda_1 \lambda_2 = -a_2$. Thus, $r<1$ for $-1<a_2<0$ and $r>1$ for $a_2<-1$. In other words, the series $\{X_t\}$ converges above the line L_3, i.e. in regions III and IV, but diverges below the line, i.e. in regions III' and IV' (Figs 2.5 and 2.6). On line L_3, $r=1$, so that the process stays at a neutral equilibrium state.

In short, the series $\{X_t\}$ converges to an equilibrium state inside the triangle of Fig. 2.5 and diverges outside. On each side of the triangle, the linear process stays at a neutral equilibrium state: its dynamics depends on the initial state (X_0, X_1). Under random perturbation, the linear process will converge to a stationary state (hence, persistent state) only inside the triangle (section 2.4).

The solution (2.16) shows that in regions under the parabola P (Fig. 2.5), X_t is a periodic function of t: its periodicity is determined by the angle w in (2.16). Eliminating λ_1 and λ_2 from (2.10) and (2.15), we obtain

$$\tan w = \sqrt{|(1+a_1)^2 + 4a_2|}/(1+a_1). \tag{2.17}$$

Thus, the periodicity of an oscillation in the series $\{X_t\}$ depends on parameters a_1 and a_2. Because we are in the regions below parabola P, $(1+a_1)^2 < 4|a_2|$ and, hence,

$$-2\sqrt{|a_2|} < 1 + a_1 < 2\sqrt{|a_2|}. \tag{2.18}$$

We see that, given a_2, $1+a_1$ can vary from $-2\sqrt{|a_2|}$ on the left half of parabola P (the left end of regions III and III') to $2\sqrt{|a_2|}$ on the other half of P (right end of regions IV and IV'). As $(1+a_1)$ so varies from left to right, given a_2, w in (2.17) correspondingly varies from π to 0. Because a sine curve has a cycle of 2π, a series $\{X_t\}$ generated by (2.16) has a cycle whose average length (c) is given by

$$c = 2\pi/w. \tag{2.19}$$

Thus, the average cycle length varies from 2 on the left half of parabola P, given a_2, to infinity on the other half, with $c=4$ at mid point, i.e. on line L_4 (Fig. 2.5).

2.2.5 Nonlinear second-order processes

Figure 2.7 illustrates three types of linear reproduction planes of model (2.6a): namely, $(a_1<0, a_2>0$; graph a), $(a_1<0, a_2<0$; graph b), and $(a_1>0,$

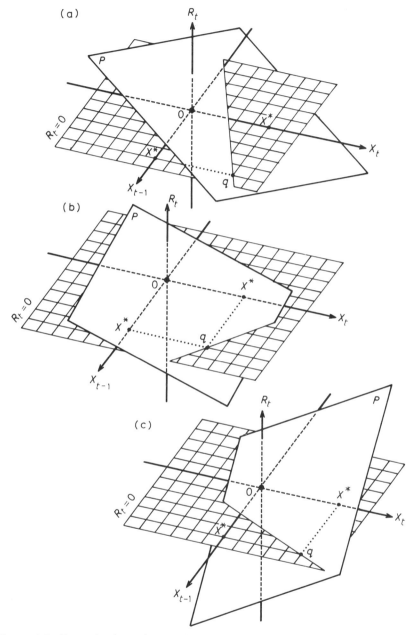

Figure 2.7 Reproduction planes of the linear second-order model (2.6a) with different combinations of a_1 and a_2: (a) $a_1 < 0$, $a_2 > 0$; (b) $a_1 < 0$, $a_2 < 0$; (c) $a_1 > 0, a_2 < 0$. P: reproduction plane. q: equilibrium point $X_t = X_{t-1} = X^*$ on the $R_t = 0$ plane marked with grid.

$a_2 < 0$; graph c). Evidently, each of these reproduction planes intersects the horizontal plane marked with the grid on which $R_t = 0$. The point q on the plane at which $X_t = X_{t-1} = X^*$ is an equilibrium point. Recall now that the log reproductive rate of an actual population must have an upper limit (R_m), a biologically realizable maximum rate. Thus, the reproduction plane of an actual population must be bent downwards (Fig. 2.8) so as to stay within the realm of the biological constraint $R_t \leqslant R_m$, analogous to the nonlinear first-order processes in Fig. 2.3.

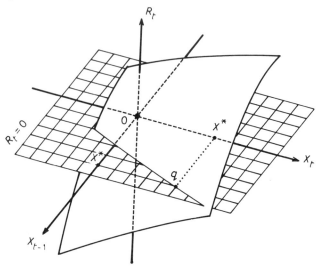

Figure 2.8 An example of nonlinear reproduction surface, corresponding to the linear reproduction plane in Fig. 2.7c, but curving downwards so as not to exceed R_m, a biologically realizable maximum log net reproductive rate.

Consider the following model to represent a nonlinear reproduction surface like the one in Fig. 2.8:

$$R_t = R_m - \exp(-a_0 - a_1 X_t - a_2 X_{t-1}), \qquad (2.20a)$$

its canonical form being:

$$R_t = 1 - \exp(-a_1^* X_t - a_2^* X_{t-1}). \qquad (2.20b)$$

These are generalizations of the first-order model (2.5a) and its canonical form (2.5b) whose dynamics were illustrated in Fig. 2.3. As a generalization of the first-order case (2.5b), the canonical form (2.20b) represents a reproduction surface which has the partial derivatives equal to a_1^* and a_2^* at the equilibrium point $q(X^*, X^*)$ as in Fig. 2.8. The exact analysis of the dynamics of model (2.20b) is difficult. However, we can make a reasonable guess by comparing it with the linear second-order scheme (2.6) and by analogy with the nonlinear first-order model (2.5b).

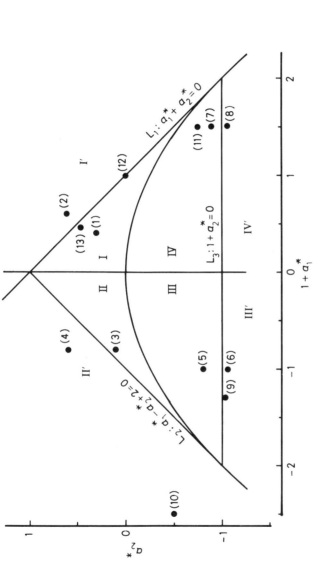

Figure 2.9 The parameter plane $(1 + a_1^*, a_2^*)$: same as Fig. 2.5 but a_1^* and a_2^* replace a_1 and a_2. The points marked by Arabic numerals in parentheses are the coordinates $(1 + a_1^*, a_2^*)$ at which the sample series in Figs 2.10 to 2.19 are generated. The coordinates of points (1) to (13) are: (0.4, 0.3), (0.6, 0.6), (−0.8, 0.1), (−0.8, 0.6), (−1.0, −0.8), (−1.0, −1.05), (1.5, −0.9), (1.5, −1.05), (−1.3, −1.01), (−2.5, −0.5), (1.5, −0.75), (1, 0), (0.47, 0.47).

For this purpose, I set up the parameter plane $(1 + a_1^*, a_2^*)$ in Fig. 2.9, analogous to the one in Fig. 2.5 for the linear scheme (2.6). Figure 2.10 illustrates the patterns of dynamics that model (2.20b) generates in the eight parameter regions at points $(1 + a_1^*, a_2^*)$ marked with Arabic numerals in Fig. 2.9. In order to compare the generated nonlinear dynamics with those of the linear process already studied, points (1) to (8) are chosen as equal to the points $(1 + a_1, a_2)$ of the linear model (2.6) that generated the eight sample series already illustrated in Fig. 2.6.

Figure 2.10 shows that the nonlinear dynamics inside the triangle, i.e. those generated at points (1), (3), (5) and (7), are similar in pattern to those of the linear scheme (Fig. 2.6). This is because nonlinear dynamics near a stable equilibrium point can be approximated by the linear model. Two noticeable differences are: (1) the nonlinear dynamics (Fig. 2.10) are no longer symmetric about the equilibrium point X^*; (2) a series produced by the nonlinear process damps down more quickly than one by the corresponding linear process (Fig. 2.6).

At point (2) in region I' outside the triangle, the nonlinear process (Fig. 2.10I') diverges in a similar fashion to that in the linear counterpart (Fig. 2.6I'). In fact, at every point above line L_1 (which includes part of region II') in Fig. 2.9, both linear and nonlinear processes diverge. However, unlike an exponential increase in the linear case, the rate of increase in the nonlinear case will, sooner or later, become constant because R_t is limited to R_m at most. On the other hand, since there is no lower limit to R_t, the nonlinear process will plunge at an exponential rate, as in the linear counterpart, if started below X^* (Fig. 2.15, p. 79).

Everywhere on line L_1, the equilibrium point X^* of process (2.20b) is neutral. This is because, on the line, $a_1^* + a_2^* = 0$ and, hence, X^* is indeterminate in (2.20b). [Line L_1 is a generalization of the point $(a_1^* = 0)$ for the first-order process (2.5b); see point (12) on line L_1.]

In both linear and nonlinear processes, every equilibrium point is unstable below line L_1 outside the triangle (Fig. 2.9). However, unlike the linear counterpart, the nonlinear process does not readily diverge. All of the series generated at points (4), (6) and (8) just outside the triangle show undamped, but stable oscillations, as expected by analogy with those in Fig. 2.3 generated by the nonlinear first-order model (2.5b) with the parameter $a_1^* < -2$. However, in region IV' towards the border to region I', the oscillation may dip down to an extreme level and, thus, is considered to be practically unbounded.

Figure 2.11 illustrates an example of deterministically chaotic behaviour (analogous to Fig. 2.3d') that the model (2.20b) could exhibit. The series $\{X_t\}$ is generated at point (9) (Fig. 2.9): this point is in region III' but only marginally outside the triangle. Here, by analogy of the first-order scheme (2.5b) (i.e. a_1^* is only slightly less than -2 as in Fig. 2.3c), I expected a limit cycle to occur. Contrary to my expectation, the generated series is quite chaotic. In fact, model (2.20b) generates a limit cycle even within region III if

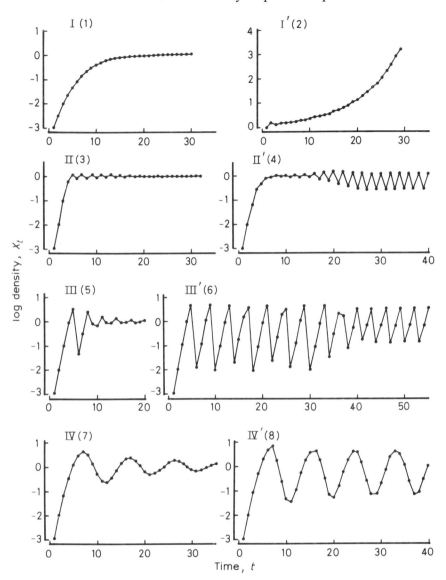

Figure 2.10 Series of log densities $\{X_t\}$ generated by the nonlinear second-order pure density-dependent process (2.20b) in the eight regions in Fig. 2.9. The Roman and Arabic numerals in each graph indicate, respectively, the region and parameter point in Fig. 2.9.

fairly close to the boundary to region III′ where a convergent oscillation would have been expected from the linear scheme of Fig. 2.6. [The reader can confirm this by computer simulation with a parameter set, e.g. $(1 + a_1^*$ $= -1, a_2^* = -0.85)$.] Thus, guessing second-order nonlinear dynamics by analogy with a linear scheme has limitations.

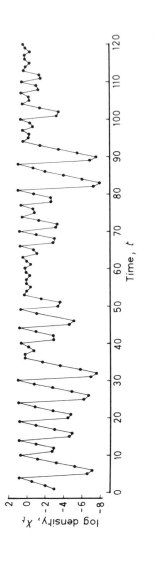

Figure 2.11 Series of log densities $\{X_t\}$ generated by the nonlinear second-order pure density-dependent process (2.20b) at point '9), region III' in (Fig. 2.9), exhibiting deterministically chaotic dynamics.

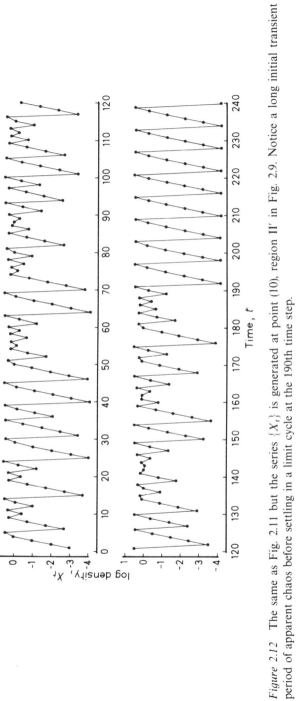

Figure 2.12 The same as Fig. 2.11 but the series $\{X_t\}$ is generated at point (10), region II' in Fig. 2.9. Notice a long initial transient period of apparent chaos before settling in a limit cycle at the 190th time step.

Another example of the dynamics of model (2.20b) is illustrated in Fig. 2.12, generated at point (10) in region II′. The generated series appears quite irregular for as long as the first 190 steps and, then, all of a sudden, converges to a regular limit-cycle pattern. The timing of convergence to the limit cycle is subject to the initial points (X_0, X_1), and, perhaps, to computer round-offs. The graph also shows that, although the series is generated in region II′, its pattern does not resemble that of its linear second-order cousin in Fig. 2.6II′. This is because the reproductive rate is bounded in the nonlinear process. Again, the analogy of the linear scheme breaks down here.

Apparent chaos (e.g. Fig. 2.11) could merely be the result of a prolonged lead time before a series converges to a limit cycle. A recent review article by Jensen (1987) and an introduction by Devaney (1990) give some insight into the nature of nonlinear dynamics without requiring advanced mathematics. Noting this, I shall not delve into this topic any further for the following reasons.

Pronounced chaotic dynamics of a nonlinear process occur only when the slope of a reproduction curve (or partial slopes of a reproduction surface) becomes sufficiently steep about an equilibrium point. This means that a population, when increasing from below the equilibrium point, rather abruptly becomes very sensitive to a slight increase in density. I feel that, though subject to observations, such an abrupt increase in sensitivity is rather unlikely to develop in an actual biological population.

On the other hand, many population processes are composed of several, sequential life-stage processes. Even if each stage process is simple and well-behaved, deterministically chaotic dynamics can readily result in the whole process. Such a complex process cannot be represented by a single reproduction surface; I shall discuss this issue in Chapter 8. Thus, I feel that chaos generated by a mathematically simple model is not all that relevant to the ecological reality. Apart from this, mathematical investigation into nonlinear dynamics is a fascinating and important subject, and I invite interested theorists to further investigate the dynamics of a nonlinear model, e.g. (2.20).

2.2.6 Higher-order processes

The second-order linear process (2.6) can be generalized to an hth-order model

$$R_t = a_0 + a_1 X_t + a_2 X_{t-1} + \cdots + a_h X_{t-h+1}. \tag{2.21}$$

Its nonlinear version as a generalization of (2.20a) is:

$$R_t = R_m - \exp(-a_0 - a_1 X_t - a_2 X_{t-1} - \cdots - a_h X_{t-h+1}). \tag{2.22}$$

For the rest of the book, I mainly deal with dynamics up to second order. In only a few cases will I refer to a higher-order process and then merely to maintain the generality of the argument. There are two reasons for avoiding the discussion of higher-order dynamics.

First is an ecological reason. As I discuss in detail in the next section, density dependence of an order much higher than 3 would, perhaps, be comparatively rare in natural populations; and, perhaps, second-order models would approximate many third-order cases.

Second is a mathematical reason. The general solution of the hth-order linear model (2.21) is:

$$X_t = k_1 \lambda_1^t + k_2 \lambda_2^t + \cdots + k_h \lambda_h^t \tag{2.23}$$

where λ_i is the ith root of the auxiliary equation

$$\lambda^h - (1 + a_1)\lambda^{h-1} - a_2 \lambda^{h-2} - \cdots - a_h = 0. \tag{2.24}$$

As is known in algebra, there exist general formulae for the roots of the polynomial (2.24) only up to the 4th degree. Therefore, we have general formulae for the solution (2.23), in terms of the coefficients a's and initial densities, only up to the 4th order; and the formulae are very cumbersome.

Differences between second and higher-order processes are not as radical as between the first two. As already shown in the preceding sections, a second-order process has a periodic solution, whereas a first-order process does not. Such is a radical difference. On the other hand, a higher-order process exhibits complex patterns as combinations of patterns that a second-order process exhibits, e.g. an oscillation superimposed upon another or on an exponential decay. Thus, the gain of insight from the analysis of cumbersome models is not great enough to be worth the effort at the level of this book.

2.3 ORIGINS OF SECOND- OR HIGHER-ORDER PROCESSES

Second- or higher-order density dependence may arise from three ecologically different, not mutually exclusive, situations: (1) when the animals concerned are comparatively long-lived, and a density effect on individuals in a past time is carried over to their present reproductive status; (2) when the realization of a density effect is somehow delayed in time; (3) when a population of a given species interacts with populations of other species, even if there is no carry-over or delayed effect.

There is a fourth situation in which an apparently high order of density dependence shows for purely statistical and mathematical reasons rather than ecological ones.

2.3.1 Carry-over and delayed density effects

Consider a species in which individual animals can live longer than one time step, e.g. one year. Suppose that the physiological condition of individuals is influenced by their density (crowding effect, e.g. malnutrition from food shortage, poisoning from environmental contamination, injuries from fighting) and that their survival and reproductive capacity are accordingly affected. Such an effect on individual animals may last more than one year or may even be passed onto their offspring. In other words, the effect of population density on the individuals' physiology more than a year ago could be carried over to still affect the overall reproductive rate of the present population. A carry-over effect for h years can be represented by a density-dependent process of order h.

However, in many natural populations, it would probably be rather rare for a density effect to be carried over more than two years or generations: those individuals that were significantly harmed by the effect would have died or recovered. In other words, a carry-over effect on the average reproductive rate of the current year would dissipate fairly quickly. If this is the case, a second-order model should be adequate to describe many natural processes.

A similar but slightly different situation arises when there is a delay in the density effect by a fraction of the regular interval of time in the discrete-time scheme; that is, the density at time $t - 1 - h'$ $(0 < h' < 1)$ influences the reproductive rate over the time interval $(t - 1, t)$. For example, potential natality in snowshoe hares (*Lepus americanus*) in summer depends on how well or poorly they fed during the preceding winter, and that, in turn, depends on the density of the animals at the beginning of the winter (section 6.6). Thus, there is a time delay in the density effect on natality of about half a year.

There are two ways of dealing with the above type of time delay. One is to approximate the density $x_{t-1-h'}$, by either x_{t-1} or x_{t-2} whichever is better correlated with $x_{t-1-h'}$. Another method is to divide the regular time interval into two subintervals, and use an auxiliary model for each subinterval. I shall discuss the use of these methods in Chapter 6.

2.3.2 Density effect in a multi-species interaction system

Consider, first, a two-species interaction system, such as a predator–prey or competing-species system, with no carry-over (nor delayed) density effect. Let the following simultaneous first-order pure density-dependent models represent the system process:

$$X_t = f(X_{t-1}, Y_{t-1}) \tag{2.25a}$$

$$Y_t = g(X_{t-1}, Y_{t-1}) \tag{2.25b}$$

where the X and Y are log densities of the two species, say x and y. Now, if we eliminate densities of one species, say the Y, from (2.25), we obtain a new equation which is second order in the X; and likewise in the Y if the X are eliminated. Thus,

$$X_t = F(X_{t-1}, X_{t-2}) \tag{2.26a}$$

$$Y_t = G(Y_{t-1}, Y_{t-2}) \tag{2.26b}$$

is another way of representing the system process (2.25).

An explicit transformation is generally difficult with a system of nonlinear equations. However, the fact that the population series of each species can be described by a single-variable time series model implies that the system representation (2.26) is possible. The principle of the transformation can be shown using the simple linear system:

$$X_t = A_1 X_{t-1} + B_1 Y_{t-1} \tag{2.27a}$$

$$Y_t = A_2 X_{t-1} + B_2 Y_{t-1} \tag{2.27b}$$

in which the As and Bs are constant parameters. This system yields, if none of the four constants are zero,

$$X_t = (A_1 + B_2)X_{t-1} - (A_1 B_2 - B_1 A_2)X_{t-2} \tag{2.28a}$$

$$Y_t = (A_1 + B_2)Y_{t-1} - (A_1 B_2 - B_1 A_2)Y_{t-2} \tag{2.28b}$$

by the following procedure. First, multiply both sides of (2.27a) by A_2 and those of (2.27b) by A_1; subtract the former from the latter; set t equal to $t-1$; and rearrange to obtain

$$A_2 X_{t-1} = A_1 Y_{t-1} - (A_1 B_2 - B_1 A_2)Y_{t-2}. \tag{2.29}$$

Further, substitute the right-hand side of (2.29) for $A_2 X_{t-1}$ in (2.27b) to obtain (2.28b), and likewise to find (2.28a).

Similarly, the three-species interaction system

$$X_t = A_1 X_{t-1} + B_1 Y_{t-1} + C_1 Z_{t-1} \tag{2.30a}$$

$$Y_t = A_2 X_{t-1} + B_2 Y_{t-1} + C_2 Z_{t-1} \tag{2.30b}$$

$$Z_t = A_3 X_{t-1} + B_3 Y_{t-1} + C_3 Z_{t-1} \tag{2.30c}$$

yields third-order density dependence in each species, if all constant parameters are not zero. In general, an h-species interaction system yields

hth-order density dependence in each species, provided that all constant parameters are not zero.

The situation in which all constants are non-zero means that all species in the system influence each other, i.e. every pair of species is directly linked by a feedback loop. An example is: in the system (2.30), x eats both y and z; y eats only z; and a predator (x or y) significantly influences the reproduction of its prey (y or z) and vice versa.

Instead, the three species may form a straight food chain, e.g., x eats y and y eats z, that is, y interacts with both x and z, whereas x and z each interact only with y. So there is no feedback loop between x and z. Thus, parameters C_1 and A_3 in (2.30) are zero. Then, after a transformation, one will find that the dynamics of y is third order and those of x and z second order. [In doing the transformation, one must be aware that $C_1 = A_3 = 0$ and avoid using them as multipliers.]

Now consider in the two-species system (2.27) that x is a trivial predator on y, that is, x depends on y but exerts little influence on y. Thus, parameter A_2 is negligible. Then, the dynamics of x in (2.27a) is first order and y (through its term Y_{t-1}) acts as a density-independent perturbation of x.

Consider further in the three-species system (2.30), x is a trivial predator on y but y interacts with z. Then, the dynamics of x is first order as in the above two-species case: x is not linked to y by a feedback loop. However, y's dynamics is now second order because of its linkage to z by a feedback loop. Thus, y acts as a second-order perturbation factor of x.

From the above argument, we conclude that the order of density dependence in a given species is equal to the number of species (including itself) with which it interacts directly; or 1 plus the number of feedback loops which involve the given species. This implies that a careful examination of the food-web structure is important to aid the determination of the order of density dependence.

We frequently encounter food webs in which two species are directly interacting with each other, i.e. they are linked to each other by a feedback loop. This means that many natural population processes would be second order.

A given species may be linked to many other species in a complicated food web (Pimm, 1982). If all of these links constitute feedback loops, the order of density dependence would be accordingly high. Then, the dynamics of the species could be very complicated. [Recall that a high order of density dependence can lead to a complicated dynamics (section 2.2.6).] Such complication may destabilize the system dynamics. If so, each species may not be involved in too many feedback loops in an existing, stable system. This would result in comparatively lower-order density dependence in species dynamics.

Some feedback loops may be weak enough to effectively form a one-way link. If so, the order of density dependence would be accordingly reduced. It is also possible that, even though the food-web linkage is complicated, the predators may act as a complex on their prey, and the prey also constitute a complex. In such a predator(complex)–prey(complex) interaction system, density dependence in each species' dynamics can be reduced to second order, and the system may exhibit a comparatively simple, stable dynamics. I shall discuss such a possibility in Chapter 6.

Note, in passing, that the two equations in (2.28) have identical coefficients in the corresponding terms on the right-hand side. This means that both species have the identical pattern of dynamics; that is, when time-plotted, the series $\{X_t\}$ and $\{Y_t\}$ exhibit an identical pattern except for a phase shift when initial densities differ between the two species. This, however, is an artificial situation arising from using the linear system. In actual nonlinear systems, the dynamics generally differ between the species involved. I shall illustrate the point with some examples in later chapters. A linear model is useful only for certain aspects of theoretical studies, and we must not overinterpret its dynamics.

2.3.3 Apparent order of density dependence

Second- or higher-order density dependence may arise when a population process happens to have a particular mathematical property, not implying any ecologically induced density dependence.

Consider the linear second-order process

$$R_t = 0.5X_t - 0.75X_{t-1}. \tag{2.31}$$

Figure 2.13a shows the reproduction plane in the (R_t, X_t, X_{t-1}) coordinate space. An inward spiral trajectory suggests a convergent oscillation of the generated series $\{X_t\}$ characteristic to region IV, point (11), of Fig. 2.9. Also shown are the projections of the trajectory onto the three coordinate planes (R_t, X_t), (R_t, X_{t-1}) and (X_t, X_{t-1}).

Now, in Fig. 2.13b, I regressed R_t on X_{t-2}. The data points line up perfectly on the straight line

$$R_t = -0.375X_{t-2}. \tag{2.32}$$

It might appear, then, that the present model is better taken as third order rather than second order. This might motivate one to seek an ecological mechanism, e.g. a density effect carried over from two time steps in the past. However, as we see below, the relationship in Fig. 2.13b is a mere mathematical correlation (or density dependence in the wide sense) and does not imply an ecologically induced two-step time-lag in the density effect.

Set $t \equiv t - 1$ in the second-order process

$$X_{t+1} = (1 + a_1)X_t + a_2 X_{t-1}$$

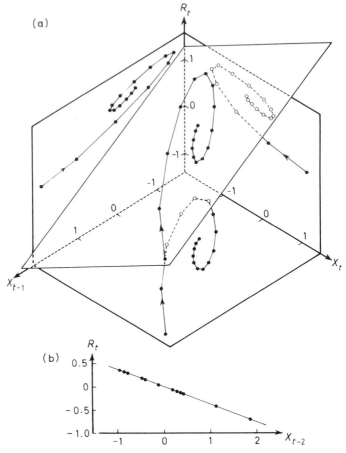

Figure 2.13 The regression of R_t on X_t and X_{t-1} (graph a) of the linear second-order pure density-dependent model (2.31), and its projections onto the three coordinate planes, (R_t, X_t), (R_t, X_{t-1}) and (X_t, X_{t-1}). The plot of R_t against X_{t-2} (graph b) exhibits a perfect regression line by (2.32).

to obtain

$$X_t = (1 + a_1)X_{t-1} + a_2 X_{t-2},$$

and, noting that $R_t = X_{t+1} - X_t$, eliminate X_{t-1} from the above two equations to obtain

$$R_t = [(a_1^2 + a_1 + a_2)/(1 + a_1)]X_t - [a_2^2/(1 + a_1)]X_{t-2}. \qquad (2.33)$$

Since $a_1 = 0.5$ and $a_2 = -0.75$ in the example (2.31), the coefficient in (2.33) for the X_t term on the right happens to vanish and the one for the X_{t-2} term turns out to be -0.375. This results in (2.32). We may take Fig. 2.13b to be a projection of the trajectory in Fig. 2.13a onto yet a fourth coordinate plane (R_t, X_{t-2}) in a four-dimensional space.

We see that a given density-dependent process can be mathematically converted to one with an arbitrarily high order of density dependence. Then, we should seek ecological mechanisms that cause density dependence in the lowest possible order, unless there is a good ecological reason to believe otherwise. I shall discuss, in section 3.2.2, how to determine the lowest order in an observed process.

2.4 DENSITY-DEPENDENT/INDEPENDENT PROCESSES

In the preceding sections, I discussed the dynamics of density-dependent processes on the condition that the effects of density-independent factors are fixed. I shall now relax this condition, allowing density-independent factors to act as perturbations of the density-dependent processes.

2.4.1 Density-dependent/independent processes as stochastic processes

In an hth-order pure density-dependent process

$$R_t = f(X_t, X_{t-1}, \ldots, X_{t-h+1}) \tag{2.34}$$

$X_t (t \geq h)$ is strictly determined when t is specified, given a set of initial states $(X_0, X_1, \ldots, X_{h-1})$. Also, no matter how irregular the generated series $\{X_t\}$ may appear when plotted against time t (e.g. Fig. 2.11), the R_t, when plotted against $X_t, X_{t-1}, \ldots, X_{t-h+1}$, are perfectly contained in an h dimensional reproduction surface defined by (2.34).

In section 1.7, I classified the effect of a density-independent factor into the three categories: vertical, lateral, and nonlinear perturbations, z, z' and z'', as in model (1.51). Let us generalize (1.51) by making these effects of the density-independent factors explicit in (2.34), i.e.

$$R_t = f[(\mathbf{X} - \mathbf{z}'), z_t''] + z_t \tag{2.35}$$

where $\mathbf{X}$ is the set $(X_t, X_{t-1}, \ldots, X_{t-h+1})$ and similarly in $\mathbf{z}'$.

Process (2.35) becomes a stochastic process in time if z, z' or z'' contains randomness. These random effects may contain a deterministic trend as well, a component that is a deterministic function of time (section 2.4.5). In the following, I mainly consider the vertical perturbation effect to investigate the basic characteristics of population processes. [Lateral and nonlinear perturbation effects will be discussed in section 2.4.5.] Thus, the log reproductive rate R_t is now a random variable: it deviates from the reproduction surface defined by (2.34) as much as z_t vertically.

2.4.2 Order of a density-dependent/independent process

The property of density effect $f(\mathbf{X})$ was investigated theoretically in the previous sections in terms of some pure-density dependent processes. Unlike the density effect, the property of the perturbation effect z_t is not a self-contained subject of theoretical investigation. The effect z is determined by whatever constitutes the set of density-independent factors involved in a process. There is no general rule that can be deduced without empirical studies.

Nonetheless, in order to maintain the generality of argument, we need to assume that z_t is somehow autocorrelated. For this purpose, I shall employ the following scheme widely used in time-series analysis:

$$z_t = u_t + b_1 u_{t-1} + \cdots + b_k u_{t-k} \tag{2.36}$$

in which the u are independent, identically distributed random numbers. The scheme (2.36) is known as the moving-average process of order k, or MA(k) for short. This is a practical, useful descriptive device (like algebraic polynomials for curve-fitting) for representing the autocorrelated effect of a density-independent factor without the need of specifying its structure.

Thus, the process we study is:

$$R_t = f(X_t, X_{t-1}, \ldots, X_{t-h+1}) + u_t + b_1 u_{t-1} + \cdots + b_k u_{t-k} \tag{2.37a}$$

or equivalently (using the relationship $X_t = R_{t-1} + X_{t-1}$),

$$X_t = X_{t-1} + f(X_{t-1}, X_{t-2}, \ldots, X_{t-h}) + u_{t-1} + b_1 u_{t-2} + \cdots + b_k u_{t-k-1}. \tag{2.37b}$$

Let us call this a density-dependent/independent process of order (h, k). When $k=0$ in (2.37), the perturbation effect becomes purely random as represented by the pure random series $\{u_t\}$. We may call this particular case a stochastic density-dependent (h) process.

The above terminology corresponds to the one referring to the linear time-series model, known as the autoregressive-moving average process of order (h, k), or ARMA(h, k) for short. When $k=0$, model (2.37b) assumes the form in which X_t is regressed on its own past terms X_{t-h} ($h=1, 2, \ldots$); hence, the term 'autoregressive process of order h' or AR(h) for short. I shall use whichever terminology is appropriate depending on the context of the argument.

A stochastic process should not be confused with a deterministic process with measurement error. If X_t' is a deterministic function of time, e.g. $X_t' = g(t)$, a measurement of X_t', say X_t, may contain random error z_t, such that

$$X_t = g(t) + z_t \tag{2.38}$$

which is not a stochastic process, even if the measurement X_t is not a deterministic function of time.

2.4.3 Examples of stochastic density-dependent (2) processes

Consider the second-order nonlinear model (2.20a) subjected to perturbation effect u_t, i.e.

$$R_t = R_m - exp(-a_0 - a_1 X_t - a_2 X_{t-1}) + u_t \qquad (2.39a)$$

its canonical form being:

$$R_t = 1 - exp(-a_1^* X_t - a_2^* X_{t-1}) + u_t. \qquad (2.39b)$$

Assuming that the mean $E(u_t) = 0$ identically, parameter R_m is now taken as the **mean** maximum log reproductive rate inasmuch as $E(R_m + u_t) = R_m$.

Figures 2.14 to 2.17 show series $\{X_t\}$ generated by (2.39b) with the parameter set $(1 + a_1^*, a_2^*)$ equal to the points (1) to (8) in Fig. 2.9: the same set of points at which the pure density-dependent series in Fig. 2.10 were generated by (2.20b). This arrangement enables us to see how the random perturbation u_t alters the deterministic dynamics of a pure density-dependent process. Also, in all examples, I use the identical sample series $\{u_t\}$, a series of independent, normally distributed random numbers with mean $= 0$ and variance 0.04 (Fig. 2.14a). So we can directly compare its effect on all examples of the series $\{X_t\}$ generated by (2.39b).

Figure 2.14 shows two series $\{X_t\}$ generated inside region I (Fig 2.9). These are compared with their pure density-dependent counterparts (smooth curves) converging to the equilibrium level $X^*(=0)$. The stochastic series I(1), generated at point (1) well inside region I, closely follows its pure density-dependent path (smooth curve). Series I(13), on the other hand, shows a considerable degree of drift away from the deterministic path (smooth curve) of its pure density-dependent analogue. This is because point (13) (Fig. 2.9), at which this series is generated, is very close to side L_1 on which every series performs a random walk.

Figure 2.15 shows series generated at point (2), region I' (Fig. 2.9), where an equilibrium point is unstable. Having started with the non-negative initial points, $(X_0 = 0, X_1 = 0.2)$, the pure density-dependent series (solid smooth curve) diverges upwards. The stochastic series with the same initial points also diverges but downwards. This is because, in this region, the direction of divergence depends (after an initial period of drift) on the first significant deviations of two consecutive points from the equilibrium point X^*. Once deviated sufficiently downwards, the diverging tendency is accelerated exponentially by the positive feedback from the density-dependent component. In the present sample series, it so happened that X_4 and X_5

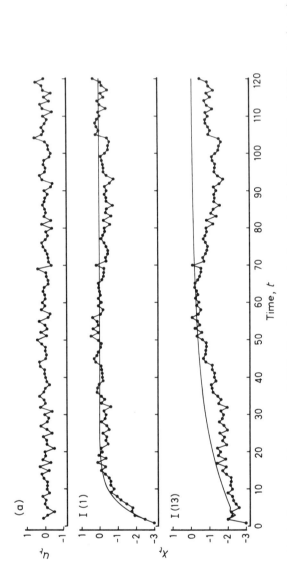

Figure 2.14 Sample series $\{X_t\}$ generated by process (2.39b). Graph a: perturbation series $\{u_t\}$. Graph I(1) and I(13): sample series of $\{X_t; X_0 = -3, X_1 = -2\}$ generated, respectively, at points (1) and (13), region I, in Fig. 2.9. Smooth curves are their pure density-dependent analogues by (2.20b).

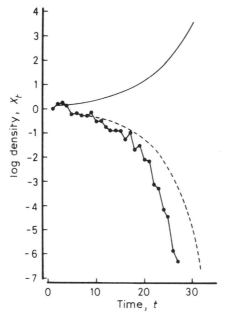

Figure 2.15   Same as in Fig. 2.14 but the series $\{X_t; X_0 = 0, X_1 = 0.2\}$ is generated at point (2), region I', in Fig. 2.9. Smooth curves are pure density-dependent analogues by (2.20b).

deviated sufficiently downwards, and the subsequent series plunged as dictated by the pure density-dependent trend (dashed curve) with the points (X_4, X_5) as its initial state.

Figure 2.16 shows two series generated at points (3) and (4), regions II and II' (Fig. 2.9), respectively. Both II(3) and II'(4) exhibit saw-tooth oscillations with no sign of drift, the tendency clearly dictated by their pure density-dependent counterparts II(3) and II'(4) in Fig. 2.10.

Examples in Fig. 2.17 are those from the four regions under the parabola p (Fig. 2.9). Series III(5), III'(6), IV(7), and IV'(8) correspond to their respective pure density-dependent series in Fig. 2.10. In these regions, vertical perturbations may cause an extreme oscillation in density. Typically, a crash decline occurred in series III'(6) immediately following an unusually high density at $t = 105$. This high density was caused by a slight but unusually high upward perturbation by u_t at $t = 104$ (Fig. 2.14a). A crash decline occurs in region III' when a very high X_t follows a comparatively high X_{t-1}; that is, when a large upward perturbation by u_t coincides with an upswing phase of a population cycle. This is because, in this region, both $1 + a_1^*$ and a_2^* are negative, so that a pair of large values in X_{t-1} and X_t can cause an extremely low R_t in (2.39b).

Series IV'(8) (Fig. 2.17) also exhibits a steep decline following the same upward perturbation as in series III'(6). However, the decline in IV'(8) occurred only after $t = 107$. It appears that there is a time delay between the

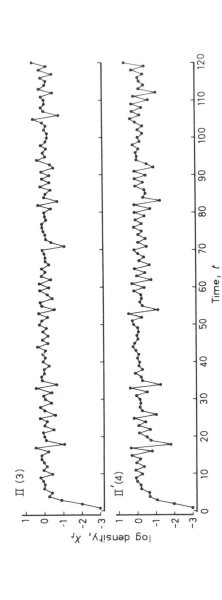

Figure 2.16 Same as in Fig. 2.14 but the two series $\{X_t; X_0 = -3, X_1 = -2\}$ are generated at points (3) and (4), region II and II', in Fig. 2.9. For their respective pure density-dependent analogues, see II(3) and II'(4) in Fig. 2.10.

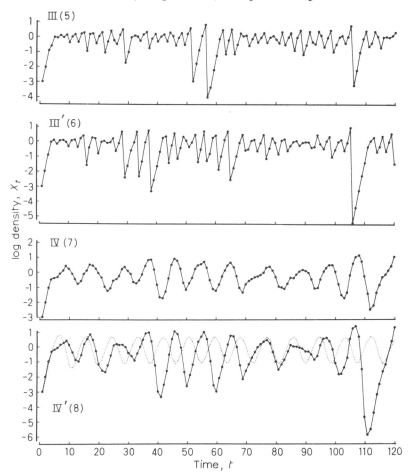

Figure 2.17   Same as in Fig. 2.14 but the four series $\{X_t; X_0 = -3, X_1 = -2\}$ are generated at points (5) to (8), regions III, III', IV and IV', in Fig. 2.9. The dotted curve in graph IV'(8) is the corresponding pure density-dependent series IV'(8) in Fig. 2.10. For the pure density-dependent analogues of the other three series, see Fig. 2.10.

large upward perturbation and the population decline. This is because, in region IV', $1 + a_1^*$ is positive, and, following the large upward perturbation by u_t at $t = 104$, density climbed up two more steps to reach an extremely high level. Only then a steep decline followed in the subsequent few steps. Thus, the effect of the large upward perturbation was delayed in time because of the density-dependent structure of the series characteristic to region IV'. Note that a series in region IV' exhibits a comparatively smooth oscillation, so that a crash decline does not occur like a relaxation oscillation in region III'.

We also see, in both Figs 2.16 and 2.17, that a distinction between damped and undamped oscillations in the pure density-dependent processes (Fig.

2.10) is obscured in the stochastic counterparts. This is because a stochastic process, even if its density-dependent component had an inclination to damp out, could not do so under constant perturbation by density-independent factors. Conversely, as exemplified by series IV'(8) (Fig. 2.17), a stochastic series may not at all follow the regular cyclic pattern (dotted curve) of its pure density-dependent analogue.

We sometimes see in the literature that a deterministic model is fitted to observed population data; hoping, perhaps, that a good fit validates the model. The above simulation suggests that a good fit may be a forced fit. A deterministic model cannot be directly compared with data. I shall discuss how to compare a model with data in Chapter 3.

2.4.4 Examples of density-dependent/independent(2, 1) processes

Consider that the vertical perturbation effect u_t in (2.39b) is generalized to MA(1), i.e. $u_t + bu_{t-1}$, so that we have the nonlinear density-dependent/independent (2, 1) process

$$R_t = 1 - \exp(-a_1^* \ X_t - a_2^* \ X_{t-1}) + u_t + bu_{t-1} \tag{2.40}$$

in which b is a nonzero constant factor and u is, as before, an independent, identically distributed random number. In the following, I shall generate sample series, using model (2.40) with the same parameter sets $(1 + a_1^*, a_2^*)$ as model (2.39b) from the regions I to IV already shown in Figs 2.14, 2.16 and 2.17.

To illustrate the effects of the perturbation $\{u_t + bu_{t-1}\}$ in (2.40), I shall use two convenient cases, i.e. $b = 1$ and -1. Also, in order to make the effects directly comparable with that of the perturbation $\{u_t\}$ in the previous simulations with model (2.39b), the distribution of the sum $(u_t + bu_{t-1})$ is made the same as u_t in (2.39b), i.e. normally distributed with mean $= 0$ and variance $= 0.04$. [This can be done by choosing the u in (2.40) to be independent, normally distributed random numbers with the mean 0 and variance 0.02, using the well-known theorems that the sum of two independent, normally distributed random numbers is again normally distributed and the variance is the sum of their variances.]

Now, $u_t + bu_{t-1}$ and $u_{t-1} + bu_{t-2}$ share the common element u_{t-1}. Therefore, they are correlated with each other. If $b = -1$, they are negatively correlated. So, low and high values tend to alternate in the series $\{u_t - u_{t-1}\}$ (Fig. 2.18a) more frequently than in the completely random series $\{u_t\}$ (Fig. 2.14a). On the other hand, if $b = 1$, $u_t + u_{t-1}$ is positively correlated with $u_{t-1} + u_{t-2}$; a high value tends to follow a high value, or vice versa. This means that the series $\{u_t + u_{t-1}\}$ (Fig. 2.18b) tends to stay on one side of the average ($= 0$) for a while before it crosses over to the other side, i.e. tends to undulate. In both cases, two data points are uncorrelated with each other if they are more than two time steps apart.

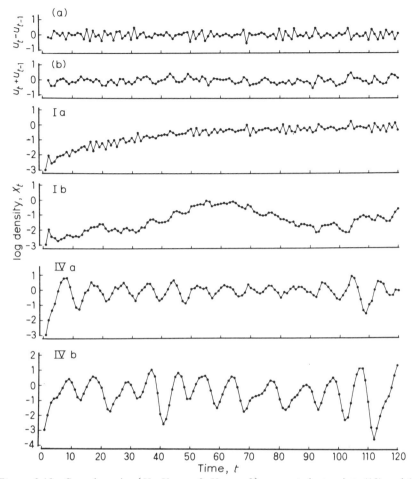

Figure 2.18 Sample series $\{X_t; X_0 = -3, X_1 = -2\}$ generated at points (13) and (7) in regions I and IV (Fig. 2.9) by process (2.40) subjected to perturbation series a, $\{u_t - u_{t-1}\}$, and series b, $\{u_t + u_{t-1}\}$.

Figure 2.18 shows four population series Ia, Ib, IVa and IVb generated by (2.40). The Roman numerals indicate, as before, the region I or IV (Fig. 2.9) and a or b indicates the series subjected to perturbation series a or b. The two series from region I are generated at point (13) (Fig. 2.9) and the two from region IV at point (7). Thus, series Ia and Ib are compared with series I(13) (Fig. 2.14), and IVa and IVb, with IV(7) (Fig. 2.17). [Remember that both I(13) and IV(7) were subjected to the uncorrelated peturbation $\{u_t\}$.] We see that perturbation series a tends to reduce drift in series Ia compared to I(13) and the amplitude of an oscillation in IVa compared to IV(7).

Perturbation series b exerts just the opposite effect. It makes population series Ib drift even more compared to I(13) and exaggerates the amplitude in IVb compared to IV(7).

Figure 2.19 shows the four series, IIa, IIb, IIIa and IIIb. Series IIa and IIb are compared with II(3) (Fig. 2.16), and IIIa and IIIb with III(5) (Fig. 2.17). Here, perturbation series a exaggerates, and series b reduces, population oscillations in both regions II and III. Thus, perturbation series a and b reverse their effects on the population series b from regions II and III as compared to their effects on those from regions I and IV.

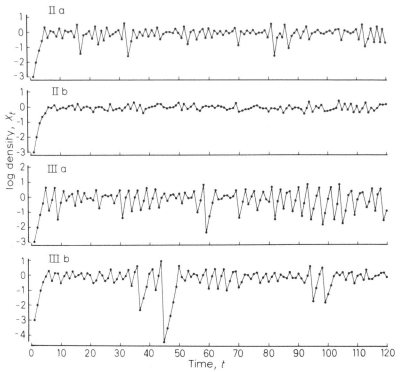

Figure 2.19   The same as in Fig. 2.18 but the four series are generated at points (3) and (5), regions II and III, subjected to the perturbation series a and b in Fig. 2.18.

The above results are explained as follows. Recall that perturbation series a tends to exhibit saw-tooth, high-frequency oscillations because two adjacent values are negatively correlated with each other. As opposed to this, the pure density-dependent population series from regions I and IV (Fig. 2.10) exhibit low frequency oscillations; one from region I can be viewed as having an infinitely long oscillation. Thus, the high-frequency tendency in perturbation series a tends to counter the low-frequency tendency of the population series from these regions. On the other hand, perturbation series b has a low-frequency, undulating tendency (due to a positive correlation between two adjacent values) which tends to 'resonate' with the similar tendency in the density-dependent component to exaggerate their oscillations.

In regions II and III, a pure density-dependent series oscillates rapidly (Fig. 2.10). Therefore, it resonates with the same tendency in perturbation series a but is countered by the opposite tendency in series b.

2.4.5 Examples of nonstationary processes

So far, I have discussed stationary population dynamics, assuming that the parameters R_m, a_1, and a_2 (or a_1^*, a_2^*) are invariant in time and, hence, that the equilibrium point X^* does not change in time. Let us relax this assumption and consider that the effects of some density-independent environmental factors progressively change in time so that the dynamics of a population becomes nonstationary. In particular, consider the following two situations.

First, progressive changes in climatic conditions favour the reproductive performance of each individual animal. Accordingly, the mean maximum log reproductive rate R_m increases with time; that is, the reproduction surface shifts upwards as in Fig. 1.6, section 1.6.5.

Second, progressive changes in the vegetation type in the habitat provides the animals with more foods, shelters or nesting sites. As a result, competition is reduced and the reproduction surface shifts laterally to the right as in Fig. 1.12, section 1.7.2.

In order to show how such environmental changes affect the pattern of population dynamics, I generalize the second-order model (2.39a) such that parameters R_m and a_0 are functions of time, i.e.

$$R_t = R_m(t) - \exp[-a_0(t) - a_1 X_t - a_2 X_{t-1}] + u_t. \tag{2.41}$$

Let us first consider the effect of $R_m(t)$ while $a_0(t)$ is fixed. For simplicity, let $R_m(t)$ be a linear function of time, e.g.

$$R_m(t) = 1 + 0.02t. \tag{2.42}$$

This means that $R_m(t)$ is 1 initially ($t=0$) but is doubled by $t=50$.

Setting, conveniently, $a_0(t)=0$, $a_1=-0.75$, $a_2=-0.5$, $X_0=-1$ and $X_1=-0.5$ in model (2.41) and using the perturbation series $\{u_t\}$ in Fig. 2.14a, I generated a series $\{X_t\}$ in Fig. 2.20b. In this series, not only does its trend increase with time, but so does the amplitude of its oscillation. The reason for the increasing amplitude is the following.

Let a_1^* and a_2^* be the partial slopes of the reproduction surface (2.41) at its equilibrium point X^*. [The partial slopes are given by the partial derivatives $\partial R_m(t)/\partial X_t$ and $\partial R_m(t)/\partial X_{t-1}$ evaluated at $X_t = X_{t-1} = X^*$.] Because $R_t = 0$ at X^* by definition, it is easy to compute that:

$$X^*(t) = -[\log R_m(t) + a_0]/(a_1 + a_2). \tag{2.43}$$

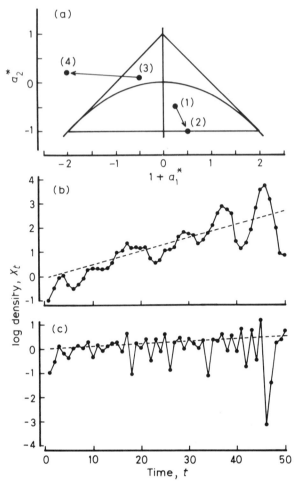

Figure 2.20 Examples of nonstationary processes generated by model (2.41) in which $R_m(t)$, and not a_0, is a function of time. So the reproduction surface shifts vertically with time.

$X^*(t)$ indicates that X^* is now a function of time. Further, it is also easy to compute:

$$a_1^*(t) = a_1 R_m(t)$$
$$a_2^*(t) = a_2 R_m(t) \tag{2.44}$$

in which $a_1^*(t)$ and $a_2^*(t)$, too, are functions of time.

Now that $R_m(0) = 1$ by the assumption (2.42), $1 + a_1^*(0) = 1 + a_1 = 0.25$, $a_2^*(0) = a_2 = -0.5$, and $X^*(0) = 0$. Thus, the series $\{X_t\}$ in Fig. 2.20b is generated at point (1) in the parameter space in Fig. 2.20a at $t = 0$ when the equilibrium point $X^*(0) = 0$. However, at $t = 50$, $R_m(50) = 2$ by (2.42), and, therefore, the equilibrium point $X^*(50)$ is, by (2.43), increased to $(\log 2)/0.25 = 2.773$. At the same time, by (2.44), $1 + a_1^*(50) = 0.5$, and $a_2^*(50) = -1$. Thus, point (1) in Fig. 2.20a has, after 50 time steps, moved to point (2). This is why the amplitude of an oscillation increases with time. Thus, an increase in $R_m(t)$ (vertical shift in the reproduction surface with time) makes the population series $\{X_t\}$ doubly nonstationary, i.e. non-stationary in equilibrium density and in the amplitude of an oscillation about the trend line. These are examples of a process in which both of the two requirements for persistence (sections 1.3) are violated: they have an unchecked trend and unregulated fluctuations about the trend.

Figure 2.20c is another example but $a_1 = -1.5$ and $a_2 = 0.1$. Thus, at $t = 0$ when $R_m(0) = 1$, the parameter set $(1 + a_1^*, a_2^*)$ is in region II at point (3) (Fig. 2.20a). After 50 time steps, $R_m(50) = \log 2$, and the parameter set moves out to point (4) in region II' (Fig. 2.20a). Accordingly, the equilibrium point X^* shifts from 0 at $t = 0$ to $(\log 2)/1.4 = 0.495$ at $t = 50$.

Now consider the second situation, i.e. the reproduction surface shifts only laterally. Here, $a_0(t)$, but not R_m, is a function of time. The equilibrium point X^* is given by:

$$X^*(t) = -[\log R_m + a_0(t)]/(a_1 + a_2). \tag{2.45}$$

Figure 2.21 shows two examples in which $a_0(t)$ increases linearly with t. Because the reproduction surface shifts only laterally, the parameters R_m, a_1^*, and a_2^* do not depend on time. Thus, these series have an upward trend dictated entirely by the trend in $a_0(t)$. However, they exhibit no systematic change in the amplitude of an oscillation about the trend. In other words, these populations are nonstationary in equilibrium density but are stationary

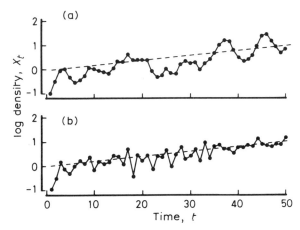

Figure 2.21   Examples of nonstationary processes generated by model (2.41) in which $a_0(t)$, and not R_m, is a function of time. So the reproduction surface shifts laterally with time.

in their fluctuations about the trend. These are examples of a process in which the first requirement in persistence is violated, whereas the second requirement is satisfied: they have an unchecked trend, but their fluctuations are regulated about the trend.

Now, we can think of a third situation in which a_1 and a_2 change so that a_1^*, a_2^*, and X^* change with time, whereas R_m and a_0 do not; that is, the environmental changes affect the dynamics of the population nonlinearly. [For causal mechanisms, see section 1.7.3.] Under the above assumptions:

$$X^*(t) = -(\log R_m + a_0)/[a_1(t) + a_2(t)] \qquad (2.46a)$$

$$a_1^*(t) = a_1(t)R_m \qquad (2.46b)$$

$$a_2^*(t) = a_2(t)R_m. \qquad (2.46c)$$

Although not illustrated, we see that a generated series would have a nonlinear trend by (2.46a), and the pattern of fluctuation about the trend would also change by (2.46b and c). Again, a generated series would be doubly nonstationary. A combination of all three situations is, of course, possible.

2.5 THE MORAN EFFECT OF DENSITY-INDEPENDENT FACTORS: INTER-REGIONAL SYNCHRONY OF POPULATION FLUCTUATIONS

To conclude the present chapter, I shall discuss an important effect of density-independent factors that Moran (1953b) suggested.

The famous Hudson's Bay Co. statistics of fur trade since the last century indicate that Canada lynx (*Lynx canadensis*) populations have not only exhibited remarkably regular 10-year cycles, but also, and more intriguingly, have been very well synchronized across all regions of Canada (Chapter 5). Spruce budworm (*Choristoneura fumiferana*) populations, too, tend to fluctuate in unison, although on a much more restricted regional scale (Chapter 9). Probably, if we paid close attention, such synchronous fluctuations among local populations, e.g. a more or less simultaneous occurrence of insect outbreaks over a wide area, would be found to be common rather than unusual in many species from diverse taxa.

Many ecologists and naturalists have sought meteorological factors as a cause of such synchrony. There seems to be one tacit assumption in their minds, i.e. that the cause of population cycles, or that of the occurrence of outbreaks, is also the cause of their synchrony. The proposition is, in other words, that there is one common extrinsic factor that governs the cyclic fluctuations in the populations and, hence, the synchrony among them. Many suggestions in the literature explaining the synchrony of the lynx cycles, for instance, typically include such factors as sunspot cycles, ozone cycles, ultraviolet-ray cycles, forest fire cycles, or even lunar cycles (Chapter 5).

However, there is no convincing evidence which, in principle, supports these suggestions.

If, on the other hand, we abandon the single-causation hypothesis, an entirely new possibility appears. As early as 1953, Moran suggested an important theorem which I restate as below:

> If two regional populations have the same intrinsic (density-dependent) structure, they will be correlated under the influences of density-independent factors (such as climatic factors), if the factors are correlated between the regions.

In particular, Moran pointed out that if the density-dependent structure is linear (as in model (2.6)) the correlation between the regional populations will be equal to that between the local density-independent conditions. This, however, would not hold exactly if the density-dependent structure is nonlinear.

In order to demonstrate Moran's theorem graphically, I compare three series of log population densities $\{X_t\}$ in Fig. 2.22. The series $\{X_t\}$ in graph a is a copy of series IV(7) from Fig. 2.17. Recall that I generated this series using the nonlinear model (2.39b) in which $a_1^* = 0.5$ and $a_2^* = -0.9$, and in which the perturbation effect u_t is an independent, normally distributed random number with mean $= 0$ and variance $= 0.04$ (Fig. 2.14a).

I now generate another series of independent random numbers, $\{v_t\}$, identically distributed as, but uncorrelated with, the series $\{u_t\}$. Then, I generate another series $\{X_t\}$ in graph c using the same model (2.39b) with the identical a_1^* and a_2^* values as in series a, but using the perturbation series $\{v_t\}$ in place of $\{u_t\}$.

I generate one more series $\{X_t\}$ in graph b, using, again, the same a_1^* and a_2^* values in model (2.39b), but using yet another perturbation series $\{w_t\}$. I generate the series $\{w_t\}$ as a blend of 0.8 part $\{u_t\}$ and 0.6 part $\{v_t\}$:

$$w_t = 0.8u_t + 0.6v_t. \tag{2.47}$$

The mixed variable w also has exactly the same distribution as its components, u_t and v_t. The mean of w_t is zero because both u_t and v_t have mean zero. The variance of w_t is also 0.04 because it is the weighted sum of the variances of u_t and v_t, i.e.

$$\text{Var}(w_t) = 0.8^2 \text{Var}(u_t) + 0.6^2 \text{Var}(v_t)$$
$$= (0.64 + 0.36) \times 0.04 = 0.04.$$

Further, w_t is normally distributed because the sum of constant multiples of independent, normally distributed random numbers is again normally distributed.

From (2.47), w_t is, as expected, 80% correlated with u_t and 60% with v_t, although the realized correlations in the simulation are 78.2% and 49.6%,

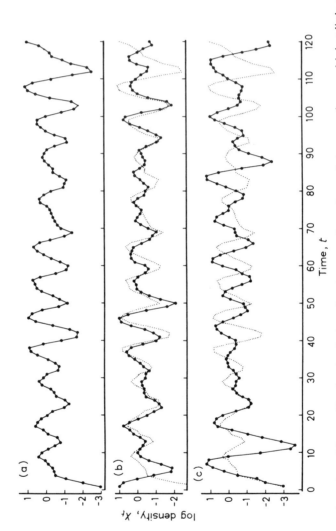

Figure 2.22 A demonstration of the Moran effect by simulation, showing a good synchrony between series a and b but little synchrony between a and c (a is reproduced in dots to compare with b and c), and a moderately good synchrony between b and c.

respectively. As a result, we see a good synchrony in cyclic pattern between series a and b; to a lesser extent between b and c; and little or no synchrony between a and c (synchrony, if any, is coincidental).

The significance of the Moran theorem is that the cause of synchrony can be completely independent of the cause of the cyclic population fluctuation. Thus, we have considerable flexibility in looking around for a possible cause of the cycles and, independently, for one of their synchrony.

3 Statistical analysis of population fluctuations

3.1 INTRODUCTION

In Chapter 2, I have shown various patterns in population dynamics generated by different types of model structures and parameter values. I shall now reverse the procedure and discuss how to guess the generating mechanism of an observed series. This is a much more demanding, challenging task, and we must admit that no method is as powerful or as widely applicable as we wish it to be. We need to utilize all available information, statistical and nonstatistical, such as life-history patterns, food-web structure and environmental influences. In this chapter I shall concentrate on statistical methods and discuss their merits and limitations. Many observed population time series may not be long enough to apply the statistical methods introduced here. Nonetheless, learning the methods will enrich perception of animal population dynamics.

I shall first deal with the analysis of density-dependent structure of a stochastic population process in section 3.2. Correlogram analysis is a useful method, if the data set satisfies certain conditions, e.g., an adequate length of data series. I use linear stationary time-series models to explain the principles of the method, then, move on to the analysis of a nonlinear structure by analogy, and to the analysis of nonstationary processes. I also introduce the method of conditional reproduction curves as complementary to the correlogram analysis.

Section 3.3 is the analysis of the effects of density-independent factors. I compare several different correlation methods and discuss their merits and limitations. The discussion of these methods reveals, in section 3.4, a widespread misconception about the role of climate in population dynamics. Inappropriate methods of data analysis could lead to misinterpretation.

Section 3.5 summarizes the various statistical characteristics of population processes to give a rough guide to diagnosing the statistical status of an observed population series. Finally, section 3.6 introduces some sampling properties of autocorrelations that should be known when carrying out a statistical analysis of time series.

In this chapter, I use rather involved computations in parts of some arguments. A nonstatistically-minded reader may ignore details of such passages at the first reading but should try roughly to grasp the principles.

3.2 ANALYSIS OF DENSITY-DEPENDENT STRUCTURE

3.2.1 Autocorrelation functions (ACFs)

In section 1.5.2, I introduced the concept of stationary stochastic processes and their autocorrelation functions. In particular, as defined in (1.15d), the autocorrelation for lag j of a stationary random process – here, generically denoted by $\{n_t\}$ – is given by

$$\rho_{nn}(j) = \text{Cov}(n_t, n_{t-j})/\text{Var}(n_t)$$
$$= \gamma_{nn}(j)/\gamma_{nn}(0) \tag{3.1}$$

where the autocovariance $\gamma_{nn}(j)$ is the covariance between two values in the series $\{n_t\}$ that are j time steps apart from each other. The Greek letters indicate that these are, as already explained in section 1.5.2, expected values (ensemble averages) of all possible series that the process $\{n_t\}$ potentially generates. The subscripts nn indicate that the random variables concerned are the n. I shall drop the subscripts whenever I see no risk of confusion.

In a stationary series already explained in section 1.5.2, γ and, hence, ρ are functions of the time-lag j but independent of the absolute (or historical) time t. The autocorrelation $\rho_{nn}(j)$ as a function of the lag j is called the autocorrelation function (ACF for short); and the set of $\rho_{nn}(j)$ arranged in the order $j = 0, 1, 2, \ldots$ is called a correlogram. Note that for $j = 0$, $\text{Cov}(n_t, n_{t-j}) = \text{Var}(n_t)$ and, therefore, $\rho_{nn}(0) \equiv 1$; also, as correlation coefficients, $|\rho_{nn}(j)| \leqslant 1$ for any j. In the following, I illustrate the basic nature of ACFs (or correlograms) and their utility in data analysis, using simple linear models discussed in Chapter 2.

Consider first the linear hth-order stochastic density-dependent process

$$R_t = a_1 X_t + a_2 X_{t-1} + \cdots + a_h X_{t-h+1} + u_t \tag{3.2a}$$

where, as before, X_t is the log population density at time t; u_t (the perturbation effect) is an independent, identically distributed random number with zero mean; and $R_t (= X_{t+1} - X_t)$ is the log rate of change in density from time t to $t+1$. [In section 3.2.4, I relax the assumption of the independent perturbation effect.] Thus, the following linear hth-order autoregressive process, AR(h),

$$X_t = (1 + a_1)X_{t-1} + a_2 X_{t-2} + \cdots + a_h X_{t-h} + u_{t-1} \tag{3.2b}$$

is an alternative expression to (3.2a). At a stationary state of a series $\{X_t\}$, the mean of R_t is expected to be zero by the first requirement for population persistence (section 1.3.1); and so is the mean of X_t for all t in (3.2b).

Now, multiply through (3.2b) by $X_{t-j}(j>0)$ and take expectations term-by-term, noticing that $E(X_{t-j}u_{t-1})=0$ and recalling the relationship (1.15b), section 1.5.2, to obtain the following relationship

$$\gamma_{XX}(j)=(1+a_1)\gamma_{XX}(j-1)+a_2\gamma_{XX}(j-2)+\cdots+a_h\gamma_{XX}(j-h). \qquad (3.3)$$

Further, dividing every term in (3.3) by the variance $\gamma_{XX}(0)$, we have the following relationship in terms of the autocorrelations ρ:

$$\rho_{XX}(j)=(1+a_1)\rho_{XX}(j-1)+a_2\rho_{XX}(j-2)+\cdots+a_h\rho_{XX}(j-h). \qquad (3.4)$$

The set of the above equations for $j>0$ is known as the Yule–Walker equations. [I shall now write $\rho_{XX}(j)$ simply as ρ_j.] The set of solutions $\{\rho_j\}$ – in terms of the parameter set $(a_1, a_2, \ldots, a_h)$ and the initial value set $(\rho_0, \rho_1, \ldots, \rho_{h-1})$ – gives the ACF of the linear process $\{X_t\}$ of (3.2b).

Notice now that (3.4) is identical in form with (3.2b) if we set $X \equiv \rho$, $j \equiv t$ and $u_{t-1} \equiv 0$. This means that, with two restrictions, the solutions for ρ_j are the same as those for X_t of the linear pure density-dependent process of order h ($h = 1, 2, \ldots$) already given in section 2.2. The restrictions are: (1) the Yule–Walker equations (3.4) are defined only when the population process $\{X_t\}$ is stationary, so that the set of parameters $(1+a_1, a_2, \ldots, a_h)$ must be within the region of stationarity (e.g. the triangle of Fig. 2.5 for $h=2$); (2) initial states $(X_0, X_1, X_2, \ldots, X_{h-1})$ in (3.2b) are equal to the set $(\rho_0, \rho_1, \ldots, \rho_{h-1})$ which is given by (3.4) when h is specified.

In particular, for the AR(1) process, i.e. $h = 1$ in (3.2b), the solution is given by (2.4), section 2.2.2, in which $X = \rho$, $a_0 = 0$, $X_0 = \rho_0 \equiv 1$, i.e.

$$\rho_j = (1+a_1)^j. \qquad (3.5)$$

The ACF evaluated on the right-hand side of (3.5) plotted against j gives the correlogram of the process concerned. As already shown in section 2.2.2, a linear AR(1) process can be stationary if, and only if, the parameter $(1+a_1)$ is within the interval $(-1, 1)$ – or a_1 is within $(-2, 0)$ – excluding the end values (Fig. 2.2c′ and d′). Thus, there are two major patterns in the ACF generated by (3.5) as the top two graphs in Fig. 3.1: (1) a monotonic exponential decay when $0 < 1 + a_1 < 1$ as in ACF I; (2) exponentially decaying oscillations between positive and negative values when $-1 < 1 + a_1 < 0$ as in ACF II. [The equivalence of (3.5) to (2.4) implies, of course, that Fig. 2.2c′ and d′ should be identical to ACFs I and II in Fig. 3.1, respectively, if the former two series were started with the initial value $X_0 = \rho_0 \equiv 1$.]

For a linear AR(2) process, the Yule–Walker equations are:

$$\rho_j = (1+a_1)\rho_{j-1} + a_2\rho_{j-2}. \qquad (3.6)$$

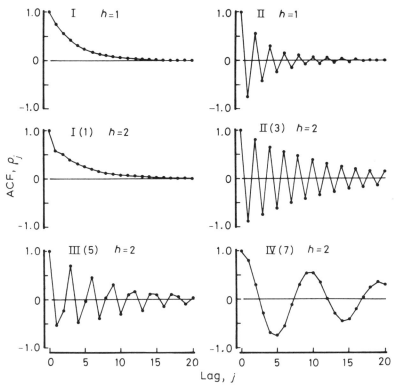

ACF, ρ_j

Figure 3.1   Typical patterns in the autocorrelation functions (ACFs) $\{\rho_j\}$ of the linear process (3.2) up to second order from the four regions (I to IV) of stationarity (Fig. 2.9, inside the triangle); $(h=1)$ and $(h=2)$ indicate the order of density dependence.

We evaluate ρ_1 as follows. Substituting $j=1$ in (3.6) to get

$$\rho_1 = (1+a_1)\rho_0 + a_2\rho_{-1},$$

and noting that $\rho_{-j} = \rho_j$ at a stationary state, we find

$$\rho_1 = (1+a_1)/(1-a_2). \qquad (3.7)$$

The relationship (3.6) is identical in form to the second-order pure density-dependent process (2.7), section 2.2.4. Therefore, the solutions for ρ_j are the same as those for X_t already given by (2.13) with the initial state (X_0, X_1) replaced by $(\rho_0 \equiv 1, \rho_1)$.

Figure 3.1 (bottom four graphs) show the ACFs $\{\rho_j\}$ by (3.6) with the parameter coordinates $(1+a_1, a_2)$ from the four regions I to IV inside the triangle of Fig. 2.9, section 2.2.5. I chose the coordinate sets to be identical with those of the four pure density-dependent processes I(1), II(3), III(5) and IV(7) in Fig. 2.6. Thus, these ACFs would be identical in pattern to the respective series $\{X_t\}$ on the left in Fig. 2.6 if the $\{X_t\}$ were started with the initial densities $X_0 = \rho_0 \equiv 1$ and $X_1 = \rho_1 = (1 + a_1)/(1 - a_2)$.

In short, the ACFs of hth-order linear stationary processes are equivalent to the patterns of their respective pure density-dependent processes $\{X_t\}$ with the initial state $(X_0, X_1, \ldots, X_{h-1}) = (\rho_0, \rho_1, \ldots, \rho_{h-1})$. Thus, a sample ACF calculated from an observed population series can extract the density-dependent structure of the population process (section 3.2.3). [The equivalence between an ACF and pure density-dependent process breaks down when a generating process is subjected to autocorrelated perturbations or if it is nonlinear in density dependence (sections 3.2.4 and 3.2.5).]

Now there is little difference in pattern between ACFs I and I(1), and between II and II(3), in Fig. 3.1. The patterns are similar because each pair originates from the same region of stationarity. Within each pair, though, one is first order ($h = 1$) in density dependence and the other second order ($h = 2$). Evidently, the ACFs do not clearly tell the difference in the order of density dependence. The following partial autocorrelation functions can reveal the difference if, again, certain conditions are met.

3.2.2 Partial autocorrelation functions (PACFs)

The exponential decay of the ACFs in Fig. 3.1 suggests that the correlation between two values in a stationary population series weakens progressively as they become further apart from each other. Nonetheless, they are still correlated with each other. On the other hand, the model (3.2b) suggests that X_t depends only on densities in the past h time steps, X_{t-1}, $X_{t-2}, \ldots, X_{t-h}$. In other words, X_t is structurally independent of the X_i for $i < t - h$ but is correlated with them. Thus, non-zero autocorrelations do not necessarily imply the structural dependence of the current density on the past densities it is correlated with. The structural dependence in the linear AR process (3.2b) will reveal itself in a **partial autocorrelation** function (PACF).

Consider the first-order ($h = 1$) case of (3.2b) in which X_t is structurally dependent only on X_{t-1}; and X_{t-1}, in turn, only on X_{t-2}. But X_t tends to be correlated with X_{t-2} through their correlation with X_{t-1}. Here, X_{t-1} acts as a medium for the correlation between X_t and X_{t-2}, even though X_{t-2} is not structurally dependent on X_t.

However, if we remove the effect of X_{t-1} as the medium, X_t should be no longer correlated with X_{t-2}. In general, X_t in an AR(h) process is not correlated with $X_{t-j}(j > h)$ if the effects of the intermediate values $X_{t-1}, X_{t-2}, \ldots, X_{t-j+1}$ are removed. We can remove the effects by calculating partial autocorrelation coefficients.

Consider the correlation coefficient between X_1 and X_2, written ρ_{12}. Likewise, ρ_{23} and ρ_{13} are the coefficients between X_2 and X_3 and between

X_1 and X_3, respectively. The partial correlation coefficient between X_1 and X_3 that removes the effect of X_2, written $\rho_{13.2}$, is then given by

$$\rho_{13.2} = (\rho_{13} - \rho_{12}\rho_{23})/[(1 - \rho_{12}^2)(1 - \rho_{23}^2)]^{1/2} \tag{3.8}$$

(Kendall and Stuart, 1968; Gottman, 1981; Priestley, 1981). For a stationary series, $\rho_{13} = \rho_2$ (lag by two time steps) and $\rho_{12} = \rho_{23} = \rho_1$ (lag by one step), so that (3.8) is reduced to

$$\rho_{13.2} = (\rho_2 - \rho_1^2)/(1 - \rho_1^2). \tag{3.9}$$

However, as already shown, $\rho_1 = 1 + a_1$ and $\rho_2 = (1 + a_1)^2$ in the stationary AR(1) process (3.5). Therefore, the numerator in (3.9) vanishes; all higher-order partials will vanish as well. Thus, in the AR(1) process, the zero partial autocorrelations reveal that density at one time step is structurally independent of densities more than one time step ago.

For $h = 2$ in the model (3.2b), $\rho_1 = (1 + a_1)/(1 - a_2)$ as already calculated in (3.7). Then, by substituting $j = 2$ in (3.6), we find

$$\rho_2 = (1 + a_1)^2/(1 - a_2) + a_2. \tag{3.10}$$

Further, substituting the above ρ_1 and ρ_2 in (3.9), we get

$$\rho_{13.2} = a_2 \neq 0. \tag{3.11}$$

But, all higher-order partials vanish.

In a stationary series, the partial autocorrelation coefficient between two values j time steps apart from each other is donated by π_j and is a function of the autocorrelation coefficients $\rho_1, \rho_2, \ldots, \rho_j$ of the series; in particular, $\rho_{13.2}$ of (3.9) is π_2, and $\pi_1 \equiv \rho_1$ by definition. In general, for the AR(h) process (3.2b), in which the parameters a_j, $j > h$, are zero,

$$\pi_h = a_h, h > 1,$$
$$= 1 + a_1, h = 1, \tag{3.12}$$

and all higher-order partials π_j, $j > h$, vanish. [For the general formula of π_j, see Jenkins and Watts (1968; section 11.1.1.) or Priestley (1981; section 5.4.5).]

Thus, we can determine the order of density dependence, h, in the linear AR process (3.2b) by observing the lag h beyond which the PACF $\{\pi_j\}$, $j > h$, truncates. [This does not apply when the perturbation effect is autocorrelated (section 3.2.4).] Figure 3.2 illustrates this attribute of PACFs of the process (3.2b) corresponding to the ACFs of the same process in Fig. 3.1.

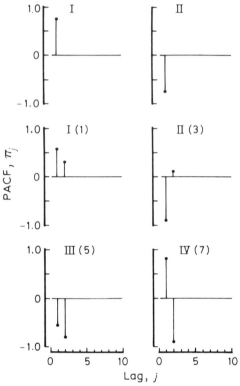

Figure 3.2   Typical patterns in the partial autocorrelation functions (PACFs) $\{\pi_j\}$ of the process (3.2), corresponding to the respective ACFs of Fig. 3.1.

The above arguments show that the ACF and PACF contain necessary information to guess the generating process (3.2b). This is not quite true with a nonlinear process (section 3.2.5). However, we still can, if not always, get some information about the density-dependent structure of a natural (non-linear) population by calculating sample correlograms.

3.2.3 Sample correlograms

Consider a stationary time series of length T, here generically denoted by $\{n_t; t = 1, 2, \ldots, T\}$. A sample autocorrelation coefficient with lag j, written $r_{nn}(j)$, is an estimator of $\rho_{nn}(j)$ and can be defined by the following two most often used formulae:

$$r_{nn}(j) = \frac{\sum_{i=1}^{T-j} (n_i - \bar{n})(n_{i+j} - \bar{n})}{\sum_{i=1}^{T} (n_i - \bar{n})^2} \tag{3.13a}$$

and

$$r_{nn}(j) = \frac{\dfrac{1}{T-j} \displaystyle\sum_{i=1}^{T-j} (n_i - \bar{n})(n_{i+j} - \bar{n})}{\dfrac{1}{T} \displaystyle\sum_{i=1}^{T} (n_i - \bar{n})^2} \qquad (3.13b)$$

where $\bar{n}$ is the sample mean $\sum_{i=1}^{T} n_i / T$. As before, so long as I see no risk of confusion, I shall drop the subscripts nn of the coefficient r and write simply r_j.

I mainly use formula (3.13a) although it tends to be more biased than (3.13b). [(If the mean of n (say μ_n) is known, formula (3.13b) in which $\bar{n}$ is replaced by μ_n, becomes an unbiased estimator.)] However, formula (3.13a) is a more efficient estimator (smaller variance) for a large T (Priestley, 1981; section 5.3.3) than formula (3.13b). Because the two formulae are asymptotically equivalent to each other, efficiency is more important than bias for large T. For a short series, r_j in absolute value larger than unity might even result with formula (3.13b), despite the fact that the theoretical correlation coefficient ρ_j will not exceed it (Kendall and Stuart, 1968); r_j in formula (3.13a) never exceeds unity in absolute value. Formula (3.13b) is historically important because it was first used in statistics for the study of sampling properties (section 3.6). I shall discuss the sampling properties of the r_j in formula (3.13b) in detail in section 3.6. A computer calculates a correlogram very quickly. Check with the software to find which formula it uses.

Now, suppose we calculate an autocorrelation coefficient with a series of length T. Then, the interval $(-2/\sqrt{T}, 2/\sqrt{T})$ – or the Bartlett band (Gottman, 1981)–approximates a 95% confidence interval for a zero autocorrelation. That is, an r falling within the above interval is considered to be not significantly different from zero (Fig. 3.3). As will be explained in section 3.6, in a series of independent, identically distributed random numbers (e.g. series a in Fig. 2.14) of length T, the sample coefficient r_j has the mean $E(r_j) = -1/(T-1)$; is asymptotically normally distributed; and is uncorrelated with r_{j+i} $(i \neq 0)$. The variance of r_j is approximately $1/T$ (Bartlett, 1946). Thus, as well known, twice the standard deviation, i.e. $2/\sqrt{T}$, on each side of the mean $-1/(T-1)$, i.e. $-1/(T-1) \pm 2/\sqrt{T}$, approximates the 95% confidence interval. Further, for large T, we can ignore $-1/(T-1)$ as compared to $\pm 2/\sqrt{T}$.

In Fig. 3.4, I generated four series $\{X_t\}$ of length 100 from regions I to IV of Fig. 2.9, using model (3.2b) for $h=2$. The perturbation series $\{u_t\}$ in the simulation is taken from Fig. 2.14a, which are independent, normally distributed random numbers with the zero mean and 0.04 variance. The parameter sets $(1+a_1, a_2)$ are chosen to be the same as those in the pure density-dependent processes I to IV in Fig. 2.6. Thus, the stochastic series in Fig. 3.4 are the respective pure density-dependent series

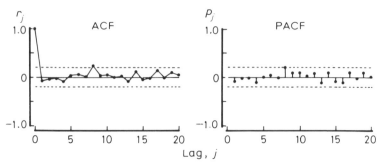

Figure 3.3 Sample ACF $\{r_j\}$ and PACF $\{p_j\}$ of a purely random series of length 100, showing their confinement within the Bartlett band, an approximate 95% confidence limit (dashed lines) for zero correlations. Note that, in a purely random series, $r_j \simeq p_j$.

(Fig. 2.6; left-hand side graphs) subjected to the perturbation series $\{u_t\}$ (Fig. 2.14a).

In Fig. 3.5, I calculated sample correlograms of the series in Fig. 3.4 – ACFs $\{r_j\}$ on the left and PACFs $\{p_j\}$ on the right – using the MINITAB ACF and PACF packages. I calculated these for the section of each series where stationarity is judged (by eye) to have been reached, i.e. $t = 20$ onwards in series I and III, and $t = 10$ onwards in II and IV.

The sample ACFs $\{r_j\}$ compare well in pattern with their respective, expected ACFs $\{\rho_j\}$ in Fig. 3.1 (bottom four graphs). We see some numerical disagreement between the sample and expected values. As will be explained in section 3.6, a sample correlogram converges to the expected one only asymptotically, and we usually do not see a good fit in a series of length less than 100. Qualitatively, however, the sample correlograms clearly reveal the pattern characteristic of the density-dependent component of the process (3.2).

The sample PACFs $\{p_j\}$ (Fig. 3.5; right-hand side) also compare well with their respective expected PACFs $\{\pi_j\}$ in Fig. 3.2 (bottom four graphs). All but PACF II in Fig. 3.5 exhibit p_1 and p_2 significantly different from zero (outside the Bartlett band) as we expect from a second-order process. In PACF II, $p_2 = 0.155$ which is inside the Bartlett band, i.e. insignificantly different from zero. This is easy to explain. As the relationship (3.12) has shown, the second coefficient p_2 is an estimate ($\hat{a}_2$) of the parameter a_2 of the generating model (3.2b) with $h = 2$. Now, when generating series II (Fig. 3.4) by (3.2b), I assigned $a_2 = 0.1$. Thus, $p_2 = 0.155$ is a reasonable estimate of a_2. However, the estimate 0.155 is small enough to fall within the Bartlett band for the data series of length 100, i.e. $\pm 2/\sqrt{100} = \pm 0.2$. In order to make the estimate $\hat{a}_2$ significantly different from zero, the length T of the sample series must be chosen to satisfy the inequality $2/\sqrt{T} < 0.1$, that is, $T > 400$. In actual population studies, it is difficult to get a data series long enough to attain such precision.

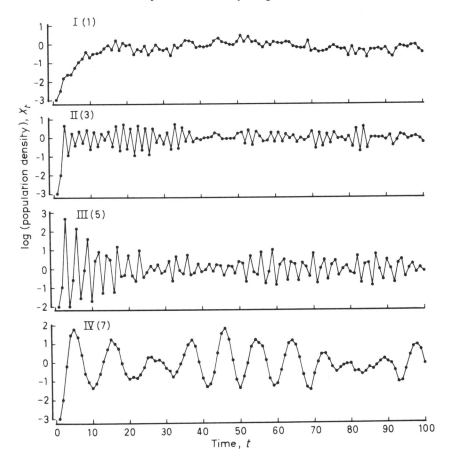

Figure 3.4   Typical examples of the series $\{X_t\}$ generated by the linear process (3.2) for $h=2$, originating at points (Arabic numerals) in regions I to IV of the triangular parameter space of Fig. 2.9.

In PACF IV (Fig. 3.5), not only the first two, but the third coefficient p_3 is outside the Bartlett band, even though p_3 is expected to be zero as an estimate of the parameter $a_3 = 0$. Again, the data series is not long enough for the sample PACF to converge to its expectation. Indeed, the second coefficient $p_2 = -0.783$, considerably overestimates the true value of $a_2 = -0.9$. A further increase in the data length to 200 (not shown) put p_3 within the Bartlett band. At the same time, p_2 was decreased to -0.88 which is now much closer to the theoretical value of -0.9. Also, we should not forget that, with the 95% confidence interval, one in 20 zero coefficients would be out of the interval.

Many population data may not be long enough for sample coefficients to converge adequately. If so, we should limit our analysis to the examination of the pattern, rather than the numerical aspect, of a correlogram. How

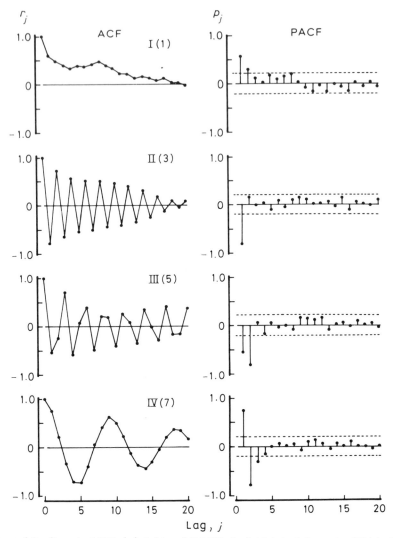

Figure 3.5   Sample ACFs $\{r_j\}$ (left) and PACFs $\{p_j\}$ (right) of the series $\{X_t\}$ in Fig. 3.4. Dashed lines in each graph on the right define the Bartlett band.

many data points are required for drawing a meaningful conclusion depends on the generating process. By analogy with a linear second-order process, a comparatively short series (say, 30 to 40 points) may be acceptable for a process originating in the central area of the triangle in Fig. 2.9. On the other hand, a much longer series would be required for those processes originating in the marginal area. Some populations, e.g. spruce budworm (Chapter 9), exhibit an unusually long cycle of 30 to 40 years. It may require ten such consecutive cycles for a meaningful statistical analysis of the generating process. Evidently, this is impractical. Then, we must resort to non-statistical analysis of such populations, paying attention mostly to qualitative aspects of the process (Chapters 6 and 9).

3.2.4 Correlograms of density-dependent population processes subjected to autocorrelated perturbation

Every linear autoregressive (AR) process discussed in the preceding section is an idealized density-dependent process subjected to a purely random perturbation effect u_t. We could determine the order of density dependence in such a process by examining the lag beyond which the PACF truncates. If, however, a process is subjected to an autocorrelated perturbation effect, this method no longer applies. To explain the principle, I shall use a mixed linear autoregressive-moving average (ARMA) process. But, first, I shall show the correlograms of a simple MA process.

In section 2.4.2, I introduced the MA(k) process (2.36) which takes the form

$$z_t = u_t + b_1 u_{t-1} + \cdots + b_k u_{t-k}. \tag{3.14}$$

Process (3.14) has the correlograms that exhibit just the opposite tendency to those of an AR(h); that is, the ACF truncates after lag k, while the PACF decays asymptotically to zero.

It is easy to see why the ACF truncates. Simply, z_t and z_{t-j} share $k-j$ common elements for $j \leqslant k$, so that they are correlated. Whereas, for $j > k$, there is no common element, i.e. no correlation. Thus, the autocovariance $\gamma_{zz}(j)$ truncates after $j > k$. For instance, in the simplest MA(1) process,

$$\gamma_{zz}(0) = (1 + b_1^2)\sigma_u^2$$
$$\gamma_{zz}(1) = b_1 \sigma_u^2$$
$$\gamma_{zz}(j) = 0, j > 1$$

and, hence, the ACF $\{\rho_{zz}(j)\}$ is:

$$\rho_{zz}(1) = \gamma_{zz}(1)/\gamma_{zz}(0) = b_1/(1 + b_1^2)$$
$$\rho_{zz}(j) = 0, j > 1. \tag{3.15}$$

[Note that the MA(1) process with the parameter equal to $1/b_1$ has the ACF identical to (3.15). To make the ACF unique, we consider the process (3.15) in which $|b_1| < 1$. In general, we only deal with an MA process that satisfies the so-called invertibility condition to make its ACF unique (Box and Jenkins, 1970; Chatfield, 1984).]

The computation of the PACF $\{\pi_{zz}(j)\}$ requires some algebra (Box and Jenkins, 1970, p. 70; Gottman, 1981, p. 173). So the result is shown to be:

$$\pi_{zz}(j) = (-1)^{j+1} b_1^j (1 - b_1^2)/(1 - b_1^{2(j+1)}), j > 0 \tag{3.16}$$

which asymptotically decays to zero. Figure 3.6 shows both ACFs and PACFs of two MA(1) processes, i.e. process (3.14) in which $b_1 = \pm 1$, $b_j = 0$, $j > 1$. [The parameter $|b_1| = 1$ apparently violates the above-mentioned invertibility condition, and the right-hand side of (3.16) is indeterminate. We can circumvent these problems by simply taking the limit $|b_1| \to 1$ by l'Hôpital's rule.]

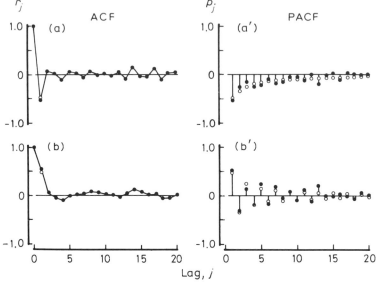

Figure 3.6 The ACFs (left) and PACFs (right) of the MA(1) process, i.e. $k = 1$ in (3.14): $b_1 = 1$ in graphs a and a′, and $b_1 = -1$ in graphs b and b′. Open circles: theoretical values given by (3.15) and (3.16). Solid circles: sample values of the MA(1) series a and b of Fig. 2.18. Notice that the theoretical ACFs in graphs a and b truncate after lag 2.

We now mix AR(h) and MA(k) in one to get ARMA(h, k):

$$X_t = (1 + a_1)X_{t-1} + a_2 X_{t-2} + \cdots + a_h X_{t-h}$$
$$+ u_{t-1} + b_1 u_{t-2} + \cdots b_k u_{t-k-1}. \tag{3.17}$$

This takes the form in which the density-dependent AR(h) process is subjected to the density-independent MA(k) perturbation. The ACF of the above process asymptotically decays to zero because its AR component so decays. The PACF of the above scheme also decays asymptotically to zero because its MA component so decays. Figure 3.7 illustrates such patterns in the correlograms of the ARMA(2, 1) process.

Note that the PACF in Fig. 3.7b′ might look like a truncation, but it is merely decaying very fast. So, we need experience and skill to read and interpret a correlogram. If the reader attempts to do a correlogram analysis, try many computer simulations. It is also important to become familiar with the sampling theory of correlograms (section 3.6).

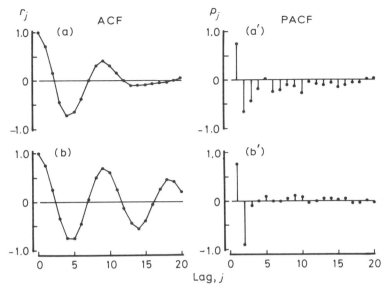

Figure 3.7   Sample ACFs (left) and PACFs (right) of the linear ARMA(2, 1) process $X_t = 1.5X_{t-1} - 0.9X_{t-2} + u_{t-1} + bu_{t-2}$; $b = -1$ in the top pair of graphs and $b = 1$ in the bottom pair. Note that the PACF in graph b′ is an example of a fast but asymptotic decay rather than a truncation after the second lag.

If the calculated PACF did not truncate, it is difficult to determine the order of density dependence. The reader will find some well-defined methods applicable to linear schemes in standard text books of time series analysis (Box and Jenkins, 1970; Gottman, 1981; Priestley, 1981). Unfortunately, their application to the analysis of population time series is limited because population processes are nonlinear in density dependence. We may carefully examine non-statistical features, e.g. food-web structure, to get some idea of the order of density dependence.

I shall now discuss the merits and limitations of the correlogram analysis of nonlinear processes.

3.2.5 Correlograms of nonlinear processes

In Figs 2.14 to 2.17, section 2.4.3, I generated some examples of population series, using the nonlinear second-order model (2.39b), i.e.

$$R_t = 1 - \exp(-a_1^* X_t - a_2^* X_{t-1}) + u_t. \qquad (3.18)$$

Figure 3.8 shows sample ACFs, $\{r_j\}$, and PACFs, $\{p_j\}$, calculated from four of those series, i.e. series I(1) in Fig. 2.14; II(3) in Fig. 2.16; III(5) and IV(7) in Fig. 2.17.

In generating these series, I used the parameter sets $(1 + a_1^*, a_2^*)$ which were equal in value to the sets $(1 + a_1, a_2)$ of the corresponding linear-process

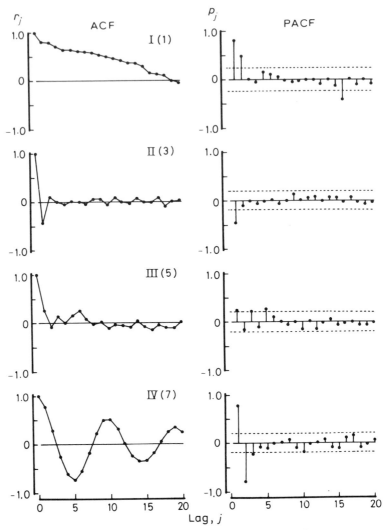

Figure 3.8   Examples of sample ACFs $\{r_j\}$ (left) and PACFs $\{p_j\}$ (right) of the series generated by the nonlinear second-order model (3.18) from regions I to IV of the triangular parameter space of Fig. 2.9.

series in Fig. 3.4. In this sense, the correlograms in Fig. 3.8 are analogous to the respective ones of the linear series in Fig. 3.5.

   The analogy is apparent in series I and IV. Both ACFs and PACFs of nonlinear I and IV (Fig. 3.8) exhibit the characteristic patterns of their respective linear analogues (Fig. 3.5). The ACF of nonlinear I, though, does not decay as fast as that of its linear analogue (Fig. 3.5).

   The analogy is also apparent in series II. However, the ACF of nonlinear II (Fig. 3.8) does not show the typical saw-tooth oscillation of its linear analogue (Fig. 3.5). This is because, as is clear in the pure density-dependent component (Fig. 2.10), nonlinear II reaches equilibrium much faster than does its linear analogue (Fig. 2.6). This property of the density-dependent component of nonlinear II shows up in its ACF (Fig. 3.8), having a significantly negative r_1 followed by a fast decay in r_2 onward. Thus, ACF of nonlinear II (Fig. 3.8) does extract the density-dependent component of its generating process.

Harder to interpret is nonlinear series III, whose ACF and PACF (Fig. 3.8) are both vaguely scattered about the Bartlett band, neither conforming to the patterns in those of its linear analogue (Fig. 3.5), nor to its pure density-dependent component (Fig. 2.10). A major reason for the vagueness is the following.

As is often the case with a nonlinear series generated in a marginal area of region III, series III (Fig. 2.17) exhibits an occasional crash decline (e.g. in the interval between $t = 50$ to 60). When this happens, it takes a while for the population to return to equilibrium. The recovery is slow because the log reproductive rate of the nonlinear process is limited to the maximum value, R_m. Thus, the process can be temporarily out of equilibrium. A sample correlogram is rather sensitive to the occurrence of such temporary non-stationary sections (or outliers) in a time series. This makes the pattern of the correlogram vague and hard to interpret.

We can delve a little more into the above issue by computing the theoretical ACF of a nonlinear system. Consider the following first-order model

$$(X_t - \mu) = (X_{t-1} - \mu) + R_m - \exp[-a(X_{t-1} - \mu)] + u_{t-1} \qquad (3.19)$$

in which μ is the mean of X_t for all t. Now, expand the nonlinear term on the right-hand side in a Taylor series, simplifying the notation $X - \mu$ to X:

$$X_t = X_{t-1} + R_m - (1 - aX_{t-1} + a^2 X_{t-1}^2/2 - a^3 X_{t-1}^3/3! + \cdots) + u_{t-1}. \quad (3.20)$$

Further, multiplying through (3.20) by X_{t-j}, taking expectations term-by-term, noting that $E(X_t) = 0$ for all t, and rearranging, we find

$$\gamma_{XX}(j) = (1 + a)\gamma_{XX}(j-1) - \sum_{i=2}^{\infty} (-a)^i E(X_{t-1}^i X_{t-j}). \qquad (3.21)$$

As compared to the Yule–Walker equations (3.4) of the linear system in which $h = 1$, the analogous equations (3.21) have extra terms comprising the product moments of orders higher than 2. In other words, the sample coefficient $r_{XX}(j)$ depends on the distribution of X.

In a process from region III, the occurrence of a crash decline, followed by a comparatively slow recovery to an equilibrium state, would make the distribution of X especially skewed as compared to that of a linear process from the same region. This might have resulted in the sum of the product moments on the right-hand side of (3.21) greatly influencing $\gamma_{XX}(j)$ on the left-hand side. This is probably why the ACF and PACF of nonlinear III are vague, differing considerably from those of its linear analogue. The involvement of such effects of higher-order moments in a nonlinear process would make the interpretation of a correlogram difficult at times.

If crash declines occur only occasionally, we may calculate a correlogram after carefully removing a nonstationary section after each crash and splicing the remaining, stationary sections together. Such an operation is justified because a nonstationary, recovering section contains little information about the density-dependent structure of the generating process. This cannot be done, of course, if the whole series exhibits a nonstationary trend.

3.2.6 Analysis of nonstationary processes

Nonstationary population processes may arise under three major circumstances: (1) lack of regulation, (2) nonstationary changes in the environment, and (3) growth of a new population on its way to a potential equilibrium state.

Circumstance 1
Random walks illustrated in Fig. 1.4, section 1.3.2, typically exemplify the first circumstance. These are realizations of the process originating at point (12) on the right-hand side of the triangle of Fig. 2.9, i.e. right on the border between region I and I'.

Figure 3.9 (open circles) shows the sample ACF (graph a) and PACF (graph a') of the random-walk series in Fig. 1.4 marked with solid circles. Graphs b and b' (open circles) show, for comparison, the sample ACF and PACF of the stationary (regulated) series I(1) of Fig. 2.14 (these are reproductions of the ACF and PACF I of Fig. 3.8). The correlograms of the two series (open circles in Fig. 3.9) look alike except that PACF(a') truncates after p_1 whereas PACF(b') does so after p_2. But that is all we can tell from the mere appearance of these sample correlograms.

A significant difference between the two processes will show in their ACFs, if we greatly extend the length, say, to $t = 500$ (see the solid circles in Fig. 3.9a and b). The ACF of the random-walk series now decays much more slowly, whereas that of the stationary series decays even faster than the ACF of the shorter version of each series.

To gain insight into the differences, let us look at the theoretical ACF of a random walk. Its generating process is, as already shown in section 1.3.2, $X_{t+1} = X_t + u_t$. Then, after a little algebra, we find:

$$X_{t+j} = \sum_{i=0}^{j-1} u_{t+i} + X_t. \qquad (3.22)$$

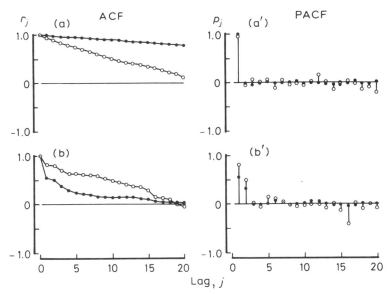

Figure 3.9 Sample ACF $\{r_j\}$ and PACF $\{p_j\}$ of an unregulated (random-walk) population series (top graphs) and those of a regulated series (bottom graphs). Open and solid circles are for the series of length 100 and 500, respectively.

[cf. (1.9).] Thus,

$$\text{Cov}(X_{t+j}, X_t) = \text{Cov}\left(\sum_{i=0}^{j-1} u_{t+i} + X_t, X_t\right) \qquad (3.23)$$

$$= \text{Var}(X_t).$$

Hence, the correlation between X_{t+j} and X_t is expected to be

$$\text{Cov}(X_{t+j}, X_t)/[\text{Var}(X_{t+j})\text{Var}(X_t)]^{1/2} = [\text{Var}(X_t)/\textit{Var}(X_{t+j})]^{1/2} \quad (3.24)$$

which converges to 1 for $t \to \infty$, given j. For small lags j, the right-hand side of (3.24) converges to 1 comparatively quickly. Because the series $\{X_t\}$ is nonstationary, the sample coefficient r_j in (3.13) is not equivalent to the expectation on the right-hand side of (3.24). Nonetheless, the r_j do converge to 1 but only very slowly.

In general, in a nonstationary process, the correlogram of a short series decays faster than a longer series. The opposite is true with a stationary process (for the reason discussed in section 3.6). [Gottman (1981) suggests that $r_j < 1/j$, at least for large j, is a sufficient condition, or a rough guide, for stationarity.] So, if we are unable to obtain a longer series, we might divide the given observed series into two or more sections, calculate ACF in each section, and see if the ACF of the whole series decays more slowly or faster than those of its shorter sections.

Another, often better way to discern a nonstationary population process is to look at the ACF of the log reproductive rates $\{R_t\}$, instead of that of log population densities $\{X_t\}$. If $\{X_t\}$ is a pure random walk, like those in Fig. 1.4, $\{R_t\}$ should be a pure random series and, hence, its correlogram should be confined within the Bartlett band 95% of the time (Fig. 3.10a). If, however, $\{X_t\}$ is a generalized random walk, like the one in Fig. 1.8b, the ACF of $\{R_t\}$ should be significantly different from zero in the first (few) lag(s)>0 and should truncate for further lags (Fig. 3.10b).

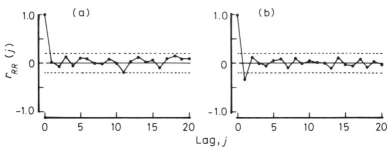

Figure 3.10 Sample ACFs $\{r_{RR}(j)\}$ of the series of log reproductive rates $\{R_t\}$. Graph a: the pure random-walk series in Fig. 1.4 (series of solid circles). Graph b: the generalized random-walk series of Fig. 1.8b.

Circumstance 2
Examples in Figs 2.20 and 2.21, generated by model (2.41), section 2.4.5, illustrate nonstationary population processes when the environment is changing with time, favouring a steady population increase. In Fig. 2.20 we assumed that the maximum log reproductive rate $R_m(t)$ increased with time, resulting in an upward shift in the reproduction surface and, hence, in an increase in the equilibrium point X^*. In Fig. 2.21, on the other hand, we assumed that the parameter $a_0(t)$ increased with time, while R_m remained constant. This resulted in a lateral shift in the reproduction surface and, as a result, X^* increased as much as $a_0(t)$. Consequently, the series $\{X_t\}$ in both processes fluctuated about a linearly increasing trend.

Figure 3.11 shows sample ACFs of these nonstationary series. The mainly positive ACFs in graphs a and b are due to the trend in the series, and their wavy pattern reflects the oscillatory population fluctuations about the trend. On the whole, however, the ACFs of the nonstationary series $\{X_t\}$ by themselves do not give much useful information.

There are two ways of dealing with a nonstationary series of the above sort: (1) detrending the series by fitting an appropriate trend line (or a curve), or (2) differencing neighbouring values in the series $\{X_t\}$. The first method requires guessing the nature of the trend and the estimation of the parameters, e.g. by least squares. The second method is more straightforward than the first. Also, it makes ecological sense because, by differencing (recalling that $R_t = X_{t+1} - X_t$), we get the series of log reproductive rates $\{R_t\}$.

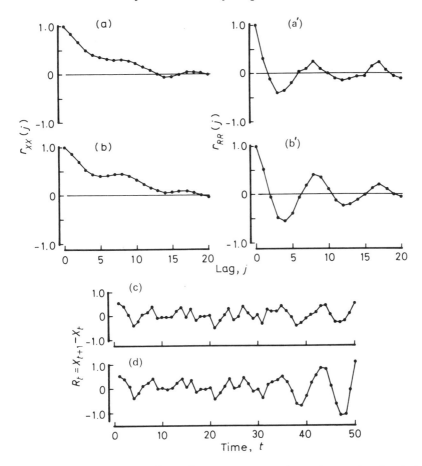

Figure 3.11 Sample ACFs $\{r_{XX}(j)\}$ (graph a) and $\{r_{RR}(j)\}$ (graph a'), of the nonstationary series $\{X_t\}$ in Fig. 2.21a. Graph c: the differenced series $\{R_t\}$. Graphs b, b' and d: the same as graphs a, a', and c but for the series in Fig. 2.20b.

In the example of Fig. 2.21a, the series $\{X_t\}$ was nonstationary because the expectation $E(R_t)$ was a positive constant. Thus, the generating process violated the stipulation (1.8) for no trend but complied with the stipulation (1.23) for regulation. Thus, detrending the $\{X_t\}$ by the differencing method, the resultant series $\{R_t\}$ (Fig. 3.11c) is stationary both in the mean and in the variance. Since the ACF of a stationary time series is, by definition, unaffected by its constant expectation, the ACF of the $\{R_t\}$ (Fig. 3.11a') would be identical to that of $\{R_t\}$ when the series $\{X_t\}$ had no trend, i.e. when $E(R_t)=0$. On the other hand, the example of Fig. 2.20a violated both stipulations. Thus, after differencing the $\{X_t\}$, the resultant $\{R_t\}$ (Fig. 3.11d) is stationary in the mean but still nonstationary in the variance.

By differencing, we can effectively detrend the series $\{X_t\}$ without in-fluencing its autocorrelation structure. [The ACF of $\{R_t\}$ maps the ACF of a detrended series $\{X_t\}$ by the relationship

$$r_{RR}(j) = \{2r_{XX}(j) - [r_{XX}(j+1) + r_{XX}(j-1)]\}/2[1 - r_{XX}(1)]. \qquad (3.25)$$

This is obtained by calculating the autocorrelation coefficients for lag j on both sides of the relationship $R_t = X_{t+1} - X_t$.]

In Fig. 3.11, ACF b' of nonstationary series d still looks much the same as ACF a' of stationary series c. This is because the length of the series ($t = 50$) is rather short. As already discussed in *Circumstance 1*, if we extend the observation, we would see that ACF a' of stationary series c damps down faster, whereas ACF b' of nonstationary series d does so more slowly.

Circumstance 3
Figure 3.12 illustrates a nonstationary series when it is growing to equilib-rium (or stationary state). Graph a is a realization of the nonlinear population model (2.39b), i.e.

$$R_t = 1 - \exp(-a_1^* X_t - a_2^* X_{t-1}) + u_t \qquad (3.26)$$

with $(1 + a_1^*, a_2^*)$ from region IV (Fig. 2.9). The population series $\{X_t\}$ in graph a starts with initial densities far below the equilibrium level which is reached around the 55th time step. Thereafter, the series maintains a stationary state. Graph c is the ACF of the growing section of the series before $t = 55$; graph c' is the ACF of the stationary section.

By differencing the population series in graph a, we obtain the series of log reproductive rates $\{R_t\}$ in graph b. The growing and stationary sections of series a correspond to two trend-free sections in series b, the first one fluctuating about the mean 1, and the second about 0. The ACF of the first section of series $\{R_t\}$ (graph d) is confined within the Bartlett bands 95% of the time, indicating that this section of $\{R_t\}$ is a series of uncorrelated random numbers. This, in turn, means that the fluctuation in series a, before it reaches the equilibrium level, is a random walk (though its drift is minimal) with an increasing trend set by the positive mean ($= 1$) reproductive rate.

[The random walk of the $\{X_t\}$ is due to the fact that until it reaches the equilibrium level, the density-dependent term in model (3.26) exerts practi-cally no influence on R_t. This is because, for a low X_t, $R_t = 1 + u_t$ practically, i.e. $\{R_t\}$ is a series of independent random numbers with the mean 1 and the variance σ_u^2. After $t = 55$, the ACF of $\{R_t\}$ (graph d') maps the oscillating ACF (graph c') of the stationary section of $\{X_t\}$ by the relationship (3.25).]

In the above simulation with model (3.26), I conveniently set the maxi-mum log reproductive rate, R_m (cf. (2.39a)), equal to 1 and the variance σ_u^2 of the perturbation effect u to 0.04. If R_m is much lower, even if positive, and the variance σ_u^2 higher, then the population, starting at a very low density

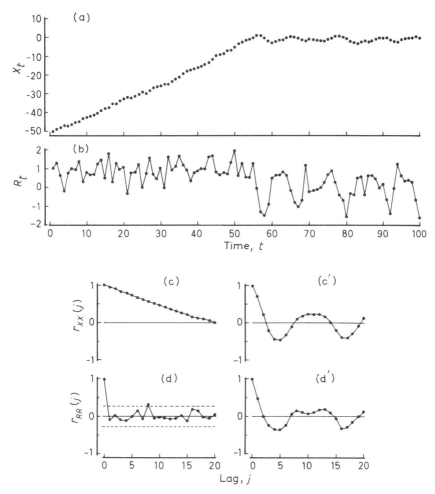

Figure 3.12   A series of log population density $\{X_t\}$ growing to an equilibrium state (graph a) and its reproductive rates $\{R_t\}$ (graph b). Correlograms of both series are calculated separately for the two sections (before and after reaching the equilibrium state): $r_{XX}(j)$ in graphs c and c', and $r_{RR}(j)$ in d and d'.

level, could perform a random walk in a typical manner and might drift further down with little sign of recovery. This implies that a population with a sufficiently low potential reproductive rate may readily become extinct once, somehow, reduced to a very low level. This could be the case of an endangered species in a marginally tolerable environment (cf. Fig. 1.11).

3.2.7 Conditional reproduction curves

The correlogram analysis I have discussed in the preceding sections plays two major roles in the analysis of density-dependent structure of an observed series. It allows us to guess both the order of density dependence and the parameter set if we know the process can be approximated by a simple model, e.g. (2.20), section 2.2.5. However, by a correlogram analysis alone, we are unable to visualize the shape of a nonlinear reproduction curve or surface. So we are unable to decide which model is appropriate to describe the observed process. This is because two series of different density-dependent relationships may have similar correlograms. The analysis of conditional reproduction curves is a second step towards the determination of the density-dependent structure of a population process.

If a sample correlogram indicates that the observed series is stationary and first order in density dependence, we may be able to choose an appropriate model of the reproduction curve by simply regressing the log reproductive rate R_t $(=X_{t+1}-X_t)$ against the log density X_t. If, on the other hand, a second-order density dependence has been indicated, we need to plot R_t against X_{t-1} as well as against X_t, that is, to construct a reproduction surface in a three-dimensional space. [The simple regression of R_t against X_t, commonly used in the literature quoted in section 1.8, does not give us much insight. It yields a projection of the surface onto the (R_t, X_t) plane (cf. Fig. 2.13), which often produces only a vague cloud of data points.] One way to get around the cumbersome task of constructing a three-dimensional graph is to estimate conditional reproduction curves in several two-dimensional graphs.

Suppose that the density-dependent structure of an observed series $\{X_t\}$ is second order and a perturbation effect is uncorrelated. Then,

$$R_t = f(X_t, X_{t-1}) \tag{3.27}$$

represents a regression surface, such as the one illustrated in Fig. 2.8. If we slice the surface vertically at a given value of $X_{t-1} = X$, say, and parallel to the (R_t, X_t) plane, then the profile of the slice represents the conditional reproduction curve

$$R_t = f(X_t | X_{t-1} = X). \tag{3.28}$$

Similarly, by fixing X_t at X and slicing the surface vertically parallel to the (R_t, X_{t-1}) plane, we have

$$R_t = f(X_{t-1} | X_t = X). \tag{3.29}$$

In an observed series of limited length, however, there would not be enough data-point sets at the given value X. So, usually we have to select an interval, say $X_a < X < X_b$, on the X_t or X_{t-1} axis. This slices the reproduction surface to the thickness $(X_b - X_a)$. If the slice is too thin, it would not contain enough data points and, if too thick, would lose resolution. Therefore, we must choose an apropriate thickness according to the sample size.

Figure 3.13 shows a family of conditional reproduction curves of the simulated series IV'(8) of Fig. 2.17. The left-hand graphs show R_t regressed on X_t given the unit interval of X_{t-1}, and the right-hand graphs, R_t on X_{t-1} given the unit interval of X_t. The theoretical curves on the extreme values X_a and X_b in each interval are calculated from the series-generating model $R_t = 1 - \exp(-0.5X_t + 1.05X_{t-1})$. We see much tighter formations of data points in the graphs on the right than in those on the left. The difference is due to differences in steepness between the partial slopes of the reproduction surface along the X_t and X_{t-1} axes.

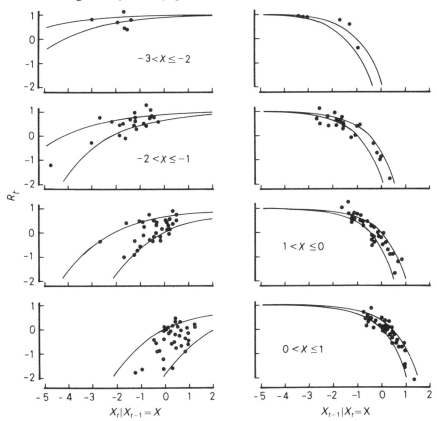

Figure 3.13   A family of conditional reproduction curves with simulated data from a nonlinear second-order density-dependent process.

If an observed series $\{X_t\}$ is nonstationary, we may not obtain a reasonably well-defined profile as it shifts from time to time. If there is a trend in the series $\{X_t\}$, remove it by fitting a trend line. [Differencing $\{X_t\}$ is obviously inappropriate here.] The profile should be more sharply defined after the trend is removed. Specially designed experiments or field observations may be useful in defining a profile (section 4.6). If an observed series is temporarily performing a random walk, as in *Circumstance 3* in section

3.2.6, we cannot determine a reproduction curve, because such a random walk contains no information about the density-dependent structure of the generating process.

3.3 ANALYSIS OF DENSITY-INDEPENDENT PERTURBATION EFFECTS

Suppose we suspect a given density-independent factor (e.g. a climatic factor) to be a source of perturbation of an observed population time series, and that we wish to test that hypothesis. Our approach depends on the type of information available.

1) Suppose that the information includes a series of yearly game statistics and meteorological records during the same time period. To see if there is any climatic influence on the game population, we measure a correlation between these two series of records or between their appropriate transformations. This method can be effective, however, only when one factor is a dominating source of the perturbation effect.

2) If there are many perturbation factors, we may look at a particular stage of the life cycle (or a period of the year) in which only one major source of perturbation operates. We measure a correlation between the survival (or mortality) and the perturbation effect during that stage or period. This requires life-table information.

3) Should there be enough information to permit a reasonably precise estimation of the reproduction surface, $R_t = F(\mathbf{X})$, we can compare the deviations of data points from the surface with the fluctuation in the suspected source of perturbation effect. In this case, even a multivariate analysis can be employed if we suspect the involvement of several different factors.

In this section, I shall discuss the first two situations, using a simple and idealized linear model to show the principles involved in the analysis. The feasibility of the third situation depends on the construction of a specific model of the density-dependent structure which I discuss in Chapter 4 as well as in Part Two.

3.3.1 Correlations between population fluctuation and perturbation effects

Consider the stationary linear second-order density-dependent process

$$R_t = a_1 X_t + a_2 X_{t-1} + z_t \tag{3.30a}$$

or (noting that $X_{t+1} = R_t - X_t$) in the form of the autoregressive AR(2) process

$$X_{t+1} = (1 + a_1)X_t + a_2 X_{t-1} + z_t \tag{3.30b}$$

in which z_t is the net perturbation effect. Consider further that z_t is largely determined by the effect, say u_t, of the factor U whose measurement is U_t. [In practice, we take u_t to be a suitable transformation of U_t, the simplest being, of course, a linear transformation.] Then, we can write

$$z_t = u_t + v_t \qquad (3.31)$$

in which v_t is the residual.

In order to discuss the principles involved in the correlation methods, I shall assume that: (1) the residual effect v_t is negligible and that u_t is an independent, identically distributed random number with the zero mean and the variance σ_u^2; (2) the log population series $\{X_t\}$ is stationary, that is, the parameter set $(1+a_1, a_2)$ is inside the triangle of Fig. 2.9.

There are several ways of detecting the influence of the perturbation factor by correlation. I shall first look at a simple correlation between X_{t+1} and u_t and one between R_t and u_t and, then, elaborate them to two other more effective methods.

1. *Correlation between X_{t+1} and u_t: $\rho(X_{t+1}, u_t)$*
The correlation coefficient is defined by

$$\rho(X_{t+1}, u_t) = \frac{\mathrm{Cov}(X_{t+1}, u_t)}{\sqrt{\mathrm{Var}(X_{t+1})\mathrm{Var}(u_t)}}. \qquad (3.32)$$

It can be readily verified that, in (3.30b) in which $z_t = u_t$,

$$\mathrm{Cov}(X_{t+1}, u_t) = \mathrm{Var}(u_t) = \sigma_u^2, \qquad (3.33)$$

$$\mathrm{Var}(X_{t+1}) = \frac{(1-a_2)\sigma_u^2}{(1+a_2)(a_1+a_2)(a_2-a_1-2)}. \qquad (3.34)$$

Substituting (3.33) and (3.34) in (3.32), we find

$$\rho(X_{t+1}, u_t) = \frac{\sqrt{(1+a_2)(a_1+a_2)(a_2-a_1-2)}}{\sqrt{1-a_2}}. \qquad (3.35)$$

[For details, see Royama (1981a).]

Figure 3.14a shows the contour lines on which $\rho(X_{t+1}, u_t) = 0.1, 0.2, \ldots,$ 0.9. We see that the correlation is perfect (i.e. 1) at the centre of the triangle, i.e. $1+a_1 = a_2 = 0$, at which, from (3.30b), $X_{t+1} = u_t$. In other words, at this point of the parameter space, the population is typically regulated, and the effect u_t reveals itself as the fluctuation in log population density X_{t+1}. The correlation deteriorates towards all sides of the triangle where the effect of

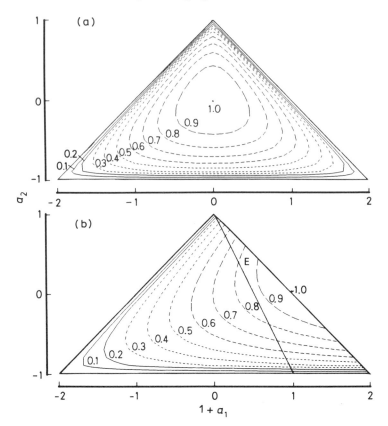

Figure 3.14 The contours of correlation coefficients $\rho(X_{t+1}, u_t)$ (graph a) and $\rho(R_t, u_t)$ (graph b) as functions of the parameters $1 + a_1$ and a_2, within the triangular region of stationarity (Fig. 2.5) of the linear process (3.30) in which $z = u$ (after Royama (1981a; figs 8 and 9)). Line E in graph b is the intersection of the graphs a and b on which the two coefficients are equal to each other.

the density-dependent component masks the effect of u_t on the population fluctuation.

Thus, if the correlogram analysis in the preceding section indicates that the observed population originates in a marginal area of the stationarity region, we cannot decide, upon finding a low correlation, whether the suspected factor has an influence on the population or is an irrelevant factor. So, we need a better indicator.

2. Correlation between R_t and u_t: $\rho(R_t, u_t)$
In model (3.30), the log population density X_{t+1} is determined by the log reproductive rate R_t, given X_t; R_t is in turn influenced by u_t in addition to the influences of X_t and X_{t-1}. So, it would seem more natural to look at a correlation between R_t and u_t rather than the one between X_{t+1} and u_t in

the first method. To do this, define the correlation coefficient $\rho(R_t, u_t)$ by substituting R_t for X_{t+1} in (3.32). Then,

$$\rho(R_t, u_t) = \sqrt{(1 + a_2)(2 + a_1 - a_2)/2} \qquad (3.36)$$

(Royama, 1981a).

Figure 3.14b shows the contour lines on which $\rho(R_t, u_t) = 0.1, 0.2, \ldots, 0.9$. We see that the correlation is perfect at the mid-point on the right-hand side of the triangle where $a_1 = a_2 = 0$, i.e. point (12) in Fig. 2.9. There, $R_t = u_t$ as is obvious from (3.30a), and the series $\{X_t\}$ is a random walk. In other words, the effect of u_t reveals itself perfectly as the fluctuation in $\{R_t\}$ when the series $\{X_t\}$ is typically unregulated. The line E in graph b marks the intersect on which $\rho(X_{t+1}, u_t) = \rho(R_t, u_t)$. On the left of E, $\rho(X_{t+1}, u_t) > \rho(R_t, u_t)$, and otherwise on the right. In other words, $\rho(X_{t+1}, u_t)$ is a better indicator than $\rho(R_t, u_t)$, on the left of line E and otherwise on the right.

Thus, by looking at both correlation coefficients, we can reveal the influence of u_t in a wider area in the parameter space. However, the problem of uncertainty still remains towards the left-hand side and the base of the triangle. We need a more effective indicator.

3. *Dichotomous nominal scale correlations*: $\delta(X_{t+1}, u_t)$ and $\delta(R_t, u_t)$
This method is a step towards better indicators which follow.

Figure 3.15 compares the series $\{u_t\}$ with $\{X_{t+1}\}$ as well as with $\{R_t\}$ of the linear model (3.30) in which I conveniently chose the parameter set to be $(1 + a_1 = 1.5, a_2 = -0.9)$. Also, in order to enhance the perturbation effect, I

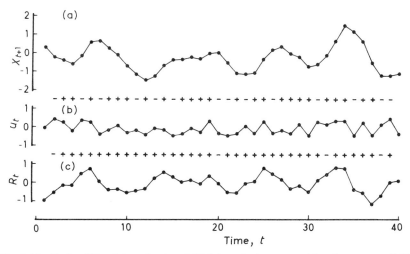

Fig. 3.15   Series $\{X_{t+1}\}$ (graph a) and $\{R_t\}$ (graph c) generated by the process (3.30) in which $z_t = u_t$ (graph b) as independent, uniformly distributed random numbers. For the plus and minus signs, see text.

chose u_t to be uniformly distributed in the interval $(-0.5, 0.5)$. From Fig. 3.14, we see that, with these parameter values, the correlations $\rho(X_{t+1}, u_t)$ and $\rho(R_t, u_t)$ are expected to be as low as 0.268 and 0.412, respectively. However, the correlations described below tend to score better.

First, look at any three consecutive points, e.g. X_t, X_{t+1}, and X_{t+2} in Fig. 3.15a and observe whether the slope of the segment $\overline{X_{t+1}X_{t+2}}$ deflects clockwise or counterclockwise from the preceding segment $\overline{X_tX_{t+1}}$. Now, look at the corresponding three points in the series $\{u_t\}$ (graph b) and observe the way the slope of segment $\overline{u_tu_{t+1}}$ deflects from $\overline{u_{t-1}u_t}$. If the direction of deflection, clockwise or counterclockwise, agrees between the two series, it is marked with a plus sign, and with a minus sign if it disagrees. The scores of the two signs are called dichotomous nominal scale data (Zar, 1984; section 19.10).

We can define the new correlation coefficient between the series $\{X_{t+1}\}$ and $\{u_t\}$ by

$$\delta(X_{t+1}, u_t) = (p-q)/(p+q) \tag{3.37}$$

where p is the number of agreements and q, disagreements (Ives and Gibbons, 1967). We can similarly define $\delta(R_t, u_t)$. As a correlation coefficient, $|\delta| \leqslant 1$; $\delta = 1$(i.e. $q = 0$) indicates perfect agreement and $\delta = -1$ ($p = 0$), perfect disagreement. If the two series are irrelevant, p is expectedly equal to q and, hence, δ is expected to be 0.

In the present example in Fig. 3.15, $p = 24$ and $q = 14$ between graphs a and b so that $\delta(X_{t+1}, u_t) = 0.26$ as compared with the sample value of $\rho(X_{t+1}, u_t) = 0.23$. Between graphs c and b, $p = 34$ and $q = 4$, so that $\delta(R_t, u_t) = 0.79$ as compared with the sample value of $\rho(R_t, u_t) = 0.37$.

Now, by making the above nominal scale data quantitative, we find even better indicators.

4. *Correlations between second differences:* $\rho(\Delta^2 X_{t+1}, \Delta^2 u_t)$ *and*
   $\rho(\Delta^2 R_t, \Delta^2 u_t)$
Let us define the first-order difference operator Δ such that

$$\Delta X_{t+1} = X_{t+1} - X_t \tag{3.38}$$

and, further, the second-order difference operator Δ^2 such that

$$\Delta^2 X_{t+1} = \Delta X_{t+1} - \Delta X_t. \tag{3.39}$$

[Although ΔX_{t+1} in (3.38) is equal to R_t, I use the difference operator Δ to make the symbolism consistent in the present argument.]

As illustrated in Fig. 3.16, if the segment $\overline{X_tX_{t+1}}$ deflects counter clockwise from the preceding segment $\overline{X_{t-1}X_t}$, then $\Delta X_{t+1} > \Delta X_t$ (upper graph), and otherwise if it deflects clockwise (lower graph). This means, of

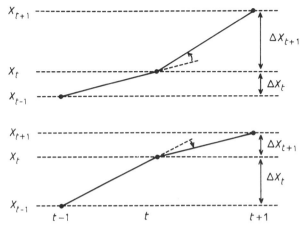

Figure 3.16   Diagrams illustrating that the sign of the second difference $\Delta^2 X_{t+1} = \Delta X_{t+1} - \Delta X_t$ indicates the slope of the segment $\overline{X_t X_{t+1}}$ being greater or lesser than the slope of the preceding segment $\overline{X_{t-1} X_t}$ (adapted from Royama (1981a; fig. 7)).

course, that the second difference $\Delta^2 X_{t+1}$ is positive in the former case and is negative in the latter case. Thus, an agreement or a disagreement in the direction of deflection between the series $\{X_{t+1}\}$ and $\{u_t\}$ is indicated by a correlation between $\Delta^2 X_{t+1}$ and $\Delta^2 u_t$. Similarly, we can take correlation between $\Delta^2 R_t$ and $\Delta^2 u_t$. The coefficients of these correlations in model (3.30) are, after some algebra (Royama, 1981a):

$$\rho(\Delta^2 X_{t+1}, \Delta^2 u_t) = (a_1^2 - 2a_1 + a_2 + 3)\frac{\sqrt{(1+a_2)(2+a_1-a_2)}}{\sqrt{12(2-a_1+a_2)}} \qquad (3.40)$$

$$\rho(\Delta^2 R_t, \Delta^2 u_t) = (a_1^2 - 3a_1 + a_2 + 6)\frac{\sqrt{(1+a_2)(2+a_1-a_2)}}{\sqrt{12[a_1^2 - (a_2+3)a_1 + 5a_2 + 6)]}}$$

$$(3.41)$$

Figure 3.17 shows the contour lines on which the above coefficients are equal to 0.1, 0.2, ..., 0.9. The two coefficients are equal on curve E in graph b; on the left of E, $\rho(\Delta^2 X_{t+1}, \Delta^2 u_t) > \rho(\Delta^2 R_t, \Delta^2 u_t)$; and otherwise on the right.

Comparing the contour lines in Fig. 3.17 with those in Fig. 3.14, we see that a combined use of the coefficients (3.40) and (3.41) detects the influence of the perturbation effect u on the population dynamics more effectively (larger coefficient values) than the simpler coefficients (3.35) and (3.36).

Table 3.1 compares the results of the above correlation methods applied to the series in Figs 2.14, 2.16 and 2.17 which were generated by the nonlinear second-order model (2.39b). In each series, I calculated the coefficients for a section (after $t = 10$ or 20) that was judged (by eye) to have

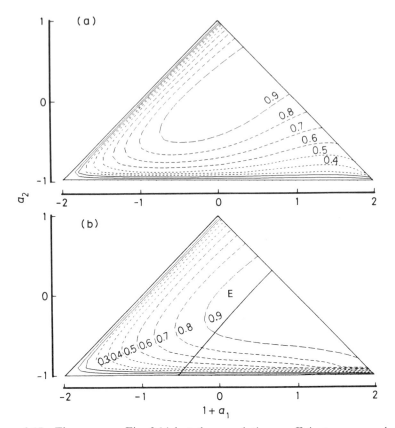

Figure 3.17   The same as Fig. 3.14 but the correlation coefficients concerned are $\rho(\Delta^2 X_{t+1}, \Delta^2 u_t)$ (graph a) and $\rho(\Delta^2 R_t, \Delta^2 u_t)$ (graph b).

reached a stationary state. With the exception of series III(5), either $\rho(\Delta^2 X_{t+1}, \Delta^2 u_t)$ or $\rho(\Delta^2 R_t, \Delta^2 u_t)$ gives the highest score in all series tested. However, the scores in the series from regions III and III′ are still comparatively low as we expect from the analogy with the linear model. Therefore, we must interpret the correlation coefficients carefully in relation to the density-dependent structure of the process concerned. Particularly important to keep in mind is that the degree of correlation does not indicate the degree of influence from the density-independent factor. The influence was the same in every one of the above examples, but the correlation varied because of the variation in the parameter values that characterize the density-dependent component of the process.

3.3.2 Utilization of life-table information

If life-table information is available, we may be able to make use of the correlation methods more effectively. By partitioning one generation into

Table 3.1 Relative effectiveness of the different types of correlation coefficients for revealing the involvement of a given density-independent factor as a major source of perturbation of an observed population series, tested on the series generated by the nonlinear second-order model (2.39b) from regions I to IV (Roman numerals) and parameter points (Arabic numerals) in the triangular parameter space of Figure 2.9.

Region Fig.	$\rho(X_{t+1}, u_t)$	$\rho(R_t, u_t)$	$\delta(X_{t+1}, u_t)$	$\delta(R_t, u_t)$	$\rho(\Delta^2 X_{t+1}, \Delta^2 u_t)$	$\rho(\Delta^2 R_t, \Delta^2 u_t)$
I(1) 2.14	0.778	0.856	0.980†	0.800	0.984*	0.936
I(13) 2.14	0.387	0.870	0.959†	0.800	0.976*	0.911
II(3) 2.16	0.603	0.453	0.694†	0.469	0.706*	0.589
II'(4) 2.16	0.717†	0.450	0.408	0.265	0.822*	0.647
III(5) 2.17	0.081	0.042	0.429*	0.265	0.326†	0.230
III'(6) 2.17	0.267	0.245	0.367	0.184	0.532*	0.444†
IV(7) 2.17	0.218	0.563	0.204	0.633†	0.359	0.805*
IV(8) 2.17	0.183	0.338	0.224	0.469†	0.214	0.577*

*highest score.
†second highest score.

several distinct life-cycle stages, we can minimize the number of density-independent factors operating in each stage. Then, we take correlation between a suspected major factor in one stage and the survival rate during that stage.

For simplicity, assume that all individuals of the population concerned breed more or less at the same time and only once a year. Consider that we divide one generation into n successive stages and determine population density at the onset of each stage. Let us also assume that the recruitment of second generation individuals (e.g. eggs) takes place at the onset of the last stage n.

Let X_{st} denote the log density at the beginning of the sth stage of the tth generation; and let H_{st} be the log survival rate during the sth stage of the tth generation, i.e.

$$H_{st} = X_{s+1,t} - X_{st}, \quad s = 1, 2, \ldots, n-1, \tag{3.42a}$$

and

$$H_{nt} = X_{1,t+1} - X_{nt} \tag{3.42b}$$

where H_{nt} is the log recruitment rate to the beginning of the $t+1$st generation. In a univoltine species, the H_s. $(s < n)$ are usually negative because mortality tends to reduce the density during these stages, except when substantial immigration occurs. The log recruitment rate H_n is, of course, positive unless substantial emigration occurs.

With the above scheme of stage division, the log reproductive rate measured from stage s of generation t to the same stage of generation $t+1$ (specified by R_{st}) is partitioned as follows:

$$\begin{aligned} R_{st} &= X_{s,t+1} - X_{st} \\ &= H_{st} + H_{s+1,t} + \cdots + H_{nt} + H_{1,t+1} + \cdots + H_{s-1,t+1}. \end{aligned} \tag{3.43}$$

Figure 3.18 illustrates the above situation.

The total effect of perturbations, say z_{st}, during the same period is also partitioned, corresponding to R_{st} in (3.43), into n components by stages, each component (say, w) influencing survival rate during the stage concerned. Thus,

$$z_{st} = w_{st} + w_{s+1,t} + \cdots + w_{nt} + w_{1,t+1} + \cdots + w_{s-1,t+1}. \tag{3.44}$$

With an appropriate stage division, we can minimize the number of influential perturbation factors operating during each stage. Then, a correlation between H_{st} and w_{st} can provide more specific information than an unpartitioned data set.

There are some important considerations to make in partitioning the life cycle. Suppose that we can divide the life cycle into distinct metamorphic

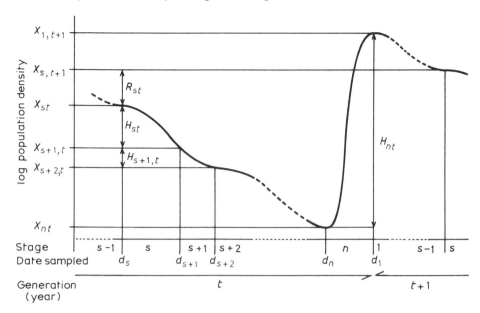

Figure 3.18 A schematic diagram illustrating the relationship (3.43) of life-table data. The curve represents seasonal changes in population density.

stages (e.g. egg, larva, pupa and adult). Because of the distinct morphological characteristics, we can often count the number of individuals that enter or exit a given stage. These metamorphic stages are then natural divisions of the life cycle.

If, however, stages are not distinct, or are difficult to distinguish in practice, the division of 'stages' could become more or less arbitrary. In such circumstances, we should take special precautions because estimates of the survival rates in two adjacent stages may no longer be mathematically independent of each other (Royama, 1984).

Suppose that we choose d_s, d_{s+1} and d_{s+2} to be the sampling dates that frame the stages s and $s+1$ (Fig. 3.18). Suppose further that d_s and d_{s+2} lie in the periods in which population changes are comparatively small. Obviously, a small shift in these dates would not much influence the assessment of the survival rate in the stages concerned. However, d_{s+1} lies in the period during which the population is decreasing rapidly. Then, a shift in d_{s+1} to the left (earlier sampling) makes an estimated log survival rate H_s high and H_{s+1} low, and vice versa if sampling is late. In other words, if the framing of the stages is arbitrary, there would be a negative correlation between the estimates of H_s and H_{s+1}.

The problem is that a negative correlation might also result from a genuine density-dependent process between the adjacent stages. A comparatively higher survival rate in a given stage, resulting in a large number of individuals entering the following stage, can in turn cause a lower survival rate there, and vice versa. An arbitrary stage division makes it difficult to distinguish between genuine and artificial relationships.

Thus, in the first place, we must make an effort, in partitioning a life cycle, to reduce any chance of getting a negative correlation due to an arbitrary stage framing. If the framing in some section of the life cycle was unavoidably arbitrary, we must be careful in interpreting that section of the life-table data.

Note finally that the relationship (3.44) may not effectively apply to those data in which the rate of change in population is partitioned into distinct types of mortality (e.g. predation, parasitism and diseases) rather than into developmental stages. This is because these types of mortality may overlap in time, and also because some individuals may be parasitized as well as diseased or preyed upon (Royama, 1981b). I shall discuss, in Part Two, various problems that we often encounter in the analysis of mortality factors.

3.3.3 Apparent density dependence of the effect of density-independent factors

In section 1.8.1, I explained that the word 'density dependence' has wide and narrow connotations. I proposed that we use the term 'density-independent factor' in the narrow sense, i.e. those factors whose measures are uninfluenced by population density (section 1.6.2). I have pointed out, however, that this usage of the term does not necessarily imply that the effect of a density-independent factor is uncorrelated with the population density.

To be precise, consider the following hth-order density-dependent process under the vertical perturbation effect z_t of the density-independent factors involved:

$$R_t = f(X_t, X_{t-1}, \ldots, X_{t-h+1}) + z_t$$

where, as before, X_t is the log population density at a given stage of the generation t, and $R_t = X_{t+1} - X_t$ is the log reproductive rate. What I have pointed out is that z_t is not necessarily uncorrelated with X_{t-h} for $h \geqslant 0$, even though they are structurally independent of each other. I have demonstrated, as one such example in section 1.6.7, that autocorrelations in the $\{z_t\}$ could result in non-zero correlations between z_t and $X_{t-h}, h \geqslant 0$.

Then, I promised in section 1.8.1 that I would explain yet another situation in which non-zero correlations could result, even if the $\{z_t\}$ are uncorrelated random numbers. To illustrate the point, I made a simulation with the following linear AR(2) process

$$X_{t+1} = 1.8X_t - 0.9X_{t-1} + u_t \tag{3.45}$$

in which u_t is an uncorrelated perturbation effect. In the simulation, $\{u_t\}$ is a series of independent random numbers uniformly distributed in the interval $[-2, 2]$ (Fig. 3.19a). In Fig. 3.19b, the series $\{X_t\}$, rather than $\{X_{t+1}\}$, generated by (3.45) is plotted to compare with $\{u_t\}$.

Now, compare the sections in series a and b in the interval $t = 13$ and $t = 28$, marked with solid circles, where the population series b has reached a plateau and somewhat stabilized temporarily. We see that the two series tend to fluctuate in opposite directions. In graph c, I regressed u_t on X_t in the interval (solid circles). We see a striking negative correlation between the two numbers, although the correlation becomes obscured when points outside the interval (open circles) are included. This is surprising because u_t was generated completely independently of X_t. This apparent paradox can be explained in rational terms.

First, recall Fig. 3.15 (section 3.3.1) in which we plotted the series $\{X_{t+1}\}$ along with $\{u_t\}$. We then compared the way the segment $\overline{X_t X_{t+1}}$ deflected from the preceding segment $\overline{X_{t-1} X_t}$ with the way $\overline{u_{t-1} u_t}$ deflected from $\overline{u_{t-2} u_{t-1}}$. We found that the direction of deflection tended to agree between the two series. I showed that the degree of agreement (or disagreement) could be indicated by a positive (or negative) correlation between the second differences $\Delta^2 X_{t+1}$ and $\Delta^2 u_t$ (cf. Fig. 3.16).

Now, look at Fig. 3.19 in which the series $\{X_t\}$, rather than $\{X_{t+1}\}$, is plotted (graph b) along with $\{u_t\}$ (graph a). We see that the directions of deflections in the corresponding segments tend to disagree between the two series as indicated by a far greater number of minuses than pluses; this is true no matter which section of the series we compare. This means, of course, that the second difference $\Delta^2 u_t$ is negatively correlated with $\Delta^2 X_t$ almost everywhere, and it is easy to show why this happens.

The covariances between u_j and X_i in the process (3.45) are expected (i.e. they are ensemble averages; section 1.5.1) to be:

$$\begin{aligned} \mathrm{Cov}(u_j, X_i) &= 0, \, i \leqslant j, \\ &= \sigma_u^2, \, i = j + 1, \end{aligned} \tag{3.46}$$

where σ_u^2 is the variance of u. It follows that, for $\Delta X_i = X_i - X_{i-1}$,

$$\begin{aligned} \mathrm{Cov}(u_j, \Delta X_i) &= 0, \, i \leqslant j, \\ &= \sigma_u^2, \, i = j + 1, \\ &= \mathrm{Cov}(u_j, X_{j+2}) - \sigma_u^2, \, i = j + 2. \end{aligned} \tag{3.47}$$

Then, for the second difference $\Delta^2 X_i = \Delta X_i - \Delta X_{i-1}$, we find

$$\begin{aligned} \mathrm{Cov}(\Delta^2 u_t, \Delta^2 X_t) &= \mathrm{Cov}(u_t - 2u_{t-1} + u_{t-2}, X_t - 2X_{t-1} + X_{t-2}) \\ &= -4\sigma_u^2 + \mathrm{Cov}(u_{t-2}, X_t) \\ &= -2.2\sigma_u^2 < 0, \end{aligned} \tag{3.48}$$

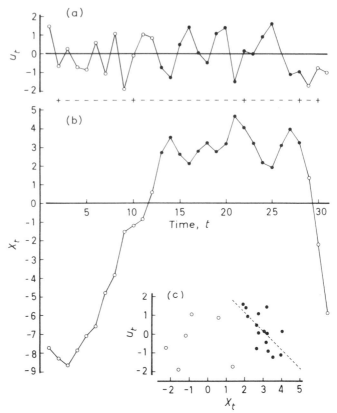

Figure 3.19 Simulation by model (3.45) showing a correlation between the uncorrelated perturbation effect u_t (graph a) and the log population density X_t (graph b) in the interval $13 \leqslant t \leqslant 28$ marked with solid circles. Graph c: regression of u_t on X_t. After Royama (1981a; fig. 6).

since $\mathrm{Cov}(u_{t-2}, X_t) = 1.8\sigma_u^2$ from (3.45). Because the covariance is negative, we see that the two quantities, $\Delta^2 u_t$ and $\Delta^2 X_t$, are indeed negatively correlated with each other. This means that, when series b in Fig. 3.19 happened to fluctuate without trend in the interval between $t = 13$ and 28, that section of the series tended to fluctuate in the opposite direction to the corresponding section of series a because the latter series, too, had no trend. Consequently, there appeared to be a negative correlation between the two series in that interval, even though u_t was generated completely independently of X_t. Obviously, if parts of the series outside this interval are included, the apparent correlation deteriorates and should vanish in the limit.

The above simulation shows that a sample correlation between population density and the effect of a density-independent environmental factor, if it is a dominant element, could happen to be fortuitously high, even if the effect is a completely uncorrelated random number. I shall show a typical example from a field study in Chapter 9.

If we take such a correlation as implying that u_t is a function of X_t, that is, u_t depends on X_t in the narrow sense, it would seriously misrepresent the structure of the population process.

3.3.4 Spurious correlations

We all know that a high correlation could be spurious, implying no causal connection between the measurements to be related. But the occurrence of spurious or nonsense correlations is particularly frequent between unrelated time series as Yule (1926) has discussed in depth. His example of correlation between a decrease in mortality in England and Wales and a decrease in the proportion of Church of England marriages, during the period between 1866 and 1911, was as high as 0.95. Another example is a 0.9 correlation between an increase in school teachers' salaries and an increase in liquor sales in France (Hoel, 1947). We understand these correlations as indicating the involvement of some common factors: during a period of good economy, both teachers' salaries and liquor sales would be likely to increase; fast progress in scientific knowledge and thinking around the turn of the century could, as Yule suggests, have contributed to decreases in mortality as well as church marriages. However, if we found a good correlation between the number of babies delivered and the stork population in Holland, we might begin to wonder if the correlation was anything but spurious.

An example of spurious correlations is one between the well-known 10-year cycle in the Canada lynx fur-trading statistics and a similar cycle in Wolfer's sunspot numbers. As I shall show in detail in Fig. 5.7, Chapter 5, the calculated correlation between log lynx fur returns each year and the sunspot numbers was 0.62 in the 30-year period beginning 1821, whereas it turned out to be -0.40 in the following 30-year period, illustrating an obvious spuriousness. The point here is, however, that with such simple correlations alone, we would not have been able to detect conclusively the spuriousness should the first 30 years of observation have been the only available information.

In many cases, spurious correlations result from similarity in trend or in the overall pattern of fluctuation (e.g. a 10-year cycle) between two time series. If a given meteorological factor in fact influences the rate of change in population, there should be a more subtle correspondence in the fluctuations between the two series than mere similarity in trend. A correlation between second differences in both time series (section 3.3.1) can greatly reduce, if not completely eliminate, a spurious correlation in trend.

The arguments in this section suggest that, by simple statistics alone, neither a high correlation promotes, nor a low correlation rejects, the hypothesis that a given factor has an influence or is an irrelevant one. We must, therefore, carefully examine the credibility of the hypothesis from the point of view of its biological implication.

3.4 FALSE INTERPRETATION OF POPULATION DATA

3.4.1 Correlation between weather and population fluctuations

Bodenheimer (1938, 1958) contended that some animal populations, particularly insects, are under the 'control' of certain climatic factors, especially temperature and humidity. His inference was based on two facts: (1) laboratory experiments show that these factors exert a dominating influence on the physiology of the animals and, hence, on their reproductive and survival rates; (2) long series of observations show a high degree of parallelism between population fluctuations and weather pattern. [Bodenheimer used the expression 'control' as synonymous with 'regulation'.]

My concern here is not so much with the validity of the 'facts' on which Bodenheimer's argument was based but with his inference leading to his climatic control (regulation) theory. As pointed out by Varley et al. (1973), his conclusion does not logically follow his observations. It is easy to show why Bodenheimer's inference is false.

I have shown in section 3.3.1 that the degree of correlation between log population density X_{t+1} and the effect u_t of the dominating density-independent factor U, say weather, depends on the density-dependent structure of the population process. In particular, I have shown, with the aid of the simple linear model (3.30), that the correlation would be highest at the centre of the triangular parameter space of Fig. 3.14a; at the centre, $a_1 = -1$ and $a_2 = 0$, so that $X_{t+1} = u_t$. In other words, a high degree of parallelism would show between the fluctuation in population density and that in weather record, only if the density-dependent component of the process perfectly regulates the population at its stable equilibrium state. At this state only the perturbation effect of weather is visible in the pattern of population fluctuation. We may say, in such circumstances, that weather 'governs' the pattern of population fluctuation, but it is misleading to say that weather 'controls' (or 'regulates') the population. This is not a matter of semantics.

If weather exerts a dominating influence – and population density has little influence – on the survival of the animals, we would be likely to see the rate of change in population fluctuating with weather, not the population itself. With the aid of model (3.30), we see that it must be the series $\{R_t\}$, rather than $\{X_{t+1}\}$, that fluctuates with $\{u_t\}$. As a matter of fact, the degree of correlation between R_t and u_t would be highest when the parameter set is at the midpoint ($a_1 = a_2 = 0$) on the right-hand side of the triangle (Fig. 3.14b). At this point, the correlation between X_{t+1} and u_t vanishes (Fig. 3.14a), and, as already explained, the population would perform an unregulated random walk.

Thus, if weather is the major factor that determines the reproduction and survival of animals, then, contrary to what he envisaged, Bodenheimer would not have seen a sign of weather controlling the population by any means, let alone the parallelism between the population fluctuation and weather pattern.

In section 1.7, I discussed the role of density-independent factors in determining population equilibria. The existence of an equilibrium density depends not only on the density-dependent structure of the population but also on the total effect of the density-independent factors involved. Among those factors some climatic factors could very well be major ones. We may say, then, that weather controls the equilibrium level of the population in conjunction with the density-dependent structure.

3.4.2 Cyclic time series as an artifact of data smoothing

An early study of the spruce budworm, *Choristoneura fumiferana*, in eastern Canada (Greenbank, 1963a; to be discussed in Chapter 9) contended that a succession of several dry, warm years tended to trigger an outbreak of this insect, and a subsequent period of wet, cool years terminated it.

The notion of such a 'cyclic weather pattern' originated from a moving-average transformation of the weather record in an attempt to extract a 'hidden cycle'. The transformation can be used to smooth a time series to effectively bring out a trend, if any. As I show below, however, this transformation also creates an artificial cyclic pattern (called the Slutzky effect) that does not necessarily exist in the original weather record. This invalidates Greenbank's analysis.

Suppose that u_t represents the perturbation effect of a major density-independent factor. Suppose further, without loss of generality, that $\{u_t\}$ is a series of independent random numbers identically distributed with the mean μ_u and the variance σ_u^2. There is neither a trend nor a systematic pattern in the series (Fig. 3.20a) as indicated by the sample correlogram not significantly different from zero (Fig. 3.20a'). Now, take the average of k consecutive values of u in the past up to time t, and make it the new random number w_t, i.e.

$$w_t = \frac{1}{k} \sum_{i=1}^{k} u_{t-i+1}. \tag{3.49}$$

While the uncorrelated original series $\{u_t\}$ shows no systematic pattern, the new k-point moving-average series $\{w_t\}$ exhibits an oscillatory motion (Fig. 3.20b). This is because in the series $\{w_t\}$, two points that are j steps apart are positively correlated with each other for $j < k$, since they share $k-j$ points of the original series $\{u_t\}$. More precisely, the autocovariance for lag j of the series $\{w_t\}$ is easily computed to be:

$$\gamma_{ww}(j) = \frac{k-j}{k^2} \sigma_u^2, 0 \leqslant j < k,$$
$$= 0, j \geqslant k. \tag{3.50}$$

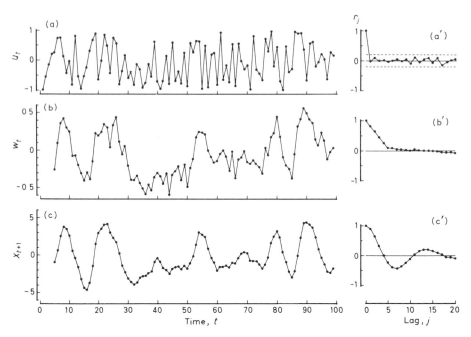

Figure 3.20   An illustration of the effect of a five-point moving-average transformation. Graph a: pure random series $\{u_t\}$. Graph b: moving-average series $\{w_t\}$ by (3.49). Graph c: log population series $\{X_t\}$ generated by (3.52). The graphs on the right: the ACFs of the series on the left. Dashed lines in graph a': the Bartlett band.

Hence, the autocorrelation for lag j is:

$$\begin{aligned}
\rho_{ww}(j) &= \gamma_{ww}(j)/\gamma_{ww}(0) \\
&= 1 - j/k,\ 0 \leqslant j < k, \\
&= 0,\ j \geqslant k.
\end{aligned} \qquad (3.51)$$

Thus, as against the identically zero autocorrelations in the original random series $\{u_t\}$, the correlogram of the moving-average series $\{w_t\}$ is positive for $j < k$, linearly decreases with increasing lags, but truncates at $j = k$ and onward (Fig. 3.20b'). The all positive autocorrelations for the first j lags less than k in the series $\{w_t\}$ indicate that the series tends to stay on one side of the average $\mu_u(=0$ in the simulation) for a while before crossing over to the other side, resulting in the oscillatory movement of graph b.

The simple moving-average transformation (3.49) is a special case of the general moving-average scheme (3.14), section 3.2.4, in which u_{t-i} is weighted by b_i. In (3.49), b_i is identically equal to $1/k$ for $i = 1, 2, \ldots, k$, and is zero for $i > k$. The moving-average scheme (3.14) is developed by Slutzky (1927), demonstrating that a wide variety of patterns in time series could be created out of a pure, uncorrelated, random series by the moving-average transformation. Slutzky shocked those people in his time who used the moving-average transformation with the intention of bringing out a 'hidden cycle' which was in fact nonexistent in the original random series. The notion of a 'cyclic tendency' in weather pattern is often a result of misinterpreting the cyclic tendency in the 'smoothed' weather data.

A subtle problem arises if one compares the smoothed weather record with a population which does tend to cycle, e.g. those generated in region IV of the triangle space of Fig. 2.5. Suppose that weather is, in fact, influencing the reproductive rate as a major perturbation factor, and that the population happens to cycle because of its density-dependent component. Then, there could be a high degree of correlation between the transformed 'cyclic' weather record and the inherently (i.e. due to the endogenous density-dependent structure) cyclic population fluctuation. It looks as though weather is controlling the population cycle, even if the original weather record is a pure random series (some examples will be shown in Chapter 9). I shall illustrate the point in the following.

Consider the linear second-order density-dependent process

$$R_t = a_1 X_t + a_2 X_{t-1} + u_t \qquad (3.52)$$

in which the pure random series in Fig. 3.20a serves as the $\{u_t\}$. Setting $a_1 = 0.6$ and $a_2 = -0.8$ conveniently, I generated the series $\{X_t\}$ in Fig. 3.20c which exhibits an oscillatory motion typical of a series from region IV of the triangular parameter space of Fig. 2.5. We see a striking similarity in the oscillatory motion between this series and the five-point moving-average series $\{w_t\}$ of (3.49) in graph b. The correlation coefficient between X_{t+1} and w_t turns out to be 0.86.

If $\{w_t\}$ was a moving-average series of weather records, such as mean daily temperatures, one might have misinterpreted that 'cyclic changes' in weather pattern was the cause of the population oscillation. The true mechanism in this model is that u_t influences X_{t+1} through (3.52), and that the oscillation in the series $\{X_{t+1}\}$ is due to the density-dependent structure of the process; whereas, the oscillation in the series $\{w_t\}$ is due to the positive autocorrelations created by the moving-average transformation (3.49). In fact, the correlogram of the series $\{X_{t+1}\}$ in graph c′ is clearly different from that of the series $\{w_t\}$ in graph b′. This reveals that the cyclic pattern in $\{w_t\}$ is not the cause of the similar pattern in $\{X_{t+1}\}$.

The high correlation between series b and c is not coincidental or spurious. A series $\{X_{t+1}\}$ originating from regions I and IV (i.e. from the right half of the triangle of Fig. 2.5) tends to be well correlated with a moving-average series $\{w_t\}$ generated from the original uncorrelated perturbation series $\{u_t\}$ if the number of consecutive values to be averaged, i.e. k, is appropriate.

On the other hand, a series $\{X_{t+1}\}$ originating from the other half of the triangle would tend to be poorly correlated with a moving-average series generated from $\{u_t\}$. Evidently, there is hardly any similarity between the slow cycle of series $\{w_t\}$ in Fig. 3.20b and the fast cycle of series II or III in Fig. 3.4.

Even if weather had little or no influence, there would still remain the danger of finding a high, spurious correlation between the population fluctuation and a moving-average transformation of the weather record. Thus, in order to detect the influence, if any, of a density-independent factor, and to interpret its role in population dynamics properly, we should use the difference transformations of section 3.3.1 rather than the moving-average transformation of (3.49).

3.5 DIAGNOSES OF OBSERVED POPULATION SERIES

Towards the end of Chapter 1, I critically reviewed the idea of detecting density dependence and population regulation. I pointed out that an attempt to detect them by means of a single statistic, such as a regression coefficient of reproductive rate on population density, would be unrewarded. I suggested that the only alternative is to look into the detail of the structure of each population process, and that such an effort inevitably develops into the study of more fundamental aspects of population dynamics. This requires analyses in many steps.

In this section, I shall summarise typical examples of the statistical characteristics that I have discussed in the present chapter. This should give us a rough guide to diagnosing the status of an observed population series. Note, however, that for a statistical analysis, time series data must be sufficiently long. There is not much we can do about a short series. [See the end of section 3.2.3 for how long is sufficient.]

First, plot an observed log population series $\{X_t\}$ as well as the log reproductive rates $\{R_t\}$ against time, calculate the correlograms of both series, if appropriate, and note the following characteristics:

1. The pattern of fluctuation in $\{X_t\}$ exhibits neither noticeable trend, nor systematic changes with time; the autocorrelation function (ACF) of $\{X_t\}$, if significantly different from zero for comparatively lower lags, decays reasonably fast. These indicate that the population is likely to be stationary and, hence, regulated (see examples in Figs 3.4 and 3.5).
2. The series $\{X_t\}$ seems to drift; its ACF decays slowly – the longer the series, the lower the speed of decay (check by dividing the series into sections, or 'jackknifing', see *Circumstance 1*, section 3.2.6); the ACF of $\{R_t\}$ is confined within the Bartlett band, or truncates after a few lower lags; the average $\bar{R}$ is not significantly different from zero (use the t-test). These indicate an unregulated random walk (examples in Figs 1.3 and 1.4).
3. The series $\{X_t\}$ has a trend; $\bar{R}$ is significantly different from zero:
   (a) $\{X_t\}$ does not change its pattern of fluctuation along the trend; $\{R_t\}$ is stationary (fast decay of its ACF). These indicate that the population is regulated about the trend which is set by a lateral shift in the

equilibrium level due to some environmental changes (examples in Fig. 2.20).

(b) $\{X_t\}$ does not drift but changes its pattern about the trend; $\{R_t\}$ is nonstationary. These indicate a gradual shift in the equilibrium state of the population due to vertical perturbation effects of environmental changes (examples in Fig. 2.19).

(c) The ACF of $\{R_t\}$ is confined within the Bartlett band, and $\bar{R} > 0$ significantly, indicating the population is growing to an equilibrium state (Fig. 3.12).

Determining the structures of population processes is an ultimate goal of studying population dynamics. The analysis of the statistical characteristics of population data discussed so far provides but a rough guide. For a good comprehension of natural population processes, we need to analyse information from systematic field studies. Such an analysis often requires a concrete working hypothesis expressed as a theoretical model. There are many alternative models depending on basic assumptions, the topic I discuss in the next chapter.

Those readers who intend to apply correlogram analysis to their data should read the following section on the sampling properties of autocorrelations before they proceed to the next chapter.

3.6 SAMPLING PROPERTIES OF AUTOCORRELATIONS

The interpretation of sample autocorrelation coefficients (or a correlogram) is not always easy. It is therefore important to familiarize ourselves with the statistical nature of the coefficients.

3.6.1 Expectations of sample autocorrelations

Moran (1948) and Orcutt (1948) found that, in a purely random series $\{u_t\}$ of length T, in which $\rho_{uu}(j)$ are identically zero for $j \neq 0$, the expectations of sample coefficients $E(r_j)$ – dropping the subscripts uu – are exactly and identically equal to $-1/(T-1)$. Marriott and Pope (1954) and Kendall (1954) independently obtained the expected values $E(r_j)$ of general case to order T^{-1} of approximation. The principle of the computation, after Kendall and Stuart (1968), is as follows.

Write the formula (3.13b) in the form $r_j = A/B$, in which

$$A = \frac{1}{T-j} \sum_{i=1}^{T-j} n_i n_{i+j} - \frac{1}{(T-j)^2} \left(\sum_{i=1}^{T-j} n_i \right) \left(\sum_{i=1}^{T-j} n_{i+j} \right) \qquad (3.53)$$

$$B = \frac{1}{T} \sum_{i=1}^{T} n_i^2 - \frac{1}{T^2} \left(\sum_{i=1}^{T} n_i \right)^2 \qquad (3.54)$$

Further, write $A = E(A) + a$ and $B = E(B) + b$ in which a and b are deviations from the expectations $E(A)$ and $E(B)$, respectively. Thus,

$$r_j = \frac{E(A) + a}{E(B) + b}. \tag{3.55}$$

Expanding the ratio on the right-hand side in a Taylor series and taking expectations term-by-term to the third term, we have an approximate value of $E(r_j)$:

$$E(r_j) \sim \frac{E(A)}{E(B)} - \frac{\mathrm{Cov}(a,b)}{E^2(B)} + \frac{E(A)\mathrm{Var}(b)}{E^3(B)} \tag{3.56}$$

in which, writing $N = T - j$, and letting the variance of the series $\{n_t\}$ be unity without loss of generality in the present arguments,

$$E(A) = \frac{1}{N}\left\{ N\rho_j - \frac{1}{N}\sum_{i=1}^{N-1}(N-i)\rho_{j+1} - \frac{1}{N}\sum_{i=1}^{j}(N-i)\rho_{j-i} \right.$$

$$\left. - \frac{1}{N}\sum_{i=1}^{N-j-1}(N-j-i)\rho_i \right\}, \ (j>0), \tag{3.57}$$

$$E(B) = \frac{1}{N}\left\{ N - 1 - \frac{2}{N}\sum_{i=1}^{N-1}(N-i)\rho_i \right\}, \tag{3.58}$$

$$\mathrm{Cov}(a,b) = \frac{2}{N}\sum_{i=-\infty}^{\infty}\rho_i\rho_{i+j}, \tag{3.59}$$

$$\mathrm{Var}(b) = \frac{2}{N}\sum_{i=-\infty}^{\infty}\rho_i^2. \tag{3.60}$$

Thus, given the true values ρ_j, we can find the expected sample values $E(r_j)$. For example, consider the following three simple cases:

Case i
When the $\{n_t\}$ is a purely random series $\{u_t\}$, so that $\rho_j = 0$, $j \neq 0$. In this case, the exact result is, as already mentioned:

$$E(r_j) = -1/(T-1), \tag{3.61}$$

though the approximate result by (3.56) is $-1/(T-j)$.

Case 2
When the $\{n_t\}$ is the MA(1) process, $n_t = u_t + bu_{t-1}, |b| < 1$, so that $\rho_1 = b/(1+b^2)$ and $\rho_j = 0, |j| > 1$, as in (3.17) and (3.18). Then, writing $\rho_1 \equiv \rho$ simply:

$$E(r_1) \sim \rho + (1+\rho)(4\rho^2 - 2\rho - 1)/(T-1), \qquad (3.62a)$$

$$E(r_2) \sim -(1 + 2\rho + 2\rho^2)/(T-1), \qquad (3.62b)$$

$$E(r_j) \sim -(1 + 2\rho)/(T-1), \ j > 2. \qquad (3.62c)$$

Case 3
When the $\{n_t\}$ is the AR(1) process, $n_t = \rho n_{t-1} + u_t$, so that $\rho_j = \rho^j$ (all j) as in (3.5). Then, for all j:

$$E(r_j) \sim \rho^j - \{[(j+\rho)(1-\rho^j)/(1-\rho)] + 2j\rho^j\}/(T-1). \qquad (3.63)$$

[Note that Kendall (1954), like Kendall and Stuart (1968), uses $(T-j)$ as the divisor in each formula. The above formulae use $(T-1)$, instead, as in Marriott and Pope (1954), which is compatible with Moran's exact result in Case 1.]

In all cases, $E(r_j)$ is a biased estimate of ρ_j to order T^{-1}; bias vanishes asymptotically, of course. As Kendall and Stuart (1968) point out, bias could be quite serious if the series is short. For example, if $\rho = 0.5$ and $T = 25$ in case 3, then, $E(r_1) \sim 0.4$, i.e. bias is as much as 20% of the true value of 0.5.

In practice, one should first decide the autocorrelation structure of the series by correlogram analysis and, then, estimate bias for correction. Iterate the procedure if necessary. However, there always remains some degree of uncertainty about the result of the correlogram analysis. [Note that a bias reduction could increase the variance of the corrected estimate; that is, it could reduce the efficiency of the estimator, thus, possibly, negating the intent to make the estimate more accurate (see below).]

Quenouille (1956) has suggested a simple and often effective method of bias reduction, popularly known as **jacknife estimation**, which does not require prior knowledge of the nature of the series for estimating the bias. This is to split the series into two halves and to calculate sample coefficients separately. Let $r_{j(1)}$ and $r_{j(2)}$ be the sample coefficients from the first and second halves. Then, the corrected coefficient r'_j is given by

$$r'_j = 2r_j - \frac{r_{j(1)} + r_{j(2)}}{2} \qquad (3.64)$$

in which r_j is the coefficient calculated using the entire length (T) of the series. The corrected coefficient r'_j is biased only to order T^{-2}. We must bear in mind, however, the variance of r'_j tends to increase; that is, r'_j is usually a less efficient estimate of ρ_j as compared to the uncorrected, hence more biased, r_j. Quenouille (1956) discusses a method to minimize the increase in the variance; see also Chatfield (1984).

3.6.2 Covariances between sample autocorrelations

Now, let us look at the covariance between the estimates r_j and r_{j+k}. The following result is due to Bartlett (1946), again using the Taylor expansion:

$$\text{Cov}(r_j, r_{j+k}) \sim \frac{1}{T} \sum_{i=-\infty}^{\infty} [\rho_i \rho_{i+k} + \rho_{i+j+k}\rho_{i-j} + 2\rho_j \rho_{j+k}\rho_i^2$$

$$- 2\rho_j \rho_i \rho_{i-j-k} - 2\rho_{j+k}\rho_i \rho_{i-j}]. \tag{3.65}$$

[The above formula applies when the time series is a Gaussian process, i.e. when the elements of the series are normally distributed. For a general, non-normal case, see Bartlett (1966) and Priestley (1981).] To obtain the variance $\text{Var}(r_j)$, simply put $k=0$ in the above formula. Again, given the theoretical values ρ_j and substituting them in the right-hand side of (3.65), we can estimate the covariances.

The two particularly important cases are:

Case 1
When the series concerned is purely random, as in the first case in section 3.5.1, so that the true autocorrelation coefficients $\rho_j(j \neq 0)$ are identically zero. Then, (3.65) is reduced to:

$$\text{Cov}(r_j, r_{j+k}) \sim 0, k \neq 0, \tag{3.66a}$$

$$\text{Var}(r_j) \sim 1/T, j \neq 0. \tag{3.66b}$$

Case 2
When the ρ_j decay to zero as j increases in such a way that they could be ignored after some j, as in comparatively lower-order AR and MA processes:

$$\text{Cov}(r_j, r_{j+k}) \sim \frac{1}{T} \sum_{i=-\infty}^{\infty} (\rho_i \rho_{i+k}) \tag{3.67a}$$

$$\text{Var}(r_j) \sim \frac{1}{T} \sum_{i=-\infty}^{\infty} \rho_i. \tag{3.67b}$$

We see that the Bartlett band (an approximate 95% confidence interval for the zero autocorrelation used in section 3.2.3) is $\pm 2\sqrt{\text{Var}(r_j)}$ of (3.66b) about the expectation $E(r_j)$ of (3.61); for large T, it is reduced to $\pm 2/\sqrt{T}$.

We also see in (3.67a) that the estimates r_j and r_{j+k} are correlated with each other, except for a purely random series as in (3.66a). This implies that biases in r_j and r_{j+k} are correlated with each other, and the shorter the length of observations (T), the higher the correlation. This illustrates why we need to be cautious in interpreting a correlogram estimated from a short series.

4 Population process models

'What we want to know is how the biological forces
of natality and mortality are so integrated and
correlated in their action as to lead to a final result
in size of population which follows this particular
curve rather than some other one.'

Raymond Pearl (*The Growth of Population*, 1927)

4.1 INTRODUCTION

In the previous two chapters, I discussed statistical properties of population processes without referring to their ecological mechanisms. In this chapter, I shall discuss several types of mechanisms and examine their interrelations. Through such comparative studies, we gain insight into the way our ideas evolve. Understanding the ecological meaning of a model promotes further improvement in its applicability through systematic generalizations.

I shall first introduce the classic logistic law and discuss its ecological mechanism in section 4.2. The logistic model is then generalized, in section 4.3, into those in a discrete-time scheme that have more descriptive power. One assumption implicit in the logistic process in discrete time is that the resource is invariant between time steps. In section 4.4, I relax this assumption, allowing for a between-step variation in the resource due to exploitation by the animals concerned. If the resource is another type of organism, the subject becomes what is usually treated as a predator–prey interaction.

In building a population process model, we wish it to be, on the one hand, simple, general and practical as an analytical tool. On the other, we may want it to be detailed, particular and realistic so that it can describe an observed process accurately. In section 4.5, I discuss how to compromise between these conflicting demands.

Often, an observed population series is inadequate to determine the reproduction curve (or surface) of a natural population. Additional observations or experiments would be necessary. I discuss this topic in section 4.6.

In the present chapter, my argument will be based mostly on deterministic, pure density-dependent processes. A stochastic component can be readily incorporated in the manner discussed in the previous chapters.

4.2. THE LOGISTIC THEORY

4.2.1 Basic idea

We all are familiar with the fact that the growth of populations of diverse organisms often exhibits a sigmoid curve. The Belgian mathematician Verhulst (1838) was the first to propose an equation that describes this type of population growth, which he (Verhulst, 1845) called 'logistique' as opposed to the 'logarithmique' of the Malthusian geometric progression (Fig. 4.1). But the work of Verhulst did not draw much attention from his peers until much later, when two Americans, Pearl and Reed (1920), without knowledge of the prior work, reinvented the logistic formula. [For a detailed

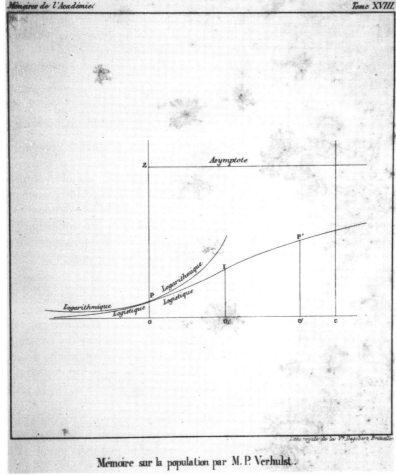

Figure 4.1   The historic graph of the logistic curve by Pierre-François Verhulst (1845).

historical account of the logistic theory, see Kingsland (1982, 1985).] The equation described many examples of population growth, as Pearl (1927) himself remarked, 'with extraordinary precision', no matter whether they were yeast cells in a culture, the *Drosophila* population in a milk bottle, or human populations in some parts of the world.

The interpretation of the logistic law is largely metaphorical, however. Given a set of environmental conditions, a population tends to grow asymptotically to a steady level. It looks as though the environment has a limited capacity to accommodate the population (carrying capacity), and, at any moment, the population increases at a rate proportional to still unutilized potential (or 'vacancy') for growth (Pearl and Reed, 1920; Gause, 1934). Chapman (1928) envisaged that the limited growth rate was due to the 'resistance' of the environment, and he coined the term 'environmental resistance'.

However, behind these ideas there is no particular ecological, or even theoretical, necessity that the logistic law must be what it is. In the following, I shall consider an ecological mechanism underlying this classic law. But, first, I shall introduce several different ways of formalizing the basic idea.

4.2.2 Classic Verhulst–Pearl equation in a continuous-time scheme

Let K be the carrying capacity of the environment at which a population is maintained in a state of equilibrium. Consider that population density at a given moment is x ($< K$) which increases to $x + \Delta x$ over an interval of time Δt. The rate of population growth per individual (per-capita growth rate) over Δt is defined by $(\Delta x/\Delta t)/x$. This rate, we assume, is proportional to the difference $(K - x)$, the 'vacancy' for further growth at this moment. Then, $\Delta x/\Delta t/x$ tends to a maximum (γ, say) as x tends to 0 and diminishes to 0 as x tends to K. The simplest form that satisfies the above tendency is:

$$(\Delta x/\Delta t)/x = \gamma(1 - x/K). \tag{4.1}$$

Parameter γ is called the 'potential rate of increase', 'innate capacity for increase', or 'intrinsic rate of natural increase', depending on the author. Note that $\gamma \geqslant 1$, or the population cannot persist. [Conventionally, the letter r is used instead of γ. However, I have been using r for denoting the reproductive rate of a population in discrete time defined in (1.3a), section 1.2.2. So, I break the convention.]

An explicit evaluation of the population density x in (4.1) depends on the population growing continuously in time or growing in a discrete fashion. Consider first that birth and death of population members occur at any moment so that there is neither a clear distinction between generations nor seasonality. The growth of a yeast-cell population in a constant environment is a typical example. In this continuous-time process, we can think of an infinitesimal interval of time $\Delta t \to 0$ and an accordingly small increment of population size $\Delta x \to 0$. Replacing Δ by the differential operator d in (4.1), and transposing x from left to right, we have the differential equation

$$dx/dt = \gamma(1 - x/K)x. \tag{4.2}$$

This, on integrating, yields the familiar Verhulst–Pearl logistic equation

$$x = x_0 e^{\gamma t}/[1 + (e^{\gamma t} - 1)x_0/K] \tag{4.3}$$

in which x_0 is an initial population density, i.e. when $t = 0$.

4.2.3 Recurrence-equation representation of the logistic growth in continuous time

In the logistic process (4.3), density x is a continuous function of time. Usually, however, we measure a population growth at regular, discrete time intervals. Therefore, it is convenient to express a continuous growth by (4.3) in terms of those discrete measurements.

This can be done by letting x_t be density at time t, which grows to x_{t+1} by the next point of time. This is equivalent to setting $t = 1$, $x_0 = x_t$, and $x = x_{t+1}$ in (4.3). Thus, we have the recurrence equation

$$x_{t+1} = x_t e^{\gamma}/[1 + (e^{\gamma} - 1)x_t/K] \tag{4.4}$$

(Fujita and Utida, 1952, 1953; Skellam, 1952; Morisita, 1965). [Notations are mine.] Note that x_t in (4.4) is a discrete measurement of the continuous time process (4.3); there is no difference in the underlying ecological mechanism between the two types of expression.

4.2.4 Logistic growth in discrete time

We now consider what we would expect if the population process concerned is truly discrete in time (e.g. those processes having non-overlapping generations or a distinct breeding season), while we retain the basic assumption (4.1) of the logistic process. There are two ways of formulating the discrete-time process analogous to the differential equation (4.2) of the continuous-time process. A simple analogue is to put $\Delta t = 1, x = x_t$, $\Delta x = x_{t+1} - x_t$ in (4.1), so that

$$(\Delta x/\Delta t)/x = (x_{t+1} - x_t)/x_t$$
$$= \gamma(1 - x_t/K) \tag{4.5}$$

and, after a rearrangement, we get

$$x_{t+1}/x_t = [1 + \gamma(1 - x_t/K)] \tag{4.6}$$

(Maynard Smith, 1968). This analogue, however, is biologically infeasible because x_{t+1} becomes negative for sufficiently large x_t.

To find a biologically feasible analogue, recall the identity $dx/x = d(\log x)$. This is a limiting form of the difference equation $\Delta x/x = \log(x + \Delta x) - \log x$; the right-hand side, after setting $x \equiv x_t$ and $x + \Delta x \equiv x_{t+1}$, is written $\log x_{t+1} - \log x_t$. Substituting this for the left-hand side in (4.1) after setting $\Delta t = 1$, we have

$$\log x_{t+1} - \log x_t = \gamma(1 - x_t/K). \tag{4.7}$$

Then, an anti-log transformation of (4.7) gives

$$x_{t+1}/x_t = \exp[\gamma(1 - x_t/K)] \tag{4.8a}$$

or, using the definition (1.3a), $r_t = x_{t+1}/x_t$:

$$r_t = \exp[\gamma(1 - x_t/K)]. \tag{4.8b}$$

This is an ecologically feasible discrete-time analogue of the classic model (4.3).

However, it is more logical to consider that formula (4.8) is a general form and classic formula (4.3) is a limiting form of (4.8) when a time step is infinitesimal. But, then, how do we derive the formula (4.8) in the first place?

Formula (4.8) may be conceived as a modification of the Malthusian geometric progression, $x = x_0 \exp(\gamma t)$. A discrete-time analogue of this equation is $x_{t+1} = x_t \exp(\gamma)$, which can be modified to (4.8) by replacing parameter γ by a density-dependent parameter $\gamma(1 - x_t/K)$ (MacFadyen, 1957; Pielou, 1977). However, this derivation, obviously inspired by the classic formula (4.3) already in mind, does not answer the question: what, ecologically, leads to the particular mathematical formulation (4.8)?

In the following, I derive, from basic principles, formula (4.8) as a general logistic formula. Through this exercise, we understand the concrete, ecological meaning of parameters γ and K. Once we know its ecological mechanism, we can systematically generalize the logistic law to formulate more flexible models that can describe and interpret a wider class of population dynamics.

4.2.5 Ecological mechanics of the logistic law

Consider a random distribution of individual organisms (represented by points in Fig. 4.2) over an available resource. Suppose further that each

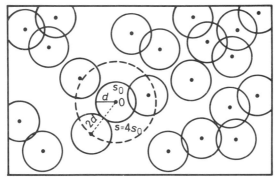

Figure 4.2   Geometric model of competition.

individual has a minimum sufficient resource requirement to survive and reproduce normally. Survival and reproduction are adversely affected if the resource falls below a certain level, whereas a surplus has no influence.

A circle of radius d around each point in Fig. 4.2 represents a minimum sufficient area (s_o) for each individual to secure its minimum sufficient requirement (n), given resource density (D) (amount available per unit habitat space); i.e. $Ds_o = n$. If several circles overlap with each other, the points involved compete for the resource. The individuals that a given individual (e.g. point o) competes with are those that fall within the concentric dashed circle with radius $2d$ and area $s = 4s_o$.

Now, let r_i be the average reproductive rate, and $Pr(i)$ be the expected proportion, of individuals having i competitors. Then, the mean reproductive rate r of the population is given by the weighted sum

$$r = r_0 Pr(0) + r_1 Pr(1) + \cdots + r_i Pr(i) + \cdots \tag{4.9}$$

[Note that r on the left is $r_t = x_{t+1}/x_t$. But, for the moment, I drop the time subscript t from all variables, so that the number subscript i would not be confused with the time subscript.]

Suppose, as the simplest case, that the points in Fig. 4.2 are Poisson distributed with mean density x. Then, the number of points within each circle of area s (with one point at the centre and excluding this point) is also Poisson distributed with the mean sx (Skellam, 1952; Morisita, 1954; Pielou, 1977). Thus, the expected proportions $Pr(i)$, $i = 0, 1, 2, \ldots$, are given by the well-known formula $[(sx)^i \exp(-sx)]/i!$. Substituting these in (4.9), we have

$$r = r_0 [\exp(-sx)][1 + (r_1/r_0)(sx) + (r_2/r_0)(sx)^2/2! + \cdots]. \tag{4.10}$$

Consider further that the reproductive rate r_i decreases as i (the number of competitors) increases. In particular, the addition of an extra competitor to the already existing i competitors within s reduces the reproductive rate of the central point by factor k, a non-negative constant not greater than 1. Thus,

$$r_i = r_{i-1}k = r_{i-2}k^2 = \cdots = r_0 k^i. \tag{4.11}$$

Parameter r_0 is a potential, biologically realizable maximum reproductive rate equal to r_m in the preceding chapters. Thus, substituting $r_i = r_m k^i$, $i = 0, 1, 2, \ldots$ in (4.10), and resuming the time subscript t, we find

$$r_t = r_m[\exp(-sx_t)][1 + ksx_t + (ksx_t)^2/2! + \cdots]. \tag{4.12}$$

By the well-known Taylor theorem,

$$1 + ksx_t + (ksx_t)^2/2! + \cdots = \exp(ksx_t)$$

so that (4.12) is reduced to:

$$\begin{aligned} r_t &= r_m[\exp(-sx_t)]\exp(ksx_t) \\ &= r_m\exp[-s(1-k)x_t], \end{aligned} \tag{4.13a}$$

or simplifying $s(1-k) \equiv c$,

$$r_t = r_m\exp(-cx_t). \tag{4.13b}$$

Further, recalling that $\log r_m \equiv R_m$, the right-hand side of (4.13b) can be written as:

$$r_t = \exp(R_m - cx_t) = \exp[R_m(1 - cx_t/R_m)]. \tag{4.13c}$$

Because (4.13c) and (4.8b) are identical in form, parameter-by-parameter comparisons reveal that

$$\gamma \equiv R_m, \text{ or } e^\gamma \equiv r_m \tag{4.14a}$$

$$K \equiv R_m/c \equiv \gamma/s(1-k). \tag{4.14b}$$

Thus, by reversing the computation from (4.13c) $\equiv$ (4.8b) to (4.7) and then to (4.1), we obtain the classic logistic formula (4.3) as a limiting form of (4.13).

The ecological meaning of parameter γ is straightforward in the relationship (4.14a). The meaning of K needs interpretation because of the complex relationship (4.14b). My interpretation is as follows.

Competition should intensify as minimum resource requirement (n) increases. Simply, the greater the demand, the higher the intensity of competition, given resource density (D). On the other hand, competition should be reduced if resource density is high, given n. Thus, the factor $(1-k)$, which measures the intensity of competition, should depend directly on n and inversely on D. Then, recalling that $s = 4s_o = 4n/D$, we see that $s(1-k)$, the denominator on the right-hand side of (4.14b), depends directly on n and inversely on D. Now, parameter r_m must have the opposite tendency: to depend directly on D and inversely on n. Then, by (4.14b), we see that K should depend directly on D and inversely on n. Thus, K and γ are likely to be correlated with each other rather than being independent parameters as some authors (Wilson and Bossert, 1971) think.

Recently, Kuno (1991) noticed that, if γ and K were treated as independent parameters, the logistic model (4.3) might exhibit anomalous behaviour. Indeed, if $\gamma < 0$ and $0 < K < x_0$, population density (x) will become discontinuous at the point in time, $t = (1/\gamma)\log(1 - K/x_0)$: approaching this point in time, x tends to $+\infty$, but as soon as t passes the point, x becomes $-\infty$. In my interpretation of the logistic law, no such anomaly exists because γ and K are bound by the relationship (4.14b): γ and K always have the same sign since c is a positive parameter.

In the above argument, parameters s_o and k (hence, c) have been treated as constants, implying that the resource is assumed to be invariant between time steps. But my theory of the logistic law is built on the assumption of competition due to a depletion of the resource during each time step. It is, therefore, implicit in the theory that the resource, being depleted during a time step, will be recovered to the same level by the onset of the next time step. I relax this assumption in sections 4.3.4 and 4.4.

4.2.6 Dynamics of the logistic models

The discrete-time logistic model (4.8) is a particular case of model (2.5), section 2.2.3, whose dynamics were already shown in Fig. 2.3. The logarithmic transformation of (4.8b) gives:

$$R_t = \gamma - \exp[\log(\gamma/K) + X_t]. \tag{4.15a}$$

For $\gamma > 0$, we can further transform (4.15a) – by setting $X \equiv \gamma X$ – to its canonical form comparable with (2.5b):

$$R_t = 1 - \exp(-\log K + \gamma X_t). \tag{4.15b}$$

Parameter-by-parameter comparisons with (2.5b) reveal that $K = 1$ and $\gamma = -a_1^*$. We see that, as γ changes in the positive domain, model (4.15b) produces all the patterns shown in Fig. 2.3. [Notice that these patterns are solely determined by γ. $K (\neq 1)$ merely controls lateral shift in log equilibrium density X^*. For $\gamma \leq 0$, no equilibrium density exists in (4.15a) since R_t cannot be positive; note that $\gamma/K = c > 0$ by (4.14b).]

The classic (continuous time) model (4.3), on the other hand, produces only one pattern of population growth, the well-known sigmoid curve converging monotonically and asymptotically to equilibrium at K.

Now, in the interval $0 < \gamma \leqslant 1$, model (4.15b) produces a growth curve monotonically increasing to equilibrium as in Fig. 2.3a'. After anti-log transformation, the curve becomes sigmoid similar to the classic logistic curve. The similarity is only superficial, however. A difference shows in the plot of X_{t+1} against X_t (Fig. 4.3), a log transformation of the so-called Ricker (1954) reproduction curves. [Call these curves $(X_{t+1}$ on $X_t)$ reproduction curves as distinguished from my $(R_t$ on $X_t)$ reproduction curves. Although these are mathematically equivalent, each has its own merit for a graphical analysis.] To make the comparison direct, I use formula (4.4) for the classic model and (4.8a) for the discrete-time analogue. An $(X_{t+1}$ on $X_t)$ reproduction curve of (4.4) always increases monotonically to a level (curve a) for all positive values of γ (and, hence $K > 0$). In contrast, a curve produced by (4.8a) always (for all positive γ) has a hump: it decreases after a peak with an ever-steepening slope (curve b). We can readily verify these features by calculating the derivative dX_{t+1}/dX_t.

Although the discrete-time logistic model can produce a variety of dynamic patterns, its descriptive ability is still limited. This is because, as noted above, its dynamics (after log transformation) are determined solely by parameter $\gamma (\equiv \log r_m)$. It means that every population having the same potential reproductive rate r_m automatically exhibits the same dynamics. This is rather unlikely in an actual situation. So, we need to generalize these basic models to gain more descriptive power.

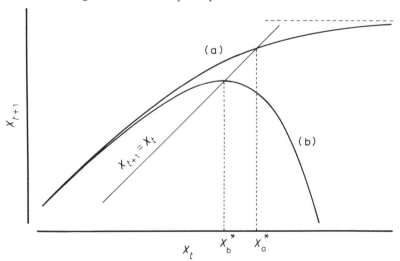

Figure 4.3 Two types of $(X_{t+1}$ on $X_t)$ reproduction curve: (a) the classic logistic process (4.4); (b) discrete-time logistic process (4.8a) and its generalization (4.19). X_a^* and X_b^* are equilibrium points, and the dashed line is an asymptote.

4.3 GENERALIZATION OF THE LOGISTIC THEORY

In this section, I discuss several ways of generalizing the logistic process and their interrelations.

4.3.1 Density-dependent parameter model

An obvious mathematical generalization of (4.15a) is model (2.5a), i.e.

$$R_t = R_m - \exp(-a_0 - a_1 X_t). \tag{4.16}$$

Parameter-by-parameter comparisons reveal that $R_m = \gamma$, $a_0 = \log K/\gamma$, and $a_1 = -1$ in (4.15a). Thus, by allowing a_1 to take an arbitrary value in (4.16), we gain an extra degree of freedom in descriptive ability over (4.15a). An ecological meaning of parameter a_1 is as follows.

Recall that in deriving the discrete-time logistic model (4.13) in section 4.2.5, we assumed that the reproductive rate of an individual was reduced by the constant factor k each time an extra individual was added in the competition area s (Fig. 4.2). Thus, if there are i competitors in the area s of a given individual, its reproductive capacity is reduced to $r_i = r_m k^i$. Let us relax this restriction and assume that the ith competitor has the reduction factor $k_i (0 < k_i < 1)$, $k_i \neq k_j$, $0 < i \neq j$, and $k_0 = 1$, such that

$$r_i = r_m k_1 k_2 \ldots k_i. \tag{4.17}$$

Incorporating expression (4.17) into (4.9) results in a cumbersome model. But we can find a convenient approximation.

Consider that parameter k_i changes monotonically as the number of competitors i changes. This means either intensifying competition:

$$1 > k_{i-1} > k_i > 0, \tag{4.18a}$$

or habituating to competition:

$$0 < k_{i-1} < k_i < 1. \tag{4.18b}$$

Usually, the number of competitors that each individual encounters will, on average, increase as population density x increases. Then, r_m would be reduced faster than the exponential reduction rate, $\exp(-cx_t)$, of (4.13b) in situation (4.18a) but more slowly in situation (4.18b). We may approximate these situations by making parameter c density-dependent. A simple way is to replace the exponent cx_t by cx_t^a in which a is a positive constant; situation (4.18a) implies $a > 1$, and (4.18b), $a < 1$. Thus, instead of (4.13b), we have

$$r_t = r_m \exp(-cx_t^a). \tag{4.19}$$

A logarithmic transformation of (4.19), with $\log c \equiv -a_0$ and $a \equiv -a_1$, yields (4.16).

An example of case (4.18a) is an increasing degree of interruption in egg-laying activity by mutual interference in some parasitoids (Hassell, 1971). A similar effect can be expected if the degree of environmental contamination increases with population density.

Conversely, case (4.18b) will be realized if an additional competitor exerts less effect than does any previous one. This fits egg-laying azuki bean weevils, *Callosobruchus chinensis*, because they tend to habituate themselves to mutual interference (Chapter 7). Some species may develop a social order to reduce conflict. This is also an example of situation (4.18b). Species, like the spruce budworm, *Choristoneura fumiferana* (Chapter 9), tend to survive overcrowding at the cost of having a reduced body size (hence, a lower potential fecundity). Whether this fits situation (4.18a) or (4.18b) depends on the balance between the gain in survival and the reduction in fecundity.

Although the generalized model (4.16) is more flexible in descriptive ability than its basic model (4.15a), it is still limited. The reproduction curve of (4.16) retains the shape of curve b of Fig. 4.3 no matter how we choose a_1: its negative slope ever steepens as X_t increases. This means that there always is a gap in between curves a and b of Fig. 4.3. To increase our descriptive capability, it is desirable to find a model which produces a family of reproduction curves that can reduce the gap. The following distinction between two types of competition leads to the discovery of such a model.

4.3.2 Population growth under scramble and contest competition

Consider that, as population density increases, an average share of the resource among the members of the population would, sooner or later, become less than the minimum resource requirement. Then, the following two extreme cases can happen.

First, the resources are equally divided among the individuals, so that, beyond a certain density, none can get enough of a share of the resource to survive and reproduce. This is the so-called scramble competition and results in the slope of a reproduction curve ever steepening as X_t increases. A typical example is curve b of Fig. 4.3. Thus, (4.19) is a model of scramble competition for all positive values of parameter a. [The reader can verify that, for all $a > 0$, the derivative dX_{t+1}/dX_t tends to $-\infty$ in the limit.]

Second, no matter how many individuals are competing, some strong (or lucky) individuals get a big enough share of the resource to survive and reproduce at the expense of the rest. This is known as contest competition. Curve a, Fig. 4.3, produced by model (4.4) of the classic logistic process is a typical example. It shows that as parent population X_t increases, offspring population X_{t+1} converges to a certain constant level.

In an actual situation, however, the production of offspring need not converge to a constant number as parent population increases even in contest competition. We only need to assume that some can outcompete others and successfully rear their offspring. That is, the asymptote of the reproduction curve need not be perfectly level but can be sloped. The slope can be negative, or even positive, so long as it does not tend to $-\infty$ (perpendicular) as in curve b (Fig. 4.3) of scramble competition. Model (4.4) acquires this property if we introduce an extra constant a such that it is generalized to either:

$$r_t = r_m/(1 + cx_t^a) \tag{4.20a}$$

(Maynard Smith and Slatkin, 1973) or

$$r_t = r_m/(1 + cx_t)^a \tag{4.20b}$$

(Hassell, 1975). In both models, a and c are positive constants. [Notations are mine.] I interpret the meaning of parameter a as the same as in the scramble competition model (4.19): $1 < a$ implies situation (4.18a), and $0 < a < 1$, (4.18b). An $(X_{t+1}$ on $X_t)$ reproduction curve by (4.20) tends to constant slope, $(1-a)$, as X_t increases (Fig. 4.4). If competition intensifies as density increases, $1 - a < 0$ as already argued; if habituation occurs, $1 - a > 0$; and in between $1 - a = 0$.

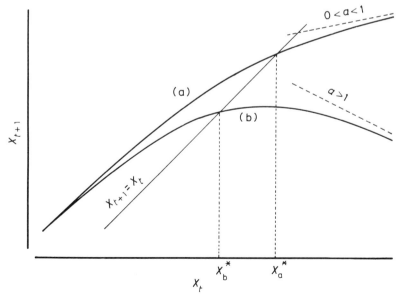

Figure 4.4 Reproduction curves of model (4.20). X_a^* and X_b^* are equilibrium points, and the dashed lines are asymptotes.

Hassell (1975) remarked that model (4.20b) with $a = 1$ represents contest competition, and $a \to \infty$, scramble competition. However, parameter a should be finite for the model to be meaningful. Also, in Hassell's perception, the difference between the two types of competition is reduced merely to one in the value of a parameter. In my perception, the difference is qualitative. Thus, I interpret (4.20) with finite parameter a to be a model of contest competition as distinguished from (4.19) as a model of scramble competition. In other words, in terms of the distinct qualitative properties of the two types of models, we can formally define the respective types of competition. [From now on, I may call (4.19) an exponential growth function and (4.20) a hyperbolic growth function.]

By log transforming the hyperbolic functions (4.20a and b), we obtain:

$$R_t = R_m - \log[1 + \exp(a_0 + a_1 X_t)] \tag{4.21a}$$

$$R_t = R_m - a_1 \log[1 + \exp(a_0 + X_t)] \tag{4.21b}$$

in which $a_0 = \log c$ and $a_1 = a$. These are comparable with (4.16), the log transformation of the exponential function (4.19). Figure 4.5 illustrates (R_t on X_t) reproduction curves of (4.16) and (4.21). Graph a shows differences, and graph b similarity, between (4.16) and (4.21) after selecting appropriate sets of parameters (R_m, a_0, a_1). The deterministic dynamics of models (4.16) and (4.21) about the equilibrium point X^* can be difficult to distinguish from each other, particularly in the graph b situation. However, when subjected to random perturbation, differences may show up in their dynamics (Chapter 7).

4.3.3 Non-Poisson distribution models

The hyperbolic growth functions (4.20) as models of contest competition are generalizations of (4.4) which is formulated in a continuous-time scheme. I have not yet been able to formulate a model of contest competition in discrete time from basic principles. Following is a hint for interested theoreticians to find an appropriate one.

So far, we have assumed that the organisms concerned are Poisson distributed over an available resource. In this pure random distribution, no individual can escape from scramble competition when population density becomes sufficiently high. If we relax this situation and redistribute competitors, there can be more individuals having a fewer number of competitors than would be expected in a completely random distribution. For example, some strong individuals can establish territories and successfully keep out competitors; early emerging insects can outcompete late emerging ones (Chapter 7).

If we find a suitable probability distribution function $\Pr(i)$ and reproductive rate r_i to represent such situations, the substitution of those in (4.9) will

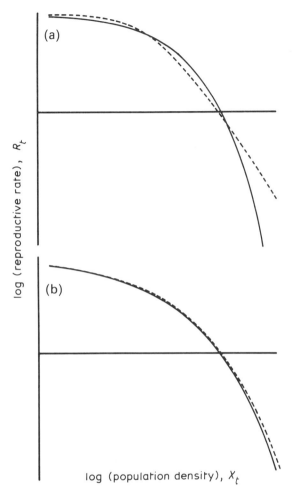

Figure 4.5   Differences (graph a) and similarity (graph b) between the (R_t on X_t) reproduction curves of model (4.19) (solid curves) and those of model (4.20b) (dashed curves).

give the model. One may very well find model (4.20) to be an approximation to the sum of such series on the right-hand side of (4.9). As far as I am aware, however, little is known about the distribution of the probability function Pr(i) for a non-Poisson distribution. In the meantime, I shall use (4.20) as empirical models of contest competition in discrete time.

4.3.4 Higher-order logistic models

The logistic model and its generalizations discussed above are all first-order density-dependent processes, meaning that x_{t+1} structurally depends only on x_t (sections 2.2.1 and 3.2.2). To increase our descriptive capability, we need a higher-order model (sections 2.2.4 and 2.2.5). For this purpose, the discrete-time logistic model (4.16) can be extended to:

$$R_t = R_m - \exp(-a_0 - a_1 X_t - a_2 X_{t-1} - \cdots - a_h X_{t-h+1}). \tag{4.22}$$

[The dynamics of the second-order ($h=2$) process was discussed in sections 2.2.5, 2.4 and 3.2.] Similarly, model (4.21a) can be extended to:

$$R_t = R_m - \log[1 + \exp(a_0 + a_1 X_t + a_2 X_{t-1} + \cdots + a_h X_{t-h+1})]. \tag{4.23}$$

An ecological interpretation of the above generalization is the following.

In the generalized logistic model (4.19), i.e. $r_t = r_m \exp(-cx_t^a)$, the parameter c represents the combined effect of average resource requirement by an individual animal, resource availability in the habitat and intensity of competition among the members of a population. These factors are assumed to be invariant between time steps in (4.19). However, the effect of population density in the previous generation (x_{t-1}) may influence any one of these factors. For example the effect of x_{t-1} on the physiology of individual animals may be carried over to reduce the average resource requirement or competitive ability of individuals at time t. The resource at time $t-1$ may be so depleted that it would not recover by time t. In other words, parameter c may depend on density x_{t-1}. A simple approximation could be to replace c by $cx_{t-1}^{a'}$ in the manner we generalized (4.13b) to (4.19) in section 4.3.1. Incorporating this approximation in (4.19), and using appropriate notations, we have

$$r_t = r_m \exp(-e^{-a_0} x_t^{-a_1} x_{t-1}^{-a_2}). \tag{4.24}$$

Further, we can extend (4.24) to order h, and after a log transformation we obtain (4.22). Likewise, model (4.23) is the log transformation of an hth-order extension of the hyperbolic growth function (4.20a)

$$r_t = r_m / (1 + e^{a_0} x_t^{a_1} x_{t-1}^{a_2} \ldots x_{t-h+1}^{a_h}). \tag{4.25}$$

One might think that (4.19) can be generalized to second order also in the following form:

$$r_t = r_m \exp(a_0 + a_1 x_t + a_2 x_{t-1})$$
$$= \exp(R_m + a_0 + a_1 x_t + a_2 x_{t-1}) \tag{4.26}$$

(Turchin, 1990). [My notations.] This is not a good model. It is hard to find a possible ecological mechanism for the form (4.26). Besides, its descriptive ability is limited compared to (4.24) as we see below.

Every population process model must satisfy the following two conditions: it must have a positive equilibrium density (x^*), and the reproductive rate r_t

must have an upper bound. Now, by definition, $x^* = x_t = x_{t-1}$ at which $r_t = 1$. Substituting these in (4.26), and by the above restrictions (assuming $a_0 = 0$ for simplicity), we find:

$$x^* = -R_m/(a_1 + a_2) > 0 \qquad (4.27)$$

$$r_t \leqslant r_m. \qquad (4.28)$$

To satisfy (4.27), the sum $(a_1 + a_2)$ must be negative because $R_m > 0$ $(r_m > 1)$ in a persistent population. Within this restriction, a_1 or a_2 can be positive. However, if a_1, say, was positive, then $|a_1 x_t| < |a_2 x_{t-1}|$, or (4.28) will be violated. It follows that both a_1 and a_2 ought to be negative to avoid all contradictions. Then, by analogy with the linear second-order model (the linearization of the nonlinear model about x^* by first approximation), we see that model (4.26) has no contradiction only in the domains in the triangle parameter space of Fig. 2.5 (section 2.2.4) where a_1 and a_2 are both negative. This severely restricts the model's descriptive capability.

Moreover, even though model (4.26) apparently has three parameters $(r_m, a_1$ and $a_2)$ to determine its dynamics (autocorrelation structure), only two of them are effective. This is because, eliminating one parameter (say, a_2) by the transformation $x \equiv a_2 x$, and writing $a_1/a_2 \equiv a$, (assuming $a_0 = 0$), (4.26) is reduced to:

$$r_t = r_m \exp(ax_t + x_{t-1}). \qquad (4.29)$$

We see that only r_m and a control the model's dynamics. This further reduces the model's descriptive capability compared to (4.24).

4.3.5 Hutchinson's time-delay logistic model

The important effect of lagged density dependence that causes population oscillations was recognized by Hutchinson (1948). In collaboration with Onsager, he incorporated the effect into the original logistic formula (4.2) and proposed the following time-delay logistic model:

$$dx(t)/dt = \gamma[1 - x(t-\tau)/K]x(t) \qquad (4.30)$$

in which $x(t)$ indicates that x is a continuous function of time t, and τ is a delay in time. This model stimulated theoreticians to further investigations into its mathematical properties (Cunningham, 1954; Wangersky and Cunningham, 1956 and 1957); see reviews by Krebs (1972), Poole (1974) or May (1981). Model (4.30) is a continuous-time analogue of the discrete-time model (4.29) in which $\log r_m = \gamma$, $a = \gamma/K$, and x_{t-1} is replaced by $x_{t-\tau}$. Thus, model (4.30) has limited applications in both theoretical and practical investigations.

4.4 PREDATOR–PREY INTERACTIONS

In section 4.3.4, we consider that the resource might vary between time steps according to differences between depletion and recovery during each time step. We developed model (4.22) or (4.23) to approximate such a situation. If the resource is a population of another organism, the process becomes a predator–prey (parasitoid–host) interaction. Then, why not incorporate the interaction explicitly into a model? [If the resource for the prey population varies as well, we have to set up a three-species interaction system model. However, I only discuss a two-species system.]

4.4.1 Conditional reproductive rates

Let x_t and y_t be, respectively, predator (parasitoid) and prey (host) densities at time step t, and let r_t and r'_t be the reproductive rates of predator and prey over the interval from t to $t+1$ time steps. Reproductive rate in either species depends, in general, on densities of both species. Thus, we write:

$$x_{t+1}/x_t = r_t(x_t, y_t) \quad \text{(Predator)} \tag{4.31a}$$

$$y_{t+1}/y_t = r'_t(y_t, x_t) \quad \text{(Prey).} \tag{4.31b}$$

The plot of r_t, and that of r'_t, against x_t and y_t in (4.31) represent the reproduction surface of predator and that of prey, respectively.

Consider the predator's reproductive rate (r_t) as a function of its own density x_t while prey density y_t is held at an arbitrary (positive) value. Write this conditional rate as $r_t(x_t|y_t)$. This is a profile of the reproduction surface (4.31a) parallel to the (r_t, x_t) plane sliced at the fixed value of y_t. Similarly, the conditional rate r_t as a function of y_t, given a fixed value of x_t, written $r_t(y_t|x_t)$, is a profile parallel to the (r_t, y_t) plane. Likewise, we define $r'_t(y_t|x_t)$ and $r'_t(x_t|y_t)$ for prey. Figure 4.6 illustrates four such profiles or conditional reproduction curves. [In the following, I drop the time subscript t because I only deal with within-time-step processes.]

Two sets of models are appropriate for describing the four curves in Fig. 4.6:

$$r(x|y) \propto \exp(-cx^a) \tag{4.32a}$$

$$r(y|x) \propto 1 - \exp(-by) \tag{4.32b}$$

$$r'(x|y) \propto \exp(-b'x) \tag{4.32c}$$

$$r'(y|x) \propto \exp(-c'y^{a'}) \tag{4.32d}$$

and

$$r(x|y) \propto 1/(1+cx^a) \tag{4.33a}$$

$$r(y|x) \propto y/(1+by) \tag{4.33b}$$

$$r'(x|y) \propto 1/(1+b'x). \tag{4.33c}$$

$$r'(y|x) \propto 1/(1+c'y^{a'}). \tag{4.33d}$$

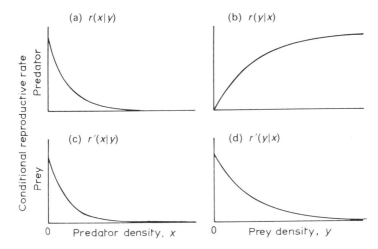

Figure 4.6 Typical conditional reproduction curves of predator and prey populations. Graphs a to d are based on models a to d, respectively, of the set (4.32).

[Models a to d in each set correspond to graphs a to d in Fig. 4.6.] The conditional rate $r(x|y)$ is the intrinsic (or self-imposed) regulation process of the predator population; and so is $r'(y|x)$ for the prey population. A logistic process is an appropriate model for these conditional rates because the resource for each species is fixed. Models (4.32a) and (4.32d) assume scramble competition after model (4.19), whereas (4.33a) and (4.33d) are after the contest competition model (4.20a).

The rate $r(y|x)$ of the predator is proportional to its functional response (predation rate) which increases as prey density y increases but, sooner or later, levels off (Fig. 4.6b). Model (4.32b) is the sum of all non-zero frequency terms of the Poisson distribution which represents prey consumption by random predation (Gause, 1934; Ivlev, 1955; Royama, 1971). Model (4.33b) assumes the Holling (1959) disc equation.

The rate $r'(x|y)$ of the prey population decreases with x due to predation. So, this rate assumes a converse of rate $r(y|x)$; i.e. proportional to the number of prey that escape predation.

4.4.2 Complete system models

We now combine the above conditional rates, resuming the time subscript t, to complete the system model (4.31). Combining the conditional reproductive rates in set (4.32), we have:

$$r_t = r_m[1 - \exp(-by_t)]\exp(-cx_t^a) \tag{4.34a}$$
$$r_t' = r_m'\exp[-(c'y_t^{a'} + b'x_t)] \tag{4.34b}$$

in which the proportionality constants r_m and r'_m are the biologically realizable maximum reproductive rates for predator and prey populations, respectively.

Alternatively, combine the conditional rates in set (4.33) to obtain:

$$r_t = r_m y_t / (1 + cx_t^a)(1 + by_t) \tag{4.35a}$$
$$r'_t = r'_m / (1 + c'y_t^{a'})(1 + b'x_t). \tag{4.35b}$$

We may even recombine conditional rates between sets (4.32) and (4.33), depending on the type of mechanism involved in each conditional rate in the system. For example, if competition among predators is of contest type, whereas it is of scramble type among prey, we may use the hyperbolic formula (4.33a) for $r(x|y)$ and the exponential formula (4.32d) for $r'(y|x)$.

4.4.3 The classic Nicholson–Bailey model and its derivatives

In this section, I shall reveal that many predator–prey models in the literature have hidden anomalies. In Table 4.1, I have selected four popular models as examples.

Table 4.1 Some predator–prey models in discrete time.

Reproductive rates		Authors
Predator, $r(x, y)$	*Prey, $r'(x, y)$*	*Authors*
$y[1 - \exp(-ax)]/x$	$b\exp(-ax)$	Nicholson and Bailey (1935)
$y^{1-c}[1 - \exp(-ax)]/x$	$by^{-c}\exp(-ax), 0 < c < 1$	Varley and Gradwell (1963)
$y[1 - \exp(-ax^c)]/x$	$b\exp(-ax^c), 0 < c < 1$	Hassell and Varley (1969)
$cy[1 - \exp(-ax)]/x$	$b\exp(-y - ax)$	Beddington *et al.* (1975)

x: predator density; y: prey density; a, b and c: constants. (Author's notations.)

The first one is the model proposed by Nicholson and Bailey (1935). This model always generates oscillations with ever-increasing amplitude and never converges to a state of persistence (section 1.2.2). This is awkward as a population process model. Suggestions have been made that the wild nature of the Nicholson–Bailey model can be 'tamed' by introducing a prey control mechanism other than predation, e.g. intrinsic regulation; or by incorporating a predator control mechanism other than prey availability (Rosenzweig and MacArthur, 1963; Murdoch and Oaten, 1975; Bulmer, 1976). The other three models in Table 4.1, proposed by Varley and Gradwell (1963), Hassell and Varley (1969) and Beddington *et al.* (1975), incorporate the control mechanism of either or both of the types suggested above. Although they apparently tamed the model, their accomplishment is only superficial because their alternative models have hidden anomalies of their own.

The conditional reproduction curves of all four models (Fig. 4.7) reveal their anomalies when compared with the corresponding curves in Fig. 4.6. The Nicholson–Bailey model has artificiality in two aspects: (1) the prey population has no intrinsic regulation mechanism since $r'(y|x)$ is flat across the prey's own density spectrum; (2) the predator population has no satiation, so that $r(y|x)$ increases unboundedly as prey density increases.

The Varley–Gradwell and Beddington *et al.* models modified $r'(y|x)$ to incorporate intrinsic regulation mechanisms in prey. The Varley–Gradwell modification, however, resulted in another artificiality, i.e. $r'(y|x)$ has no upper bound for $y\to0$. In both models, $r(y|x)$ is still unbounded as in the Nicholson–Bailey model. The Hassell–Varley model leaves $r'(y|x)$ unchanged (no intrinsic regulation) but modifies $r(y|x)$ which now does not have an upper bound: $r(x|y)\to\infty$ as $x\to0$. Also, as in other models, $r(y|x)$ of this model has no upper bound.

When formulating the system models (4.34) and (4.35), section 4.4.2, I carefully eliminated all anomalies exhibited by the above four models. Nonetheless, my models are still a simplistic idealization of an actual process. I have ignored certain other features that also influence system dynamics. In constructing a model, we always encounter the problem of how to compromise between realism and simplification, which topic I shall discuss below in some detail.

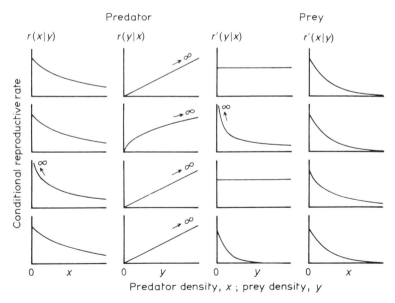

Figure 4.7 Typical conditional reproduction curves of the predator–prey models in Table 4.1. From top to bottom rows: the Nicholson–Bailey, Varley–Gradwell, Hassell–Varley and Beddington *et al.* models.

4.5 SIMPLE vs DETAILED MODELS

4.5.1 General principles

An ideal model might be one which represents the components of the system process in as much detail as we can conceive. Sooner or later, however, such a model would become too cumbersome to handle. Even with the simplest predator–prey interaction conceivable, an exact representation of its system components would lead to a complicated nonlinear model that is mathematically intractable. Besides, the incorporation of too many details makes it difficult to measure its parameters with precision and would, therefore, negate the sole purpose of building a detailed model. Further, as the model becomes more detailed, it tends to lose generality. We must somehow find a compromise between the ideal and practical and between the particular and general. From this point of view, a simple model has obvious merit.

In building a simple model, we should consider two major points: (1) the elimination of infeasibilities in basic biological properties such as unlimited satiation and reproductive capacity; (2) a parsimonious structure, such that the model is, at least, able to simulate most of the qualitative aspects of the system's dynamics, even if it is quantitatively not so accurate. In the following, I shall discuss these points with some examples.

4.5.2 Partitioned and unpartitioned models

In section 3.3.2, I partitioned one generation (of a univoltine species) into n successive stages, such that the overall reproductive rate $r_t = x_{t+1}/x_t$ is partitioned into stage survival rates, h_{st} $(s = 1, 2, \ldots, n-1)$, and natality (oviposition rate), h_{nt}, i.e.

$$r_t = h_{1t} h_{2t} \ldots h_{nt}. \tag{4.36}$$

In general, the rate h_s depends on density x_s at the onset of stage s. This is because some kind of competition among the members of the population is conceivable at any stage. If we assume scramble competition after (4.13b), survival rate h_{st} is given by

$$h_{st} = \exp(-c_s x_{st}); \quad s = 1, 2, \ldots, n-1$$
$$h_{nt} = r_m \exp(-c_n x_{nt}). \tag{4.37}$$

Substituting (4.37) in (4.36), we have

$$r_t = r_m \exp(-c_1 x_{1t} - c_2 x_{2t} - \cdots - c_n x_{nt}). \tag{4.38}$$

[Never confuse (4.38) with (4.26). We are dealing with within-generation processes here.]

In formulating a partitioned model such as above, we should be aware that: (1) the density-dependent structure in each stage is an approximation, and the estimation of a stage constant will contain some error; (2) error induced by an approximation tends to accumulate rather than cancel between stages; (3) therefore, the incorporation of too many stages will produce a poor result. Thus, there must be an optimal number of components we should incorporate into a model.

On the other hand, an unpartitioned model, as a regression model, has the advantage of being able to predict output x_{t+1} directly by input x_t (or $x_{t-1}, x_{t-2}, \ldots$, if necessary), bypassing all intermediate stages that may be black boxes. Moreover, such a regression model may describe an observed process even better than does a partitioned one, a point we should seriously consider in choosing a model. I shall discuss this subject again using a concrete example in sections 7.4 and 8.5.1.

4.5.3 Log linearization of a predator–prey interaction model

Every predator–prey system model discussed in section 4.4 assumes the form:

$$r_t = r_m f(x_t)g(y_t) \qquad (4.39a)$$
$$r'_t = r'_m j(x_t)k(y_t). \qquad (4.39b)$$

Each of the right-hand sides, when log transformed, becomes a linear combination of the functions of X and Y:

$$R_t = R_m + F(X_t) + G(Y_t) \qquad (4.40a)$$
$$R'_t = R'_m + J(X_t) + K(Y_t) \qquad (4.40b)$$

in which a capital letter represents the logarithm of the corresponding small letter.

The above structure does not accurately represent an actual predator–prey interaction system. In actual systems, the nonlinear relationship between x and y cannot usually be linearized by a log transformation. Nonetheless, the structure (4.39/40) is often capable of reproducing many aspects of observed dynamic patterns (Part Two). The structure is a compromise between retaining the model's ability to simulate some essential features of the observed system with some loss of descriptive power and the practicality of fitting the model to data.

The following is an aspect of nonlinear structure that the log linear model (4.40) cannot describe. In models (4.32) and (4.33), the conditional rate $r'(y|x)$ for prey was assumed to be a monotonically decreasing function of prey density y (Fig. 4.6d). This is not necessarily true (Bulmer, 1976; Rosenzweig, 1977). Given predator density x, the proportion of prey killed (y'/y) increases as y decreases as is obvious in Fig. 4.8a. There-fore, $r'(y|x)$ could be lower when y is low than when y is high. This could result in positive dependence of $r'(y|x)$ on y towards the lower end of y (see the dashed curve in Fig. 4.8b). However, sooner or later as y increases further, the intrinsic regulation of the prey population overrides the predation-induced positive density dependence. Therefore, for suffi-ciently large y, $r'(y|x)$ will have to decrease (solid curve, Fig. 4.8b). In order to describe such an effect of predation, we need a nonlinear model. But, then, we may lose quantitative accuracy.

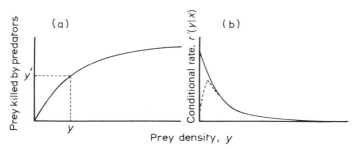

Figure 4.8 Number of prey killed (y') by predation, given prey density (y) (graph a) and its influence on the prey's conditional reproductive rate $r'(y|x)$ (graph b).

Under these circumstances, no single model may be satisfactory. We may have to use several types of model to describe different aspects of an observed process (Chapter 8).

4.5.4 The number of constant parameters required in a model

A stationary population time series is characterized by the mean and autocovariances (Chapter 3). In a deterministic model, these are the equilib-rium density and dynamic pattern. A practical model must have a minimum number of constant parameters to determine adequately the above statistical or mathematical characteristics.

Consider the basic, two-parameter logistic model

$$R_t = R_m - \exp(-a_0 + X_t). \qquad (4.41)$$

[cf. (4.15a).] The (log) equilibrium density X^* is determined by both parameters: since $R_t = 0$ at $X_t = X^*$, it is given by

$$X^* = \log R_m + a_0. \qquad (4.42)$$

The $(R_t$ on $X_t)$ reproduction curve shifts vertically and laterally as R_m and a_0 change, respectively, and, accordingly, X^* changes (see Figs 1.6 and 1.12, sections 1.6 and 1.7).

Now, the dynamics of a first-order model are determined by the slope of the reproduction curve at X^* (cf. Fig. 2.3, section 2.2.3). In (4.41), the slope (the derivative at $X = X^*$) is given by:

$$dR_t/dX_t|_{X=X^*} = -R_m. \tag{4.43}$$

We see that the dynamics of (4.41) is determined by R_m. In other words, once X^* was determined from observation, no more degree of freedom (flexibility) is left for model (4.41) to describe the dynamics. For example, suppose we used the values of R_m and a_0 determined by independent observation or experiment, and model (4.41) failed to generate the observed dynamic pattern. Then, the model is judged inadequate, and we may try the generalized model (4.16). The addition of parameter a_1 provides an extra degree of freedom. If this model failed again, we may move on to a higher-order model. A careful examination of disagreement and agreement between the observed and the simulated is an important step towards a better understanding of the observed process.

The predator–prey models discussed in section 4.4.3 have only two or three constant parameters. These are obviously inadequate for describing four (potentially independent) conditional reproduction curves. There ought to be a minimum of four parameters – at least one for each conditional rate – to construct a two-species system model.

The system model (4.34) or (4.35) has eight constant parameters, much in excess of four. However, two are concerned with the equilibrium levels, and another two for determining the variances, in both species. The remaining four parameters (two for each species) determine the autocorrelation structures of both species.

In fitting a model, it is often practical first to determine the autocorrelation structure by correlogram analysis (Chapter 3). [Always transform the data into logarithms.] The mean and variance can then be adjusted to their observed values by the transformation

$$X \equiv nX + m \tag{4.44}$$

in which n and m are constant adjustment factors (section 5.7.2). A transformation by (4.44) does not influence autocorrelations as is obvious from (3.13), section 3.2.3.

Many iterations may be needed to find a set of parameter values to describe observed dynamics to a reasonable degree. An adequate yet parsimonious use of constant parameters is the key to a useful model.

4.6 EXPERIMENTAL APPROACH TO THE DETERMINATION OF REPRODUCTION CURVES AND SURFACES

So far, I have been assuming that we determine a reproduction surface from time-series data. It would be difficult to do so, however, if the data series is comparatively short or if the population process is very stable. Unless the generation span is very short, adding more data points is a major problem. A researcher working for as long as 30 years on a univoltine species can generate no more than 30 data points. If the observed series fluctuates little or exhibits a stable, regular pattern, we would be unable to see more than a fraction of the reproduction surface. Often, we have little or no control over the variation of data points in a time series. A laboratory or field experiment may supplement inadequate parts of the time-series data.

The laboratory experiment is straightforward in principle, if the process concerned is first order in density dependence. Prepare first-generation individuals (parental stock) at an appropriate life stage. Introduce them into the experimental space in varying densities with replicates. Maintain the uniformity of environmental conditions as required. Count the number of offspring surviving to the same stage as their parents at the beginning of the experiment. Then, calculate the reproductive rate

$$r = x_2/x_1 \tag{4.45}$$

in which the numerals 1 and 2 stand for the parental and progeny generations, respectively. The regression of log r on log x_1 will estimate the reproduction curve (Chapter 7).

With a predator–prey system, we need to determine a reproduction surface for each species:

$$x_2/x_1 = r(x_1, y_1) \quad \text{(Predator)} \tag{4.46a}$$

$$y_2/y_1 = r'(x_1, y_1) \quad \text{(Prey).} \tag{4.46b}$$

If we need conditional reproduction curves for n fixed values on each (x_1 or y_1) axis for each species, there will be, without replicates, $4n$ experimental sets ($2n^2$ data points) altogether. We see that an experiment with a two-species system is much more laborious than one with a single species.

Although it may be laborious, the above type of experiment saves time if we can conduct all experimental sets simultaneously. Once a reproduction curve or surface is determined, we can deduce its population dynamics in time. However, the experiment must be carefully designed to replicate the conditions under which the time-series data are taken; or the deduced dynamics may not match the time-series data. For example, one might use an experimental population made up of individuals of uniform age. Then, scramble competition might result because they are equally strong. On the other hand, the population, from which time-series data are taken, may be composed of individuals of different ages at any moment. Then, older ones may have advantages over younger ones. So they compete in the contest fashion. I show some examples in Chapters 7 and 8.

In the field, one may look for many locations to cover a desired range of variation in density. But one must be aware that such a set of spatial data can be equivalent to time-series data only when the following assumptions hold. All local populations are supposed to be random samples from the same ensemble (section 1.5). That is, all sample populations have the same density-dependent/independent structure, and their parameters do not depend on their relative positions in the space – equivalent to 'stationarity' in time series.

If the food supply, predator–parasitoid composition, or weather influence is radically different between sample populations, they would have different equilibrium densities. If predators or parasitoids move between sample plots according to the local abundance of the prey, the density-dependent structure of either species will be affected. To make sure whether or not the above assumptions hold, one needs good knowledge of the ecosystem composition in each sample plot as well as knowledge of the life history of the species concerned and its position in the food web it belongs to. If it is decided that the assumptions hold reasonably well, proceed to the following.

If the structure is judged first order, calculate the reproductive rate (4.45) and draw a reproduction curve. If it is judged second order due to the interaction with a second species, then one must measure population densities of other species to construct conditional reproduction curves in the manner of the laboratory experiment already discussed.

If no information is available on the population of the second species, or the second-order structure is caused by a delayed effect (section 2.3.1), one may construct a reproduction surface in terms of a time-lag, e.g. $R_t = f(X_t, X_{t-1})$. In practice, one plots conditional reproduction curves $R_t = f(X_t | X_{t-1})$ and $R_t = f(X_{t-1} | X_t)$ as in the example in Fig. 3.13.

PART TWO

Analysis of classic cases

'And I may say that though I have now arrived at
what I believe to be the true solution of the case,
I have no material proof of it. I know it is so, because
it must be so, because in no other way can every
single fact fit into its ordered and recognized place.'

> Hercule Poirot (in *Murder in Mesopotamia* by
> Agatha Christie)

For the study of population dynamics, the material used must, at least,
satisfy either of the following two qualifications: (1) the observation must be
sufficiently long so that some time-series aspects of a population process can
be analysed with a reasonable degree of credibility; (2) information on the
demography and ecology of the population must be adequately detailed, so
that a suitable model can be built to explain some qualitative, if not
quantitative, aspects of the population dynamics.

In the following five chapters, I shall discuss several classic examples of
population dynamics. The first example, the lynx (*Lynx canadensis*), provides
an unusually long series of fur-trade records which typically depict the
widely occurring wildlife's 10-year cycle along the North American boreal-
forest zone (Keith, 1963). The records are suitable material for a statistical
analysis. The lynx, however, is an extremely shy animal so that only a very
little is known about its ecology.

In the second example, the snowshoe hare (*Lepus americanus*), the
fur-trade records are not as long and as extensive geographically as those
on the lynx. However, its ecology has been studied intensively in recent
years by some ecologists and naturalists. As the lynx's chief prey, its
population dynamics should be treated as an integrated part of the lynx
10-year cycle.

In the subsequent two chapters, I shall discuss the oft-cited work of Utida
on the azuki bean weevil (*Callosobruchus chinensis*). His work comprises two
separate series of experimental studies under the series titles: 'Studies on
experimental populations of the azuki bean weevil' and 'Host–parasite
interaction in the experimental population of the azuki bean weevil'. These

laboratory studies provided both very long observations and details of the ecology of the species concerned.

The final example, the spruce budworm (*Choristoneura fumiferana*), is a devastating pest of spruce-fir forests in eastern Canada and the adjacent part of the United States. The famed Green River Project of the Canadian Forest Service conducted extensive life-table studies of it. The discussion is largely based on the analysis of these life-table data.

In order to read Part Two, it is not absolutely essential to have full knowledge of Part One. Whenever necessary, I shall refer the reader to certain sections of Part One for a better comprehension of my argument.

5 Analysis of the lynx 10-year cycle

5.1 INTRODUCTION

Nothing has so intrigued and fascinated many naturalists and ecologists as the persistent 10-year cycle of Canada lynx, *Lynx canadensis*. The well-known archives of the fur trade between the Hudson's Bay Company and the Canadian trappers in the past two centuries have been a rich source of speculation about the cause of the cycle. As a result, diverse theories exist in the literature. However, many of these ideas were not substantiated by observations, and some statistical analyses ignored ecological mechanisms.

The ultimate aim of my analysis here is to pave the way to a comprehensive, coherent interpretation of the lynx 10-year cycle. However, because the lynx is so much dependent on its chief prey, the snowshoe hare (*Lepus americanus*), a full comprehension of the lynx population process is difficult without knowledge of hare ecology, the topic of Chapter 6. In the present chapter, I shall concentrate on the analysis of the fur-return statistics and on developing a model that later assists in interpreting the relationship between the snowshoe hare and its predator complex of which the lynx is a constituent.

I shall first examine, in section 5.2, the nature of the fur-return statistics and discuss how they can be related to changes in lynx populations. This will be followed by: a brief review of early theories (section 5.3); the autocorrelation structure of the cycle (section 5.4); and a review of models proposed by several statisticians (section 5.5). All of these statisticians' models assume linear dependence of the animal's reproductive rate on its own population density. However, the lynx data show nonlinear dependence (section 5.6). In section 5.7, I apply the two types of nonlinear models developed in Chapter 4: the second-order logistic and predator–prey models. In these sections, I deal with stationary series of fur-return records, i.e. those in which mean and covariances (of fur returns) do not systematically change in time. Section 5.8 discusses nonstationary series.

5.2 FUR-RETURN STATISTICS AS AN INDEX OF LYNX ABUNDANCE

Elton and Nicholson (1942) have compiled the most complete and consistent table of the numbers of lynx pelts collected each year from 1821 to 1939 (Fig. 5.1) in ten different districts across Canada (Fig. 5.2). The figures from

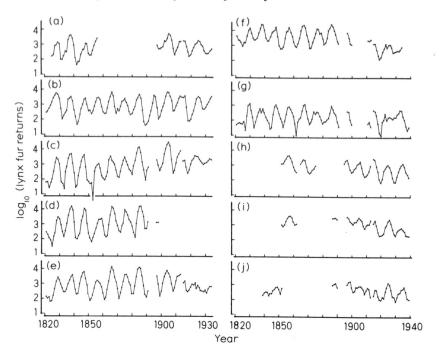

Figure 5.1 Hudson's Bay Company records of lynx fur returns from ten districts across Canada; after Elton and Nicholson (1942; table 4). Names and locations of Districts (a) to (j) are in Fig. 5.2.

1821 to 1913 are based on the London archives of the Hudson's Bay Company, and those from 1915 to 1939 are from the books kept at the Company's Fur Trade Department in Winnipeg; for details, see section 4, p. 226, of Elton and Nicholson (1942). The 10-year cycle still persists in more recent records of pelts harvested across Canada compiled by Statistics Canada (Fig. 5.3).

Many people have questioned whether the fur-return statistics are a reasonable index of the actual abundance of the animal. Since my argument in this chapter relies on the statistics, I shall first examine their reliability.

Gilpin (1973), Weinstein (1977) and Winterhalder (1980) held the most pessimistic view. They suggested that the cycle in fur returns did not reflect the actual abundance of the lynx, but it was a similar cycle in the snowshoe hare abundance that determined trappers' effort. However, the old notes kept by some resident traders and local naturalists clearly indicate that the abundance and scarcity of the 'cats' tend to agree well with the peaks and troughs of the fur-return cycle (Elton and Nicholson, 1942). More recent information on lynx demography (Nellis *et al.*, 1972; Brand *et al.*, 1976; Brand and Keith, 1979) has provided direct evidence, although incomplete, that the cyclic fluctuation in lynx abundance is genuine, and that the pelt harvest does cycle in parallel to the census data (Fig. 5.4).

Figure 5.2 Approximate locations of the ten districts across Canada where lynx pelts were collected for Hudson's Bay Company (Fig. 5.1), based on Elton and Nicholson (1942; figs 1–6). Districts are: (a) West, (b) Mackenzie River, (c) Athabasca Basin, (d) Upper Saskatchewan, (e) West Central, (f) Winnipeg Basin, (g) North Central, (h) James Bay, (i) Lakes and (j) Gulf.

Nonetheless, the fur-return records do seem to exaggerate the cyclic changes in lynx populations. For example, Brand and Keith (1979) observed that the lynx population in Rochester, Alberta, increased 4.3-fold over 5 years, while the pelts harvested during the same period showed a 20-fold increase (Fig. 5.4). Similarly, a peak rate of increase in fur returns in Fig. 5.1 often exceeds a biologically probable maximum reproductive rate of the lynx.

We can roughly estimate the highest possible rate of increase using the demographic data of Brand and Keith (1979): The data show that: the sex ratio was about 1:1; the largest average *in utero* litter size of an adult female (excluding yearlings) was 5.0 (4.6±0.4); the highest pregnancy rate in adults was 0.73; and the highest survival rate (May to November) was 0.93 in adults (including yearlings) and 0.35 in kittens. Then, the maximum average rate of population increases could not be much higher than 1.57 because:

(Proportion adult females in population) × (pregnancy rate) × (*in utero* litter size) × (survival of kittens) + (survival of adults)
$$\leqslant 0.5 \times 0.73 \times 5.0 \times 0.35 + 0.93 = 1.57.$$

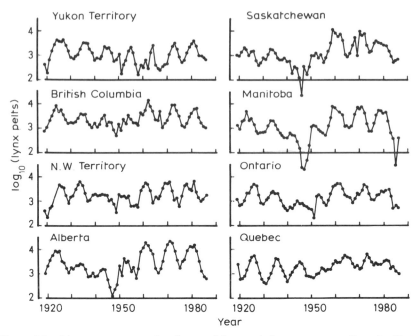

Figure 5.3 More recent records of annual lynx pelt harvest across Canada. Data taken from Dominion Bureau of Statistics (1965) and Statistics Canada (1983).

The above calculation ignores winter mortality and breeding by yearlings. If these are included in the calculation, the maximum rate of increase would be further reduced. The yearlings do breed when food is plentiful, but their average pregnancy rate, litter size and survival rate are significantly smaller than those of adults (Brand and Keith, 1979). On the other hand, the pregnancy rate of adults used in the above calculations might be an underestimate because of the small sample size. But even a 100% pregnancy would enhance the maximum rate of population increase only slightly to 1.81. The maximum rate is probably not higher than 2. Figure 5.5 shows, however, that the peak rate of increase in a pelt-harvest cycle often exceeds the probable maximum rate of 2, the upper horizontal line in each graph. [I do not consider the effect of migration on the rate of change in a local population because the effect should be negligible if we consider a sufficiently large geographical area as in Fig. 5.2.]

A probable lower limit in the rate of increase is difficult to establish. The lowest rate of change observed by Brand *et al.* (1976) was 0.45 from 1964 to 1965 (Fig. 5.4). The combination of the lowest pregnancy rate (0.33), litter size (3.1) and survival rate (0.62) in Brand and Keith (1979) yields a minimum of 0.68. Probably, the average lower limit is not much lower than 0.4. But the minima in the cycles in Fig. 5.5 are often below 0.4, the lower horizontal line in each graph. Thus, I suspect that the fur-return statistics tend to exaggerate the amplitude of the lynx cycle in both directions.

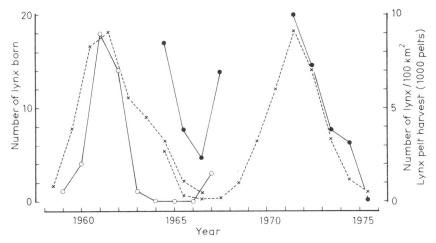

Figure 5.4 Yearly changes in some indices of lynx abundance in a study area near Rochester, Alberta. Open circles: number of lynx born which survived to first capture (Nellis *et al.*, 1972; fig. 1). Solid circles: number of lynx censused per 100 km² during winter near Rochester, Alberta (Brand *et al.*, 1976; table 1). Crosses: number of lynx pelts harvested in Alberta; earlier half from Nellis *et al.* (1972; fig. 1), and later half from Brand and Keith (1979; fig. 4).

Three factors – trap bias, sampling (harvest) error and trappers' effort – can conceivably cause a discrepancy between fur returns and the actual abundance of the lynx.

A lynx is more readily trapped when its chief prey, the snowshoe hare, is scarce, and vice versa (Ward and Krebs, 1985). Because the trough and peak of a snowshoe-hare cycle is only slightly ahead of the corresponding lynx cycle (section 6.8), this type of trap bias could reduce the amplitude of a fur-return cycle. Also kittens are less likely to be trapped than adults (Brand and Keith, 1979). Since the proportion of kittens in a population is higher during the increasing phase of a cycle (Brand and Keith, 1979), this trap bias, too, could reduce the amplitude.

Figure 5.1c shows an unusually low harvest in 1853 in Athabasca Basin District. This caused an unusually low rate of change in fur harvest from the 1852 to 1853 seasons and, conversely, an unusually high rate from 1853 to 1854 (Fig. 5.5c). With a sufficiently large number of pelts harvested, this type of error would have a minor influence.

Perhaps the most consistent cause of the exaggerated amplitude of the fur-return cycle is the trappers' effort. Apparently, the company traders paid the trappers a fixed price for more than a century (Newman, 1985). Then, for a unit trap effort, the net profit would be low at the trough of a lynx cycle and high at a peak. Thus, the trappers must have been discouraged to catch the animals when scarce and vice versa when abundant. [Brand and Keith (1979) thought that the trappers put more effort into catching when

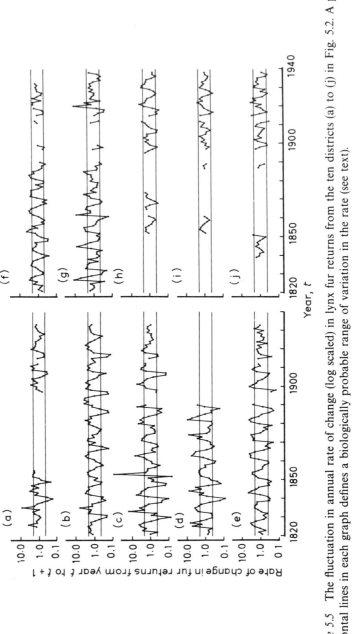

Figure 5.5 The fluctuation in annual rate of change (log scaled) in lynx fur returns from the ten districts (a) to (j) in Fig. 5.2. A pair of horizontal lines in each graph defines a biologically probable range of variation in the rate (see text).

the pelt price was high, and vice versa. However, as mentioned above, the trappers were paid a fixed price. Also, there is no clear relationship between the number of pelts traded and the market price in recent years (Fig. 5.6).]

Another important source of discrepancy is changes in the boundaries of some districts (Elton and Nicholson, 1942). For instance, in the Athabasca Basin District, one post (Fort St. Johns) was not in operation between 1824 and 1857, resulting in a probable reduction in the number of pelts bought in the District. In 1871, a new post was established to extend the District's area of coverage. In 1881, and again in 1899, there were further extensions of the District. An apparent increasing trend in fur returns in the District (Fig. 5.5c) from 1855 onwards could very well be the result of such changes (section 5.8).

Interestingly, an uninterrupted record from the Mackenzie River District, which underwent the minimum degree of changes among the ten districts, shows the most stable cyclic fluctuation (Fig. 5.1b). If we reduce the variance of the series in Fig. 5.1b in such a way that the rate of change in Fig. 5.5b is

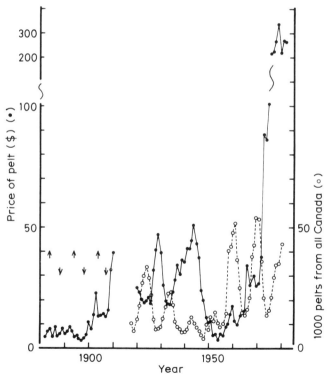

Figure 5.6 Comparison between lynx fur returns (arrows and open circles) and price of pelt (solid circles). Fur data: peaks and troughs before 1910 are indicated by upward and downward arrows (cf. Fig. 5.1); after 1918 are pooled data from all regions in Fig. 5.3. Price data: up to 1910 are the London market price (Jones, 1914; p. 216), and from 1920 on, Canadian price compiled by Dominion Bureau of Statistics (1965) and Statistics Canada (1983).

accommodated within the biologically probable range, the fur-return series should reasonably approximate the dynamics of the lynx population in the District. Thus, this series is most suitable for the analysis of basic characteristics of the lynx cycle. Some of the districts show a deterioration in their cyclic pattern from 1915 onwards (Fig. 5.1). I discuss probable causes in section 5.8.

5.3 EARLY THEORIES OF THE 10-YEAR CYCLE

The intriguing features of the lynx 10-year cycle (Fig. 5.1) are its remarkable regularity and synchrony across all regions of Canada. Many speculations about the cause of the cycle have been published in the past half-century. Table 5.1 gives a summary of typical theories proposed before 1965; many variations of these are reviewed in Keith (1963). There are two major categories, namely, extrinsic (density-independent) and density-dependent causes.

Table 5.1 Early theories on the lynx cycle.

	Ultimate cause	*Proximate cause*	*Effect on:*	*Authors*
I.	Extrinsic			
	Sunspot cycle			Elton (1924)
	Lunar cycle	Moonlight quantity	Reproductive potential	Siivonen and Koskimies (1955)
	Ozone cycle			Huntington (1945)
	Ultraviolet-ray cycle	Plant nutrient	Nutritional imbalance in hare	Rowan (1950)
	Weather cycle	Plant succession	Changes in hare habitat	Grange (1949)
	Forest-fire cycle	Plant succession	Changes in hare habitat	Grange (1965)
	Plant-nutrient cycle	Inherent to plants	Nutritional imbalance in hare	Lauckhart (1957)
II.	Density-depenent			
	Changes in prey population	Overpopulation in hare	Epidemics	MacLulich (1937)
			Shock diseases	Green and Evans (1940)
			Physio-logical stress	Christian (1950)
	Grazing by hare	Plant–hare oscillation		Lack (1954)
	Predation	Hare–lynx oscillation		Moran (1953a)

On the whole, the theories of extrinsic causes have little merit. They are entirely based on correlations, poorly supported by evidence for causal relationships. The sunspot and related theories (e.g. the ozone and ultraviolet-ray cycle) have long been rejected since MacLulich (1937) because of their poor correlations with the lynx cycle (Fig. 5.7); see the comment in section 3.3.4. Note, however, that a correlation might have been judged 'good' if one had data only for the first 30 years. Only after an extended observation, we know that the correlation is spurious.

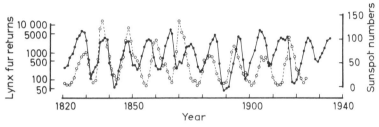

Figure 5.7 Comparison between the lynx series (solid lines) from Mackenzie River District (Fig. 5.1b) and Wolfer's sunspot numbers (dashed lines) taken from Yule (1927; table A).

As long as a 'good correlation' exists, the idea of extrinsic causes seems to remain attractive as in the more recent resurrection of the theories based on the lunar cycle (Archibald, 1977), forest-fire cycle (Fox, 1978) and weather-pattern cycle (Arditi, 1979). However, the occurrence of spurious (or nonsense) correlations between unrelated time series is frequent (section 3.3.4). Conversely, a poor correlation between population fluctuation and some extrinsic factor does not necessarily imply that the factor has little influence (section 3.3.1). Therefore, correlation means little unless we see a plausible causal relationship. The suggested causations in the above theories are rather farfetched.

One might believe that the factor responsible for the regularity of cyclic fluctuations must also be the cause of their synchrony. Thus, one seeks an extrinsic cyclic phenomenon as a possible cause. However, the Moran (1953b) effect (summarized below) explains the interregional synchrony of population fluctuations without assuming any extrinsic cyclic causes:

> If several regional populations have the same intrinsic, density-dependent structure, they will be correlated under the influence of some extrinsic factor (e.g. a climatic factor), if the factor is correlated between the regions.

[See section 2.5 for details and Fig. 2.22 for demonstration.] Under this effect, the factor that causes cyclic population fluctuations need not be the one that causes their synchrony. Thus, we have a considerable flexibility in looking for a solution. It is unfortunate that this important theorem has been almost completely ignored in the history of population ecology for more than 30 years since its proposal.

From the Moran effect point of view, each theory in the density-dependent category has merit. For example, a second-order density-dependent process (e.g. a predator–prey interaction) could readily generate a cyclic population fluctuation (section 5.7.2). Under the Moran effect, those regional populations cycling independently of one another could be brought into synchrony. However, none of those early theories in the density-dependent category is adequate in its original form to explain the lynx cycle. I shall examine and incorporate these ideas in appropriate places as I develop my argument in this and the following chapters. But, first, let us look at some statistical characteristics of the lynx cycle.

5.4 STATISTICAL CHARACTERISTICS OF THE LYNX CYCLE

Let X_t be the log population density in year t and R_t be the log rate of change in the population (or, simply, reproductive rate) from year t to $t+1$, i.e.

$$R_t = X_{t+1} - X_t. \tag{5.1}$$

If R_t depends on densities in the h most recent years, it is said to be an hth-order density-dependent process, written as

$$R_t = f(X_t, X_{t-1}, \ldots, X_{t-h+1}) + z_t \tag{5.2a}$$

or alternatively by (5.1),

$$X_t = X_{t-1} + f(X_{t-1}, X_{t-2}, \ldots, X_{t-h}) + z_{t-1} \tag{5.2b}$$

in which z is the net effect of perturbation by density-independent factors. Setting $z \equiv 0$, the relationship (5.2a) represents a reproduction surface (regression of R_t on $X_t, X_{t-1}, \ldots$) in an h dimensional space, and z_t is a vertical deviation of an observed R_t from the surface. [In this sense, z is called a vertical perturbation effect on the reproductive rate (section 1.7). Other types of perturbation effects – lateral and nonlinear – will be considered later.] Our ultimate goal is to determine the reproduction surface (5.2a) and to find an appropriate model which reasonably approximates to it.

The lynx series from the Mackenzie River District (Fig. 5.1b) is characterized by a stable cyclic pattern without any noticeable trend in the mean and in the amplitude of a cycle (variance). When a given statistical parameter, e.g. the mean, the variance or a covariance, is practically constant and does not systematically change with time, it is considered to be stationary (section

1.3). The lynx series from the District is practically stationary in the mean and in the variance. Then, we can extract its dynamic pattern by calculating correlations between two data points in the series that are j $(= 1, 2, 3, \ldots)$ years apart from each other. These correlations are called autocorrelations of the time series (section 3.2).

We calculate an autocorrelation coefficient in much the same way as the usual correlation between measurements of certain objects, such as the weight and height of people. We simply assemble pairs of data points in the series that are j years apart from each other, e.g. $(X_1, X_{1+j}), (X_2, X_{2+j}), \ldots, (X_{T-j}, X_T)$, in which T is the total length of the series. Then, we calculate the coefficients for $j = 1, 2, 3, \ldots$ by the formula (3.13), section 3.2.3. In practice, we use computer software, such as the MINITAB ACF package. The number j is called the lag, and the autocorrelation as a function of the lag is called the autocorrelation function (or ACF for short). The plot of the autocorrelation coefficients against lags $1, 2, 3, \ldots$ is called the correlogram.

The graphs on the left in Fig. 5.8 show (sample) correlograms calculated from the six series (b to g, Fig. 5.1). [A parameter value calculated from data is called a sample value as distinguished from a theoretical (or expected) value.] In the calculations, I used the whole length (114 years) in series b (Mackenzie River) and the first 71 years in the other five series. Series a, h, i, and j are too fragmentary for a meaningful correlogram analysis.

All correlograms exhibit much the same cyclic pattern. The correlation coefficients peak nearly every ten lags. The first peak comes mostly at lag 9 although the coefficients at lag 10 are very close (the peak is at lag 10 in the Mackenzie River series). The second peak is at lag 19 in every series; and the third is mostly at lag 29 (at 28 in the Athabasca Basin series). The pattern indicates that the average cycle length of all lynx series is slightly less than ten years. Also, the correlograms damp down so slowly that the correlation even at lag 29 is still quite high. This means, of course, that the number of fur returns this year would still be correlated with those of 29 years ago, indicating a highly stable cyclic fluctuation. However, such a persistently high correlation does not imply that the lynx abundance today is governed by some density-dependent factor whose effects has persisted for such a long time. It does not necessarily imply that the order (h) of the density-dependent process (5.2) is as high as 10, 20 or even 30. A population process in which $h = 2$ can generate a cyclic pattern like the lynx cycle (section 5.5.1).

One way to determine the order (h) of density dependence is to look at the partial autocorrelation function (PACF) (section 3.2.2). Suppose X_t is determined only by X_{t-1} and X_{t-2}, i.e. $h = 2$ in (5.2b). Even so, X_t can still be correlated with X_{t-3} and backwards because X_{t-1} structurally depends on X_{t-2} and X_{t-3}. The partial autocorrelation between X_t and X_{t-3} and backwards should become insignificant if there is no structural dependence.

The PACFs calculated on the right-hand graphs in Fig. 5.8 show that the coefficients for lags 1 and 2 are always high in absolute values, whereas those for the higher lags are much less, even if some of them are still

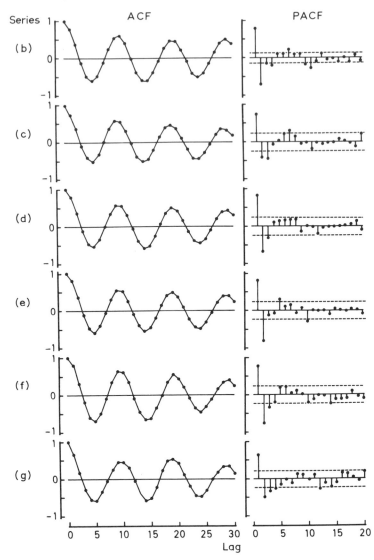

Figure 5.8 Autocorrelations (ACF: left graphs) and partial autocorrelations (PACF: right graphs) of the six lynx fur-return series (b)–(g) in Fig. 5.1. A pair of dashed lines in each PACF-graph defines an approximate 95% confidence interval for the zero correlation (section 3.2.3).

significantly different from zero. Thus, the density-dependent structure of the lynx population process is at least second order but not much higher. In practice, we can consider order 2 to be a good approximation.

With the above statistical characteristics of the lynx cycle in mind, let us look at the models that several statisticians proposed to simulate the lynx series.

5.5 REVIEW OF STATISTICAL MODELS OF THE LYNX CYCLE

Theoretically, two types of population process can generate cyclic fluctuations (section 2.4). One is a second- or higher-order density-dependent process with uncorrelated perturbation effect, e.g. $h \geqslant 2$ in (5.2a) and z is an uncorrelated random number, say u. The other is a first-order density-dependent process ($h = 1$) with autocorrelated perturbation effect, e.g. z has a periodic component in it. I shall use my notations in this review.

5.5.1 Linear second-order autoregressive model

Moran (1953a), probably the first person to attempt an analysis of the statistics of some game birds and fur-bearers, conceived that the lynx cycle was the first type of population process. He applied the simplest, linear second-order autoregressive model, AR(2) for short, i.e. the model (5.2b) in which $f(X_{t-1}, X_{t-2}) = a_1 X_{t-1} + a_2 X_{t-2}$. Thus:

$$X_t - m = (1 + a_1)(X_{t-1} - m) + a_2(X_{t-2} - m) + u_{t-1} \qquad (5.3)$$

where X is the log number of pelts, m is the mean of X, a_1 and a_2 are constants, and u is an independent and identically distributed random number. He fitted this model to the lynx series from the Mackenzie River District (Fig. 5.1b). His estimates of the parameters are: $m = 2.9036$, $a_1 = 0.4101$, $a_2 = -0.7734$, and the residuals (u) have the mean $= 0$ and variance $= 0.04591$. [These values are based on X being the common (base 10) logarithm.] Thus, (5.3) becomes

$$X_t = 1.0549 + 1.4101 X_{t-1} - 0.7734 X_{t-2} + u_{t-1}. \qquad (5.4)$$

In Fig. 5.9a, I generated a series using model (5.4) with the initial numbers (X_0, X_1) taken from the lynx series (Fig. 5.1b) and the series $\{u_t\}$ generated by MINITAB – base 1, normal, mean $= 0$, standard deviation $= \sqrt{0.04591}$. Despite its simplicity, the model appears to mimic the cyclic pattern of the lynx series reasonably well. However, the sample correlogram (Fig. 5.10a) of the simulated series does not quite match the sample ACFs of the lynx series (dashed graph).

The sample PACF (Fig. 5.10a′) of the simulated series appears to be basically similar to that of most lynx PACFs (Fig. 5.8, right graphs); the coefficients for the first two lags are large in absolute values compared to those for the higher lags. Yet, there is a subtle difference compared to the lynx PACFs. While the coefficients for lags higher than 2 in Fig. 5.10a′ are all insignificantly different from zero (as expected from the theoretical PACF of the model which truncates after lag 2 – section 3.2.2), the lynx PACFs in Fig. 5.8 for the higher lags are not all insignificant. Moreover,

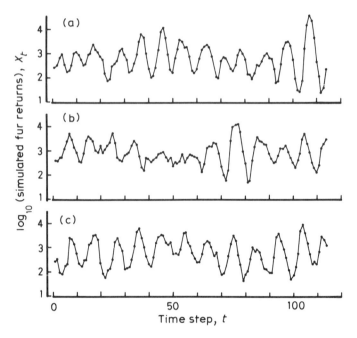

Figure 5.9 Simulations of lynx cycle with three linear models: (a) Moran's AR(2); Tong's AR(11); (c) Campbell and Walker's AR(2) with periodic perturbation.

they appear to have a somewhat systematic trend which is not apparent in the simulation.

Although the result of simulation is not quantitatively satisfactory, Moran's model provides a useful insight into two important aspects of the lynx cycle: the order of density dependence and the synchrony of cycles among the regional populations.

As I have discussed in section 2.3, a second- (or higher-) order density-dependent structure may originate from two major possible sources. One is the carried-over density effect: a second-order structure would result if the effect of population density on the physiology of individual animals in one year is carried over to the following year so as to affect the reproductive rate of the population in the following year. The other source is the effect of interaction with other species: if two species (or two groups of species) interact with each other, the density-dependent structure of either species (or group) should be second order. It is not known whether the carry-over density effect is a source of the second-order density dependence in lynx. However, the lynx, as a predator, must interact with its prey. Therefore, a second-order model is an appropriate and parsimonious choice from the ecological point of view.

The second insight of Moran's model is the effect of the density-independent perturbation factor on the synchrony among the regional populations (section 5.3). We may assume that local lynx populations are fluctuating cyclically, but independently of one another, due to the interaction with prey

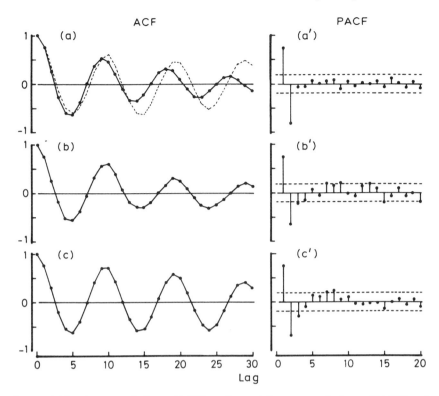

Figure 5.10 Autocorrelations (ACF) and partial autocorrelations (PACF) calculated from the three simulation series a, b and c of Fig. 5.9. The dashed curve in (a) is the ACF of the lynx series (Fig. 5.8b). A pair of dashed lines in each PACF-graph defines an approximate 95% confidence interval for the zero correlation.

in each locality. These local interaction systems are subjected to a certain extrinsic perturbation factor which is correlated among the localities (e.g. similar climatic influences). Then, by the Moran theorem (section 5.3), we would expect synchrony among the local population cycles. Some climatic influence in winter is a likely source of perturbation, and the climate does tend to be correlated among regions (Chapter 6).

Despite its importance, Moran's work has been largely ignored in the ecological literature. Recently, however, some statisticians have revived interest in Moran's statistical model and have further elaborated it. Let us look at their efforts.

5.5.2 Eleventh-order autoregressive models

The fact that the PACFs of the lynx series do not seem to truncate completely after lag 2 suggests three types of possible mechanisms or their combinations. Referring to the general model (5.2), these are: (1) the order of density dependence (h) is higher than 2; (2) the perturbation effect z is not an independent random number; and (3) the density-dependent structure (f) is nonlinear.

Addressing the first possibility, Tong (1977) examined the optimal (or most parsimonious) order for a linear autoregressive model to describe the lynx series, using the Akaike Information Criterion. [The A.I.C is a widely-used method in time-series analysis to determine the order of linear autoregressive scheme (Priestley, 1981; Chatfield, 1984; Brockwell and Davis, 1987).] Tong found order 11 to be optimal, i.e. model (5.2b) in which f is a linear function of $X_{t-1}, X_{t-2}, \ldots, X_{t-11}$:

$$X_t - m = (1 + a_1)(X_{t-1} - m) + a_2(X_{t-2} - m) + \cdots$$
$$+ a_{11}(X_{t-11} - m) + u_{t-1} \tag{5.5}$$

where $m = 2.9036$ is the mean of the $\log_{10}$ lynx fur returns as before; the estimates $\hat{a}_h$ for $h = 1$ to 11 are: 0.13, -0.51, 0.23, -0.29, 0.14, -0.14, 0.08, -0.04, 0.13, 0.19 and -0.3; and the random number u has the mean $= 0$ and variance $= 0.0437$.

Figure 5.9b is my simulation using (5.5) with the $\hat{a}_h$; the $\{u_t\}$ are the MINITAB random numbers – base 1, normal, mean $= 0$, standard deviation $= \sqrt{0.0437}$. To initialize the simulation I used the first 11 values of the lynx series (Fig. 5.1b); these are omitted in the illustration. The sample ACF and PACF are given in Fig. 5.10b and b'.

We see a little improvement in the correlograms in Tong's AR(11) model over Moran's AR(2). The sample PACF damps out more slowly (because it truncates only after lag 11) than that of the AR(2) in Fig. 5.10a' (which truncates after lag 2). However, there is no visible improvement in the dynamic pattern of the simulated series (Fig. 5.9b) over that of the AR(2) series (Fig. 5.9a).

Such a small improvement of the AR(11) model cost a large number of lagged terms. How can we interpret its 11-year time-lag in ecological terms? It can be argued that second-order density dependence should be common among wildlife populations, but much more than third order is unlikely (section 2.3). It is hard to imagine any ecological reason for the effect of density 11 years ago to influence the population today. The AR(11) model is like a polynomial curve fitting and provides no insight into the ecological structure of the lynx population process.

More recently, Haggan and Ozaki (1981) proposed a modification of the linear AR(h) scheme. They modified the ith autoregressive parameter a_i, which is constant in the linear scheme, into a function of X_t, say $a_i(X_t)$, such that

$$a_i(X_t) = \alpha_i + \beta_i \exp[-\gamma(X_t - m)^2], \quad i = 1, 2, \ldots, h. \tag{5.6}$$

where α_i, β_i and γ are constants, and m is the mean of the X as before. [The scheme (5.6) can be considered to be a particular type of nonlinear perturbation effect (section 1.7.3), which in this case, is correlated with density.] When the constants are appropriately chosen, the process can oscillate. In particular, for a small deviation of x_t from the mean m, the coordinates

$$[a_1(X_t), a_2(X_t), \ldots, a_h(X_t)] \tag{5.7}$$

can be outside the stationarity region for the AR(h) scheme. Then, the process exhibits a divergent oscillation: it oscillates with an increasing amplitude. However, as X_t deviates further away from the mean m, the parameter coordinates (5.7) return to the stationarity region, where the process tends to damp down. [The above behaviour can be understood easily by the analogy with the second-order scheme ($h=2$) whose stationarity region is defined by the triangle in Fig. 2.5, section 2.2.4; the parameter coordinates $[1 + a_1(X_t), a_2(X_t)]$ oscillate between region IV inside the triangle and region IV′ outside.] The alternation between divergent and convergent oscillations produces a limit cycle.

Using the Akaike Information Criterion, Haggan and Ozaki also found $h = 11$ to be an optimal order, as had Tong. It is rather surprising that this elaborated limit-cycle model still requires an 11-year lag to simulate the lynx cycle optimally. The Haggan–Ozaki model is based on an equation describing random mechanical vibrations and does not illuminate the ecological mechanics of animal population dynamics.

5.5.3 Linear autoregression with periodic perturbation

In addition to his AR(2) model, Moran (1953a) considered a model in which the lynx cycle is governed by some meteorological or other terrestrial phenomenon which itself exhibits periodic oscillations. The model is:

$$X_{t+1} - m = \alpha \sin(\beta t - \gamma) + u_t \tag{5.8a}$$

in which the parameters α, β and γ are constants representing, respectively, the amplitude of an oscillation, the periodicity and the initial phase.

We can write (5.8a) in the form of (5.2a): by subtracting $X_t - m$ from both sides of (5.8a) and substituting R_t for the difference $X_{t+1} - X_t$ on the left by the relationship (5.1), we have

$$R_t = -(X_t - m) + \alpha \sin(\beta t - \gamma) + u_t. \tag{5.8b}$$

Thus, (5.8) is a linear first-order density-dependent process subjected to the periodic perturbation due to the harmonic term.

[One might consider that (5.8a) represents a density-independent process because population density on the left-hand side is determined solely by a density-independent element. The alernative expression (5.8b) reveals that it is, in fact, a density-dependent process. The expression (5.8a) means that the equilibrium state of the process is so stable – because of its density-dependent structure (5.8b) – that only the effect of density-independent perturbation shows up in population fluctuation. For details, see section 3.4.1.]

Model (5.8) can be generalized to:

$$R_t = a(X_t - m) + [\alpha \sin(\beta t - \gamma) + u_t] \tag{5.9}$$

(Bulmer, 1974). Campbell and Walker (1977) generalized (5.9) even further by adding a second-order density-dependent term such that:

$$R_t = a_1(X_t - m) + a_2(X_{t-1} - m) + \alpha \sin(\beta t - \gamma) + u_t \tag{5.10}$$

and fitted the model to the lynx series from the Mackenzie River District (Fig. 5.1b). Their estimates of the parameters are: $a_1 = 0.0302$, $a_2 = -0.3228$, $\alpha = -0.2531$, $\beta = 2\pi/9.5$ and $\gamma = -(5\pi - 4.04)/19$; the u have the mean$=0$ and variance$=0.0440$. Figure 5.9c is my simulation using their parameter estimates. Its sample ACF and PACF are given in Fig. 5.10c and c′. We see that the simulation closely resembles the lynx series in most aspects.

One may interpret the periodic terms in (5.9) and (5.10) to represent the effect of some cyclic meteorological factors. Moran wrote, however, that he knew of no such cyclic meteorological phenomena that could possibly and significantly influence the animals' survival. On this basis, Moran rejected this type of model. Even today, I know of no such factor, and I believe it unlikely that one will be found in the future.

However, a periodic, trigonometric polynomial function can be used to simulate an autocorrelated perturbation effect. [For probable origins of autocorrelation in perturbation effects, see section 2.3.] In fact, many types of time series can be expressed by the sum of some trigonometric functions just as many types of graph can be approximated by algebraic polynomials. Indeed, with such a mathematical device, a lynx fur-return series can be simulated almost perfectly (see Chapter 20 of BMDP Statistical Software edited by Dixon, 1981). This subject is usually treated under 'periodogram and Fourier analyses' in time-series analysis. However, I shall not pursue this topic in this book: it does not provide insight into the ecological mechanism of population dynamics.

All of the statisticians' models reviewed above are linear in density dependence. However, as shown below, the density-dependent structure of lynx process is nonlinear. A population process model should incorporate this property.

5.6 THE LYNX CYCLE AS A NONLINEAR DENSITY-DEPENDENT PROCESS

5.6.1 Asymmetry in the lynx cycle

A series of log population densities generated by any of the above models exhibits a symmetrical fluctuation both about the mean and about the peak (or trough) of a cycle. This means that if those series in Fig. 5.9 are inverted or turned over, their pattern of fluctuation would still look the same.

The lynx cycle, however, is not symmetrical (Moran, 1953a; Williamson, 1972). In the Mackenzie River series (Fig. 5.1b), it takes an average of 5.78

Table 5.2 Asymmetry about the peak in the lynx cycle from Mackenzie River District (Figure 5.1b).

Cycle No.	1	2	3	4	5	6	7	8	9	10*	11	Mean	S.d.
No. years to:													
increase	7[†]	6	6	5	5	6	6	6	6	—	6	5.78[‡]	0.441
decrease	4	4	4	4	3	4	4	3	4	—	4	3.80	0.422
Total	11[†]	10	10	9	8	10	10	9	10	11	10	9.70[§]	0.823

*Peak is indeterminable. [†]Minimum number. [‡]Excluding cycles No. 1 and 10.
[§]Excluding cycle No. 1.

Table 5.3 Asymmetric oscillation about the mean in the (log) lynx fur returns from Mackenzie River District (Figure 5.1b).

Cycle No.	Trough	Peak
1	2.430*	3.774
2	1.991	3.533
3	1.653	3.404
4	2.352	3.458
5	2.373	3.827
6	2.407	3.352
7	2.303	3.647
8	1.591	3.605
9	2.021	3.845
10	2.537	—
11	1.903	3.553
12	2.686	—
Mean	2.1652[†]	3.5998
S.d.	0.3573	0.1735

*Could be less.
[†]Excluding cycle No. 1.

years from a trough to reach a peak but only 3.80 years to decrease to the next trough (Table 5.2). Also, the lynx (log) numbers at troughs are slightly more variable than those at peaks (Table 5.3). In other words, after a log transformation, the lynx cycle is asymmetrical about the peak (or trough) and about the mean. This means that the reproduction surface of the lynx process is curved.

5.6.2 Nonlinear reproduction surface

The reproduction surface of a second-order density-dependent process, i.e. $h=2$ in (5.2a), is a regression of R_t on X_{t-1} and X_t. But a three-dimensional regression is difficult to see. It so happens, however, that we can see the curved surface of the lynx process in a much simpler way.

A linear second-order process – in which R_t ($=X_{t+1}-X_t$) is determined by both X_t and X_{t-1} – can be transformed, utilizing its recurrence property, into a form in which R_t depends on X_t and X_{t-2}. Such a transformation of (5.3) yields:

$$R_t = [(a_1^2 + a_1 + a_2)/(1 + a_1)](X_t - m)$$
$$- [a_2^2/(1 + a_1)](X_{t-2} - m) + u_t - [a_2/(1 + a_1)]u_{t-1}. \qquad (5.11)$$

[cf. Equation (2.33); see section 2.3.3, for details.]

Now, the parameters a_1 and a_2 may be such that the coefficient of X_t in (5.11) could be small compared to that of X_{t-2}, i.e. R_t could practically depend on X_{t-2} alone. This means that if R_t is regressed on X_{t-2}, we would find, practically, a linear relationship between them. Figure 2.13b illustrates an ideal case.

Moran's AR(2) model (5.4) happens to be one such case. Because $a_1 = 0.4101$ and $a_2 = -0.7734$, the coefficient of X_t in (5.11) is only -0.1384 compared to that of X_{t-2} being -0.4242. Thus, R_t regressed on X_{t-2} scatters closely about the regression line (Fig. 5.11a) with a -0.81 correlation. The other models (5.5) and (5.10) exhibit a similar tendency (Fig. 5.11b and c).

[The least-square estimate of the regression coefficient in the AR(2) example is -0.5706, much lower than the expected coefficient of -0.4242. This is largely because: the residual $\{u_t - [a_2/(1 + a_1)]u_{t-1}\}$ in (5.11), being a moving-average, is positively autocorrelated; this causes a negative correlation between the residual and the regressor X_{t-2}; and this correlation, in turn, shifts the slope of the regression downwards. For general discussions of bias in least-square estimates, see Kennedy (1984).]

A strong correlation between R_t and X_{t-2} also exists in the lynx series, but the regression is distinctly curved (Fig. 5.11d). This suggests that the lynx reproduction surface, $R_t = f(X_t, X_{t-1})$, must be curved like that illustrated in Fig. 2.8. Thus, as Moran has remarked, 'the [lynx] process would be better represented by some kind of nonlinear model'.

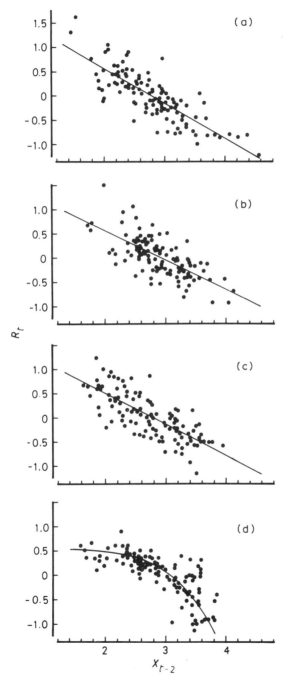

Figure 5.11   Regressions of $R_t (= X_{t+1} - X_t)$ on X_{t-2}. Graphs a, b and c: three simulation series a, b, and c in Fig. 5.9. Graph d: the lynx series b, Figs 5.1 and 5.5 (the curve is drawn by eye).

5.7 APPLICATION OF NONLINEAR DENSITY-DEPENDENT MODELS TO THE LYNX CYCLE

A major reason for the nonlinearity is that every animal has a limited reproductive capacity (section 2.2.3). As Fig. 5.11d shows, the regression curve has an upper bound (say, R_m). This implies that the reproduction surface is asymmetric about the equilibrium point. This, in turn, causes an asymmetric cycle in a lynx fur-return series.

The following two types of model, second-order logistic and predator–prey models developed in Chapter 4, have these characteristics. Let us see how well these models can describe the lynx series from the Mackenzie River District (Fig. 5.1b).

5.7.1 Second-order logistic model

A straightforward nonlinearization of Moran's AR(2) model is the second-order logistic model (2.39a) (sections 2.4.3 and 4.3.4):

$$R_t = R_m - \exp(-a_0 - a_1 X_t - a_2 X_{t-1}) + u_t. \tag{5.12a}$$

There are four parameters in (5.12a) which can be reduced to two in the canonical form:

$$R_t' = 1 - \exp(-a_1^* X_t' - a_2^* X_{t-1}') + u_t' \tag{5.12b}$$

by the transformations:

$$X = R_m X' - (a_0 + \log R_m)/(a_1 + a_2)$$
$$a_1 = a_1^*/R_m$$
$$a_2 = a_2^*/R_m$$
$$u_t = R_m u_t'. \tag{5.13}$$

The general model (5.12a) has the same ACF as that of (5.12b) (section 2.2.5).

In order to describe the lynx series with (5.12), we must find an appropriate set of parameter values with which we can simulate: (1) a dynamic pattern in terms of autocorrelation function (ACF), (2) equilibrium point (X^*) and (3) amplitude of cycle (variance of X). I know of no well-defined method to estimate satisfactorily parameter values from data. So I shall do it by trial and error: first roughly guess their values and refine them by repeated fitting to the data set.

In Fig. 5.11d, we see that $R_m \sim 0.6$ (the upper asymptotic level of the reproduction curve) and $X^* \sim 3$ ($X_{t-2} = X^*$ at which $R_t = 0$). In the meantime, the equilibrium point X^* of (5.12a) is given by

$$X^* = (a_0 - \ln R_m)/(a_1 + a_2) \sim 3. \qquad (5.14)$$

[To calculate X^*, set $X_t = X_{t-1} = X^*$, $R_t = 0$, and $E(u_t) = 0$. Note that in the canonical form (5.12b), $X'^* = 0$.]

Now, model (5.12b) generates a cycle similar to lynx when the parameter set $(1 + a_1^*, a_2^*)$ is in region IV or IV' of the parameter space in Fig. 2.9. A good example is the simulation series IV(7) in Fig. 2.17 with $a_1^* = 0.5$, $a_2^* = -0.9$ and $\{u_t'\}$ in Fig. 2.14a. Then, we can determine a_1, a_2 and u_t from the relationships in (5.13), and a_0 by (5.14).

After a fine tuning, I chose:

$$R_t = 0.597 - \exp(-2.526 - 0.838X_t + 1.508X_{t-1}) + u_t. \qquad (5.15)$$

To see whether or not this model fits the lynx series, we can use the method of conditional reproduction curves (section 3.2.7).

A conditional reproduction curve is a vertical slice of a three-dimensional (R_t on X_t and X_{t-1}) regression surface. Suppose we slice the surface parallel to the (R_t, X_t) plane at the value of $X_{t-1} = X_a$, say. The profile of the slice is a conditional reproduction curve, written: $R_t = f(X_t | X_{t-1} = X_a)$. Similarly, we can define $R_t = f(X_t | X_{t-1} = X_b)$. On the left-hand side of Fig. 5.12 are shown five pairs of such conditional curves of model (5.15). The fixed values X_a and X_b are moved progressively from graph to graph; $|X_a - X_b| = 0.4$ conveniently. Similarly, on the right-hand side are another five pairs of conditional curves $R_t = f(X_{t-1} | X_t)$.

Now, we sort the data (the Mackenzie River series) by pairing each X_t with its previous point X_{t-1}. We assemble all those pairs (X_t, X_{t-1}) with the X_{t-1} falling in the interval (X_a, X_b). Then, we plot $R_t(= X_{t+1} - X_t)$ against X_t (left-hand side graphs, Fig. 5.12). Similarly, on the right-hand side, we have the plot of R_t against X_{t-1}, given the X_t falling within (X_a, X_b). We see that the model reproduction curves fit the data points reasonably well. To see how reasonable the goodness of fit is, compare with an example in Fig. 3.13 (section 3.2.7) in which the conditional reproduction curves and data points were both generated by model (2.20) – same as (5.12) – so that the curves fit data points ideally.

Model (5.15) generates an asymmetric cycle (Fig. 5.13a), an improvement over the linear models. However, its pattern is not quite the same as a lynx series. The sample ACF (Fig. 5.13b) of the model damps down much faster than lynx correlograms (Fig. 5.8). Also, the sample PACF (Fig. 5.13c) truncates much too abruptly after lag 2 compared to the lynx. Probably, we need a more concrete process model to describe the lynx dynamics.

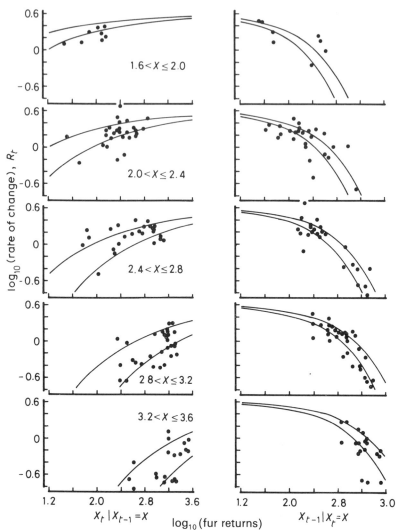

Figure 5.12 Family of conditional reproduction curves of the second-order logistic model (5.15) compared with the lynx data points from the Mackenzie River series. Left graphs: plot of R_t on X_t, given $X_{t-1} = X_a$ or X_b. Right graphs: plot of R_t on X_{t-1}, given X_t. The range $X_a < X \leqslant X_b$ is given in each pair of graphs.

5.7.2 Predator–prey interaction model

Because the lynx is a predator, we should investigate how much of its dynamics can be attributed to its interaction with prey.

Let x_t and y_t be, respectively, predator and prey densities at time step t, and let r_t and r'_t be their reproductive rates from t to $t+1$. Each species'

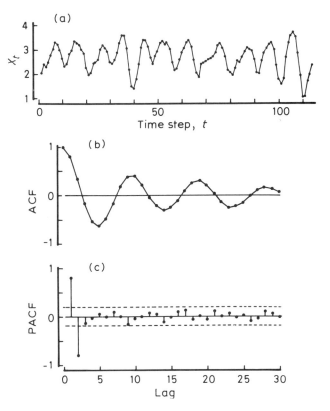

Figure 5.13 A simulation with the second-order logistic model (5.15) (graph a), its sample autocorrelations (ACF) (graph b), and partial autocorrelations (PACF) (graph c).

reproductive rate depends on the densities of both species. Thus, their interaction is represented by the system of two reproduction surfaces:

$$x_{t+1}/x_t = r_t(x_t, y_t) \qquad \text{(for predator)} \qquad (5.16a)$$

$$y_{t+1}/y_t = r'_t(y_t, x_t) \qquad \text{(for prey).} \qquad (5.16b)$$

[cf. (4.31), section 4.4.1.] This system yields four conditional reproductive rates, $r(x|y)$ and $r(y|x)$ for the predator, and $r'(y|x)$ and $r'(x|y)$ for the prey: $r(x|y)$ and $r'(y|x)$ are the intrinsic regulation mechanisms (e.g. logistic mechanism) of the predator and prey; $r(y|x)$ represents the functional response of the predator; and $r'(x|y)$ is the prey's survival from predation, a converse of $r(y|x)$. The plot of each rate against x or y forms a conditional reproduction curve. Figure 4.6 illustrates the typical shape of each conditional curve.

The following system model (developed in section 4.4) incorporates the above features:

$$r_t = r_m[1 - \exp(-by_t)]\exp(-cx_t) \tag{5.17a}$$

$$r'_t = r'_m \exp[-(b'y_t + c'x_t)] \tag{5.17b}$$

in which r_m and r'_m are the maximum reproductive rates for predator and prey, respectively, and b, b', c and c' are constant parameters (cf. (4.34)). These parameters represent the effects of the resource requirement and competition (given the resources in the habitat) in each species (sections 4.2.5 and 4.4).

After a log (base 10) transformation of (5.17), adding random perturbation effects u_t and u'_t, we have

$$R_t = R_m + \log_{10}\{[1 - \exp(-by_t)]\exp(-cx_t)\} + u_t \tag{5.18a}$$

$$R'_t = R'_m + \log_{10}\{\exp[-(b'y_t + c'x_t)]\} + u'_t \tag{5.18b}$$

in which $R \equiv \log_{10} r$ and $R' \equiv \log_{10} r'$. The lynx cycle can be simulated with the system model (5.18) in two steps: we choose a first set of parameter values to simulate the dynamic pattern and then choose a second set to match the mean and variance.

Although there are six constant parameters in (5.18), only four of them are essential in determining the system's dynamics (autocorrelation structure). This is because, by transformations $cx \equiv x$ and $b'y \equiv y$, we can effectively reduce the parameters to R_m, R'_m, b/b' and c'/c without affecting the autocorrelation structure of the system.

Now, the mean maximum rate for the lynx (r_m) is probably about 1.6 and, almost certainly, not larger than 2 (section 5.2). As for r'_m, the maximum rate for the lynx's chief prey, snowshoe hare (*Lepus americanus*) discussed in the next chapter, is probably three to four times as high as the lynx's. We do not know about the other two constants, b/b' and c'/c. Therefore, with r_m and r'_m chosen from the above range, and by iteration, we determine values of b/b' and c'/c with which the system model can simulate the lynx cycle.

The final set of values I chose is ($r_m = 1.86$, $r'_m = 6.05$, $b/b' = 5.00$, $c'/c = 6.00$). Figure 5.14 shows a simulation with (5.18) in which the perturbation effects u_t and u'_t are chosen to be independent, uniformly distributed random numbers in the interval (-0.065, 0.065). As it stands, the simulated predator series does not quite resemble the lynx series. However, the simulation is not directly comparable with the data because the lynx series (Fig. 5.1) are not census data but trapping data. And we do not know what an actual population cycle looked like. Therefore, to make a simulation comparable with the data series, we should 'sample' the simulated series in Fig. 5.14a as if the trappers harvested the pelts. [Recall my argument in

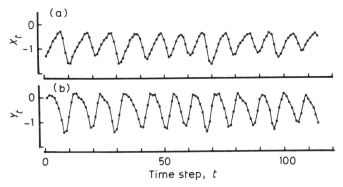

Figure 5.14 A simulation with the predator–prey model (5.18); X and Y are, respectively, predator and prey ($\log_{10}$) populations. The parameter values used in the simulation are given in the text.

section 5.2 that the trapping practice tends to stretch the amplitude of a pelt-harvest cycle disproportionately in both directions as compared to that of an actual (and unknown) lynx cycle.]

Let x be lynx density in the model and x' be the number to be trapped in the simulation. The simplest relationship between x and x' would be:

$$x' = mx^n \qquad (5.19)$$

where m is the trapping effort and n the trapping bias by which the amplitude is magnified. By log transformation and using the simplified notations, $X \equiv \log x$, $X' \equiv \log x'$, and $M \equiv \log m$, (5.19) is written:

$$X' = nX + M.$$

Further, we should add a random number (s) as a sampling error, such that, resuming the time subscript t:

$$X'_t = nX_t + M + s_t. \qquad (5.20)$$

Taking the expectations on both sides, and assuming the mean of s to be zero:

$$E(X') = nE(X) + M. \qquad (5.21)$$

Similarly, the variance of X' is:

$$\mathrm{Var}(X') = n^2 \mathrm{Var}(X) + \mathrm{Var}(s). \qquad (5.22)$$

To make X'_t in (5.20) directly comparable to the lynx data, we choose n, M and s_t such that $E(X')$ and $\mathrm{Var}(X')$ equal the mean and variance of the lynx series in Fig. 5.1b. For $E(X)$ and $\mathrm{Var}(X)$ on the right-hand side of (5.21) and (5.22), we substitute the mean and variance of the simulation series in Fig. 5.14a.

Figure 5.15 shows a result of trapping simulation by the above procedure in which $n=0.65$, $M=4$ and s_t is a random number uniformly distributed in

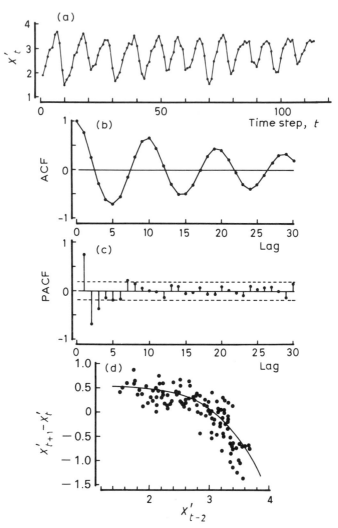

Figure 5.15 A simulation of lynx (log) fur returns $\{X_t'\}$ (graph a) after a simulated 'trapping' of the series $\{X_t\}$ in Fig. 5.14a by the scheme (5.20); parameter values are given in text. Graphs b and c are autocorrelations (ACF) and partial autocorrelations (PACF) of the series $\{X_t'\}$. The log rate of change, $X_{t+1}' - X_t'$, in the series is regressed on X_{t-2}' in graph d to compare with the same regression in the lynx series in Fig. 5.11d. To assist visual comparison, the regression curve in Fig. 5.11d is reproduced here.

$(-0.11, 0.11)$. The sample mean and standard deviation of the simulated series (graph a) are 2.7776 and 0.5578 compared with 2.9036 and 0.5584 of the lynx series. Autocorrelations and partial autocorrelations are shown in graphs b and c. Graph d shows the log rate of change in series a, i.e. $X'_{t+1} - X'_t$, regressed on X'_{t-2}, revealing the curved reproduction surface comparable to the same regression in the lynx series (Fig. 5.11d). Clearly, the statistical characteristics of the simulation closely resemble those of the lynx cycle.

5.8 NONSTATIONARY ASPECTS OF THE LYNX CYCLE

In the foregoing arguments, I have mainly considered the stationary series of (log-transformed) fur-return records from the Mackenzie River District (Fig. 5.1b). In this series, mean and cycle amplitude do not appear to change systematically with time; hence, the series is considered to be stationary. However, the series from the Athabasca Basin District (Fig. 5.1c) exhibits an increasing trend in the mean from the mid 19th century, while no noticeable changes occurred in the amplitude of a cycle. Conversely, the series from the James Bay District (Fig. 5.1h) exhibits a decreasing trend, but, again, little change in the amplitude.

Two different mechanisms are conceivable for the above features. First is changes in the 'sample' size. Fur returns as samples of a lynx population must be influenced by changes in the area covered by the district where the pelts were harvested. As already mentioned in section 5.2, the Athabasca Basin District was reduced around 1850 but was expanded towards the turn of the century on several occasions. Provided that there was no change in the lynx population process in the area, changes in the fur-trading scheme could have influenced parameter m, i.e. the 'trapping effort' in (5.21), which progressively shifts the mean of the (log) pelt harvest without influencing the amplitude of its cycle.

The second possibility is that some environmental conditions changed in such a way that the reproduction surface of the lynx process itself (not the fur-harvest scheme) shifted 'laterally'. The principle of the mechanism can be explained using the second-order logistic model (5.12a). Suppose that some environmental changes influence food supply, say, which in turn influences the intensity of competition. As a result, parameter a_0 in model (5.12a) would change its value. This would shift the mean of the series of log densities $\{X_t\}$ without influencing its amplitude. [See the illustration in Fig. 2.21 and discussion in section 2.4.5.] No information for testing this possibility is currently available, however.

Consider finally that environmental changes influenced the average reproductive performance of the animals so that the maximum (log) reproductive rate was changed. Such a change in the rate would shift the amplitude as well as the mean of a cycle. We can see this effect by changing R_m in model

(5.12a) as illustrated in Fig. 2.20, section 2.4.5. An apparent deterioration in the cyclic pattern, associated with a decreasing trend, towards the end of the series from the West Central and Winnipeg Basin Districts (Fig. 5.1e and f) could have been a consequence of such environmental changes, although I have no suggestion of what could have actually happened.

5.9 DISCUSSION

The fact that the system model (5.17) is capable of simulating the detail of the lynx cycle does not imply that the cycle is simply a result of the lynx's interaction with its chief prey, the snowshoe hare. An apparent problem here is the fact that the lynx is not always a major predator of the hare; the lynx alone rarely takes a substantial toll of the hare. However, many other predators, which tend to cycle with the lynx, prey on the hare as well as other types of prey. Thus, we must consider an interaction between a predator complex and a prey complex, rather than one between a single predator and a single prey species. Some clues lie in the ecology of the hare, which I shall discuss next.

6 Snowshoe hare demography

6.1 INTRODUCTION

When Keith (1963) published his 'Wildlife's 10-year cycle', available information on the theme was minimal. Many theories were no more than conjectures. In 1961, realizing that further theorizing would get him nowhere, Keith and a team of researchers from the Wisconsin school of wildlife ecology, launched a long-term field study on snowshoe hare (*Lepus americanus*) populations near Rochester, Alberta. A number of important papers from this study have appeared since then, including the monograph (Keith and Windberg, 1978) that provides a nearly complete 15-year set of demographic data. I shall call this work 'the Rochester study'.

Stimulated by the Rochester study, a group of zoologists from the University of British Columbia began field investigations in 1976 in the Kluane Lake Region of the Yukon Territory. These have produced equally important results; major ones were published in the series 'Population biology of snowshoe hares' (I. Krebs *et al.*, 1986; II. Smith *et al.*, 1988; III. Sinclair *et al.*, 1988). I shall refer to their work as 'the Yukon study'. Many experimental and field investigations by other independent researchers have offered valuable information on the ecology of this species.

In this chapter, I attempt a coherent interpretation of the results of these works. I begin by assembling some statistics of hare abundance since 1850 from various independent sources to verify the already well-known 10-year cycle (section 6.2).

The demographic data from the Rochester study (section 6.3) reveal the annual recruitment of young animals to be the major component of the cyclic population process. The recruitment rate is further partitioned into natality and subsequent survival of the young. Natality is largely determined by the overwintering physiological state of breeding animals, indicated by their relative body-weight changes during the previous winter (section 6.4). A major factor determining the physiological condition is winter food supply: not only its quantity but quality (nutritive value and digestibility) influences hares' body-weight (section 6.5).

Section 6.6 shows that hares' natality depends on their density through the depletion of woody browse during the preceding winter. After heavy browsing by a large number of hares in one winter, the recovery of quality vegetation takes more than one growing season. As a result, the density dependence of natality tends to lag by one winter. Section 6.9 discusses the significance of such lagged density-dependent natality in hare population dynamics.

Radio telemetry (monitoring individual hares by signals emitted from radios attached by collars) in both the Rochester and Yukon studies reveal mortality factors and their impact on hare populations. Predation and, to a lesser extent, starvation are major factors. Many predators prey on snowshoe hares as well as on several species of grouse (Tetraonidae) in many localities. I argue that an interaction between the predator and prey complexes must be a major cause of the ubiquitous 10-year wildlife cycle in North American boreal forests (sections 6.7 and 6.8).

To conclude this chapter, I synthesize the results (including those from Chapter 5) to explain the persistence and regional synchrony of the robust 10-year cycle along the boreal forest zone (sections 6.9 and 6.10).

6.2 EVIDENCE FOR THE 10-YEAR CYCLE

The Hudson's Bay Company fur-return statistics between 1849 and 1904 (Fig. 6.1a: first series), taken from MacLulich (1957; table 1), provide early evidence of the 10-year cycle in snowshoe hare abundance. Each data point is plotted, as in the lynx data (Fig. 5.1), against 'the year of the outfit' in which the harvest season ended. In the company's books, one harvest season started on 1 June of year $t-1$ and ended on 31 May of year t; the year t, in which the furs were shipped to the London headquarters, is called 'the year of the outfit' (Elton and Nicholson, 1942) or 'the year of end of biological production' (MacLulich, 1937; 1957). Poland (1892) and Seton (1909), who originally compiled the statistics from the company's books, used the year $t+1$ in which the furs were sold at the London market.

Unlike lynx furs which were gathered across Canada, only those hare pelts brought by the trappers to trading posts along the shores of Hudson Bay (Fig. 6.1c) reached the London market and, thus, were recorded in the books. Also, unknown proportions of hares harvested were consumed locally as food or were utilized for blankets and clothing (MacLulich, 1937). Because hare skins are not as valuable a commodity as lynx furs, the early hare-pelt statistics are neither as extensive geographically, nor as reliable an index of the abundance of the animal as in the lynx. Yet, the pelt statistics show evidence of the cyclic fluctuation in hare populations. In more recent years, however, the number of hare pelts harvested in Canada, for instance in Alberta (the second series in Fig. 6.1a: Statistics Canada, 1983), shows little correlation with actual abundance of the animal (see below). This is probably because of the steady decline of the value of hare pelts in the market since World War II.

On the other hand, being much more abundant and easier to observe than the lynx, snowshoe hares have been the object of more detailed ecological studies. MacLulich (1937) gives an index of hare abundance in Ontario between 1906 and 1935, using replies to the questionnaires sent out by the National Parks Bureau of Canada (now Parks Canada); the inquiries covered the hatched area in Fig. 6.1c. Figure 6.1b (solid circles) shows the

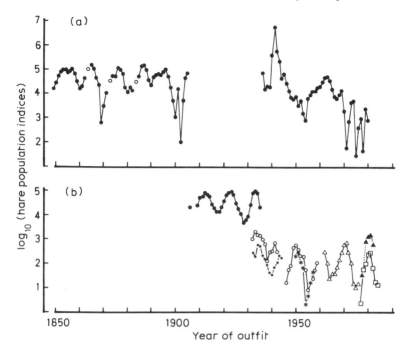

Figure 6.1   Annual snowshoe hare population indices from various areas since 1849. Graph a: pelt-harvest statistics from Hudson's Bay Company (first series; an open circle is an average of two successive years) and from Alberta (second series). Graph b: Inquiry and census data from the locations marked in graph c. Solid circles: Ontario questionnaires (hatched area). Two series of open circles: Lake Alexander and Cloquet, Minnesota (locations 1 and 2). Small asterisks: Oba, Ontario (location 3). Large asterisks: Anzac, Alberta (location 4). Open and solid triangles: Rochester, Alberta (location 5). Squares: Kluane Lake, Yukon Territory (location 6).

result which provides direct evidence of the cyclic fluctuation in hare populations. Since MacLulich did not publish the numerical data, I read off that part of his figure 3. [MacLulich's (1937) figure 3 shows the Hudson's Bay Company statistics and the Ontario inquiry result as one continual graph. It has since been reproduced in many text books without mentioning the fact that the data come from two quite different sources.]

Even more reliable indices of hare abundance since the early 1930s in much more restricted areas are available. In his table B, Keith (1963) compiled the data from six different sources taken between 1932 and 1959. I plotted the four of these that are based on direct counts (Fig. 6.1b). These are: two sets of census data (open circles) from the Lake Alexander area (Green and Evans, 1940; Chitty and Nicholson, 1943) and Cloquet Experimental Forest (Marshall, 1954), both in Minnesota; a series of trapping records (small asterisks) from Oba, Ontario (Hess, 1946); and Keith's (1963) own record in Anzac, Alberta (large asterisks).

Finally, the two most recent and intensive studies conducted near Rochester, Alberta (Rochester study) and in the Kluane Lake region of Yukon Territory (Yukon study) provide the most reliable census data by capture–recapture techniques. The series of open triangles (Fig. 6.1b) is the result of the Rochester study (Keith and Windberg, 1978; the estimated numbers per 100 ha are given in Cary and Keith, 1979; table 8). The subsequent series of solid triangles is the result of surveys by helicopter (Keith et al., 1984; fig. 3); I adjusted the vertical scale to make the series continuous from the census series. The open squares (Fig. 6.1b) are annual averages of the five populations from the Yukon study – 'Silver Creek', 'Beaver Pond', 'Kloo Lake', '1050 control' and 'Gribble's control' (Krebs et al., 1986; table 4) expressed as number per 100 ha.

Although the sources of information are diverse and fragmentary, the data in Fig. 6.1 depict the persistence of the popularly-acknowledged hare cycles for more than a century and their synchrony over large areas.

6.3 ANNUAL CHANGES IN DEMOGRAPHIC PARAMETERS

Suppose that there are y_t adult hares in a study plot at the onset of the breeding season of year t, of which s_t proportion survives to the next breeding season. Suppose also that each female produces, on average, n_t young (as realized natality) of which s'_t proportion survives to the next breeding season. Then, at the beginning of the next breeding season, there would be $y_t s_t$ adults carried over from the previous spring and $(y_t/2)n_t s'_t$ newly recruited individuals; $(y_t/2)$ is an approximate number of breeding females. Thus, the next breeding population y_{t+1} is the sum of the above individuals, i.e.

$$y_{t+1} = y_t(s_t + n_t s'_t/2). \tag{6.1}$$

[This relationship ignores the effect of hare movements across the border of the study plot.]

Keith and Windberg (1978) and Cary and Keith (1979), as part of the Rochester study, provide the demographic data for the right-hand side of the relationship (6.1). Figure 6.2 shows the annual hare density (y_t) (number per 100 ha in April), the adult survival rate (s_t) and the recruitment rate ($n_t s'_t/2$). [There are four distinct litter groups in each breeding season in Rochester (section 6.4.2). The number n is the sum of the mean litter size of the four groups, and s' is the average survival rate of all litter groups combined.]

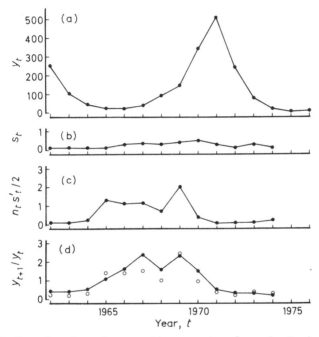

Figure 6.2 Snowshoe hare demographic parameters from the Rochester study. Graph a: y_t, hares per 100 ha in April (Cary and Keith, 1979; table 8). Graph b: s_t, April-to-April adult survival (Keith and Windberg, 1978; table 7). Graph c: $n_t s'_t/2$, recruitment rate – realized natality (n_t) from Cary and Keith (1979; table 8), and birth-to-spring survival of young (s'_t) from Keith and Windberg (1978; table 10). Graph d: y_{t+1}/y_t, annual rate of change in population; solid circles, calculated from the series $\{y_t\}$ in graph a; open circles, sum of the parameters in graphs b and c.

The series of solid circles in Fig. 6.2d is the rate of population change (y_{t+1}/y_t) calculated directly from the series $\{y_t\}$ of graph a. In the same graph, I also plotted the sum ($s_t + n_t s'_t/2$) (open circles) by combining graphs b and c. The sum is an estimate of the rate of change y_{t+1}/y_t by relationship (6.1). The directly calculated rate and the estimated rate do not precisely agree with each other. Probably, a major cause of the disagreement is the effect of local hare movement across the border of a study plot because relationship (6.1) does not take that effect into account. [I shall discuss the effect later.] Allowing for the effect of hare movements, Fig. 6.2 reveals that the variation in the recruitment rate (graph c) is the dominant component of the variation in the rate of change in population (graph d) and that adult survival (graph b) has only a minor influence.

The recruitment rate per female ($n_t s_t'$) can be further partitioned, after log transformation, into log n_t and log s_t' (Fig. 6.3). Clearly, the major source of annual changes in the recruitment rate (graph a) is those in juvenile survival (graph b), while natality (graph c) has a comparatively minor influence. In other words, juvenile survival is the major component that determines the cyclic hare population process. However, because knowledge of the natality process will help us understand the hare population dynamics, I shall discuss it in some detail.

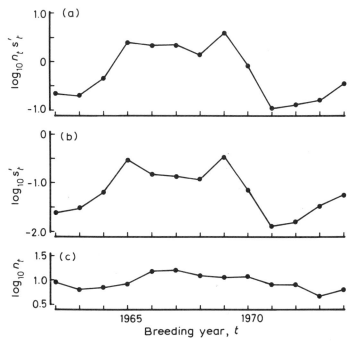

Figure 6.3   The recruitment rate per female ($n_t s_t'$) after log transformation (graph a) is partitioned into log s_t' (juvenile survival: graph b) and log n_t (realized natality: graph c).

6.4 ECOLOGICAL PROCESS DETERMINING NATALITY

The Rochester study reveals that the physiological condition hares have acquired during winter (indicated by their relative body-weight changes) governs their reproductive activity and therefore their natality in the following summer.

Parameter *n* in Fig. 6.3c is 'realized natality' which takes into account the effect of mortality among the pregnant and nursing females. In this section, however, I consider 'potential natality' (*n'*) which excludes the effect of the female mortality to avoid complication.

6.4.1 Dependence of natality on relative body-weight changes during winter

Figure 6.4 shows the annual potential natality (graph a) and percent losses or gains in body-weight in both sexes (graph b) during the preceding winter (from November to April). The correlation between the two parameters (graph c) reveals that the greater the weight loss during the winter, the lower the natality in the following summer. [In graph c, I substitute weight changes in males for those in females. This is primarily

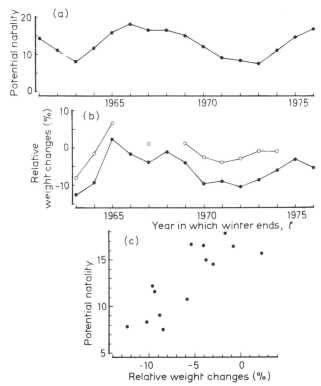

Figure 6.4 Annual changes in: (a) potential natality (Keith and Windberg, 1978; table 23); (b) relative body-weight changes (open circles for females and solid circles for males) during the preceding winter (Keith and Windberg, 1978; table 20); and (c) the correlation between (a) and (b), modified from the bottom graph in Keith and Windberg (1978; fig. 8).

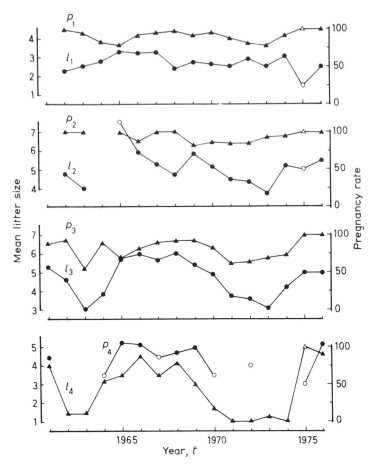

Figure 6.5   Annual changes in: mean litter size l_i (circles, scaled on the left; from Cary and Keith, 1979; table 4) and pregnancy rate p_i (triangles, scaled on the right; from Cary and Keith, 1979; table 2) in the four ($i=1$ to 4) litter groups. Open marks: samples with less than five females.

because the data for females are incomplete. The similar trend between the two sexes justifies the substitution. However, male weight changes influence their sexual activity and are also relevant to the process concerned.]

I now partition potential natality into pregnancy rate and litter size to reveal their dependence on body-weight changes during the winter.

6.4.2 Pregnancy rate and litter size

Individual female hares usually have several (mostly up to four) litters a season around Rochester. Because the majority of females tend to raise their litters more or less at the same time, there tend to be four distinct litter groups each season (Meslow and Keith, 1968). Cary and Keith (1979)

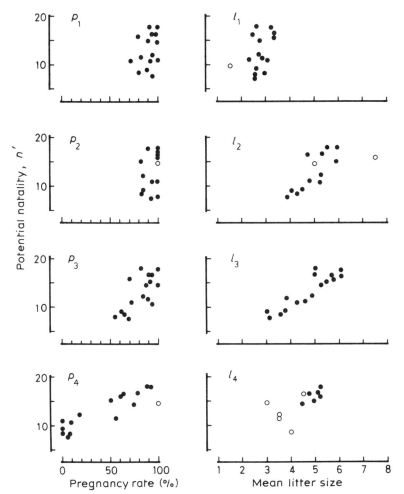

Figure 6.6 Regressions of potential natality (Fig. 6.4a) on pregnancy rate (p_{it}, Fig. 6.5) and on mean litter size (l_{it}, Fig. 6.5) in each litter group ($i = 1$ to 4). Open circles: samples with less than five females.

calculate potential natality (n'_t) in year t by the formula

$$n'_t = \sum_{i=1}^{4} p_{it} l_{it} \tag{6.2}$$

in which p_{it} and l_{it} are, respectively, mean pregnancy rate and litter size (mean number of viable embryos) of the ith litter group in year t.

Figure 6.5 shows the annual estimates of p_i and l_i ($i = 1$ to 4). Regressions of n'_t on p_{it} and l_{it} for each litter group ($i = 1$ to 4) (Fig. 6.6) clearly show that the annual changes in potential natality depend mostly on the changes in p_3, p_4, l_2, l_3 and l_4. There are some orderly changes in pregnancy

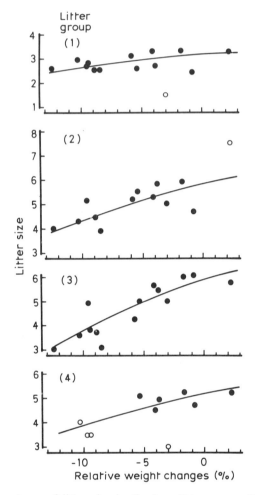

Figure 6.7 Dependence of litter size in the four litter groups (Fig. 6.5) on relative body-weight changes during the preceding winter (Fig. 6.4b). Open circles: samples with less than five females. Regression curves are drawn by eye, ignoring apparent outliers (all of them from small samples) in the litter groups (1), (2) and (4).

rate between litter groups: p_1 and p_2 are least variable between years and mostly higher than 80%, whereas p_3 and p_4 tend to be spread out over lower percentages. As for litter size, l_1 is least variable between years but smallest in the mean, while l_2, l_3 and l_4 are spread out over higher ranges. The variabilities in pregnancy rate and litter size in each litter group depend, to different degrees, on the females' overwintering conditions.

6.4.3 Effect of overwintering condition on litter size in summer

Figure 6.7 shows the influence of winter body-weight changes (an index of overwintering condition) on mean litter size. It reveals that after a poor winter (heavy loss of weight), the mean litter size does not exceed four in all litter groups. After good winters (little loss of weight), l_2 and l_3 tend to be

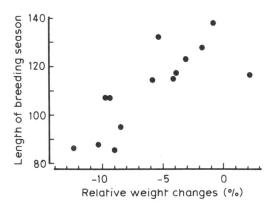

Figure 6.8 Correlation between length of breeding season (Fig. 6.9a) and males' relative body-weight changes during the preceding winter (Fig. 6.4b).

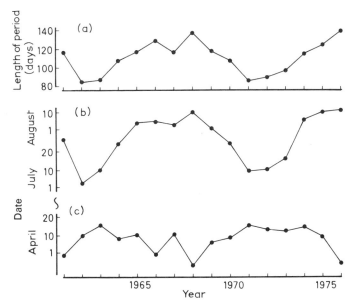

Figure 6.9 Length of breeding season (graph a) measured from mean date of first-litter conception (graph c) to mean date of testis regression to 4 g (graph b). Numerical data from Keith and Windberg (1978; table 23).

larger; after a very good winter, l_3 is almost doubled compared to the size after very poor winters. Size l_4 does not increase even after a good winter as much as l_2 and l_3.

Size l_1 is always small (about three) regardless of the overwintering condition. Keith and Windberg (1978) find that earlier in the spring, litter size tends to be even smaller within the group. But they find that the young from the first litter group tend to be heavier and to survive better than those from the later groups. Evidently, it is advantageous to raise young as early in the spring as possible but in small numbers.

6.4.4 Effect of overwintering condition on pregnancy rate in summer

The Rochester study shows that the length of breeding season depends on relative body-weight changes during winter (Fig. 6.8), meaning that the

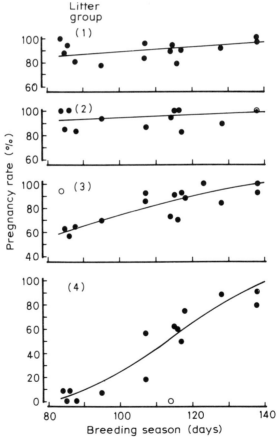

Figure 6.10 Dependence of pregnancy rates in the four litter groups (Fig. 6.5) on the length of breeding season (Fig. 6.9a). Samples with less than five females are omitted. The regression curves are drawn by eye, ignoring the two apparent outliers (open circles) in the last two litter groups.

season tends to be short when sexual vigour is low after a poor winter. The length of breeding season (Fig. 6.9a) is measured by the interval between the mean date of first-litter conception (Fig. 6.9c) and the mean date on which average testis weight had regressed to 4 g (Fig. 6.9b); this marks the end of males' sexual activity (Keith and Windberg, 1978).

It is clear in Fig. 6.9 that the length of the season is determined mainly by the time of termination rather than the onset of the season. This suggests that sexually vigorous hares tend to remain active, whereas less vigorous ones tend to curtail their activity, towards the end of the season. If more individuals curtail their activity in a summer following a poor winter, the pregnancy rate in that year would be more heavily reduced towards the later part of the breeding season as we see in Fig. 6.10: pregnancy rate is substantially low in a short season in the last two litter groups.

Now, it appears in Fig. 6.9 that the onset of breeding (graph c) is somewhat negatively correlated with the time of termination (graph b). This does not necessarily imply that more vigorous individuals start breeding earlier than less vigorous ones. Rather, the onset of breeding depends on the earliness of the spring, while most individuals start more or less simultaneously (Keith and Windberg, 1978). Probably, an early arrival of spring tends to relieve the hares from winter stress. This tends to enhance the sexual vigour of the hares, resulting in a prolonged breeding season.

6.5 CAUSE OF BODY-WEIGHT CHANGES DURING WINTER

I now show that body-weight changes are governed by the quantity and quality of foods (woody browse) which, in turn, depend on hare density.

6.5.1 Winter foods

The chief winter foods for snowshoe hares are woody browse (terminal twigs) of various shrubs and trees. Bryant and Kuropat (1980) list some abundant and commonly eaten species roughly in the order of the hares' preference: willow (*Salix*), aspen (*Populus*), larch (*Larix*), birch (*Betula*), pine (*Pinus*), fir (*Abies*), spruce (*Picea*) and alder (*Alnus*). Least preferred species like *Picea* nevertheless comprise a large fraction of hares' diet (Smith, personal communiction). On the other hand, a 'juvenile form' shoot in *Betula, Salix, Populus*, etc. usually deters hares from browsing. [The juvenile form is one- or two-year-old sprouts growing straight from stumps; it is distinct from older 'mature form' shoots. Current-year twigs growing on a mature shoot are palatable to hares.]

Feeding experiments by Bookhout (1965) and Pease *et al.* (1979) have shown that an individual hare, in order to maintain its body-weight during winter, requires daily roughly 300 (fresh) g (although depending on temperatures) of mixed-species browse, no more than 3–4 mm in basal diameter. The

300-g requirement includes a lot of discarded material. Wild snowshoe hares usually clip off a twig at a point where it is less than 5 mm in diameter. Pease *et al.* (1979) found that hares in captivity did not maintain weight and died if fed with thicker browse, even much in excess of 300 g. Evidently, not only the quantity of foods but quality is essential for the hares to maintain their body-weight. Let us look at the quantity first.

6.5.2 Quantity of food

For six winters beginning in 1970, Pease *et al.* (1979) measured the standing crop of browse and hare density in two study plots (the 40-ha and 28-ha plots near Rochester) three times each winter (November, January and March). Only those twigs utilizable and accessible to hares (i.e. up to 60 cm above the ground or snow cover) were counted. Then the authors calculated, for each plot, the ratio of the browse available in the habitat to the total quantity required by all hares to maintain body-weight from the day on which the measurement was taken until the arrival of spring (1 May) (Fig. 6.11). [Although a hare requires about 300 (fresh) g of terminal twigs less than 5 mm in basal diameter daily, the authors measured the standing crop of terminal twigs up to 15 mm in diameter. This is because the authors found that their experiment went well when they supplied hares with 3 kg of such thick twigs and let the hares eat, at their discretion, the required amount of choice browse less than 5 mm in basal diameter.]

All measurements in Fig. 6.11 are plotted on a log scale so that we can see the rate of change between measurements as the difference between the data points. We see that the stocks of woody browse (graphs a and a') were heavily depleted from November (open circles) to March (triangles) when the hares were numerous in the first two winters (graphs b and b'). For example, in the winter of 1970–71 (graph a), 75% of the November stock had been depleted by March. Nevertheless, the annual (November-to-November) changes in the browse stock (graphs a and a'; open circles) are comparatively small, indicating a quick replenishment by new growth during the summer months.

We also see in Fig. 6.11 that the food available-to-required ratios in the first two Novembers (open circles) in graph c are below unity (dashed line), indicating an absolute inadequacy of food supply for the hares to maintain body-weight until spring. Despite the heavy depletion of food supply already noted, the ratio had increased to about unity by March (triangle) in each year. This is because many hares did not survive these winters because of heavy predation (section 6.7). As hare density continued to decline in the subsequent winters, the ratio increased above unity, indicating an 'apparent' sufficiency of food.

In Fig. 6.12, I regressed the mean relative weight change in males during a given winter (Fig. 6.4b) on relative food supply at the beginning of the winter

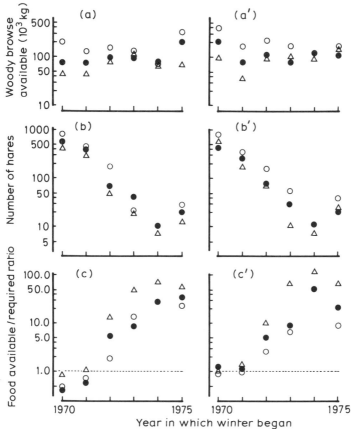

Figure 6.11   Annual variations in: standing crop of woody browse (less than 15 mm in basal diameter) per 100 ha of habitat area (top graphs); hares per 100 ha (middle graphs); and the browse available/required ratio (bottom graphs) in two study plots (the 40-ha plot on the left and the 28-ha plot on the right) near Rochester. Measurements taken in November (open circles; no measurement in 1974), January (solid circles), and March (triangles). After Pease *et al.* (1979; tables 1 and 2).

(November) (graph a). We see an apparently good correlation between them. Now, recall that the annual food available-to-required ratio (Fig. 6.11, bottom graphs) was almost completely dependent on hare density (Fig. 6.11, middle graphs). This is because, as also noted, the standing crop of utilizable browse in the habitat (Fig. 6.11, top graphs) did not change as much from year to year. Consequently, body-weight changes, which are positively correlated with the relative food supply (Fig. 6.12a), are negatively correlated with hare density (Fig. 6.12b). [Note that the data points for the two plots (open and solid circles) are not independent sets because body-weight changes were not measured separately.]

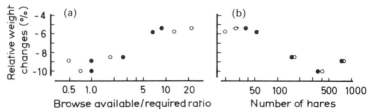

Figure 6.12 Dependence of relative body-weight changes in males during winter (Fig. 6.4b) on: (a) woody browse available/required ratio in November (Fig. 6.11; bottom graphs); and (b) hares per 100 ha (Fig. 6.11; middle graphs). Open and solid circles are for the 40-ha and 28-ha plots, respectively.

An additional experiment by Vaughan and Keith (1981) shows a similar tendency. In three winters beginning 1974, they built several enclosures in the 40-ha plot used by Pease *et al.* (1979). These were divided into four categories of treatment according to high and low hare densities and scarce and abundant food supplies, denoted by the obvious symbols, HS, LS, HA, LA. In the scarce-food (S) treatments, in which the hares ate natural woody browse, the food available-to-required ratio in November was less than unity, indicating an absolute shortage of food, in all replicates. The ratio was, on the average, 0.37 in HS and 0.78 in LS.

In the abundant-food (A) treatments, natural foods were supplemented by commercial rabbit food. Then the authors weighed sample hares in January, in February and in March. Their tables 5 and 6 summarize relative weight changes in the same individual between measurements. Weighing of the same individuals restricted the sample size, and the results are somewhat variable. Nevertheless, if an average is taken in each treatment by pooling the data for

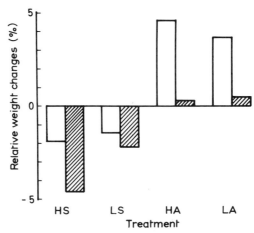

Figure 6.13 Relative body-weight changes in hares between January and February (open bars) and between February and March (hatched bars) in four experimental treatments in the Rochester study. H and L = high and low hare densities; S and A = scarce and abundant food supply. After Vaughan and Keith (1981; tables 5 and 6).

males and females and for all three winters, obvious effects appear (Fig. 6.13).

No doubt, the total quantity of foods in the habitat is a major factor which governs relative weight changes. However, the way foods are supplied (distributed) in the habitat is crucial.

6.5.3 Effect of food distribution on hares' foraging

A glance at Fig. 6.11c and c′ reveals that food available-to-required ratios are well above unity when hare density declined after 1971. Nonetheless, on the average, the hares were still losing weight (Fig. 6.12a). Only those hares in the food-supplemented treatments in the Rochester study (Vaughan and Keith, 1981) gained weight (Fig. 6.13). How is it that apparently abundant natural food, indicated by browse available-to-required ratios being 5, 10 or even 20 in Fig. 6.12a, was still not enough for the hares?

A similar phenomenon was observed in the Yukon study. Sinclair *et al.* (1988) evaluated relative body-weight changes indirectly in terms of crude protein content in faeces collected in these plots. [The crude protein value in the faeces is proportional to the amount digested and, thus, indicates the maintenance of body-weight.] In their experimental plots with ample supply of commercial rabbit food from September to May, the hares maintained weight better than those in the control plots without the commercial food. More hares died of starvation in the control plots than in the experimental ones (section 6.7).

Nonetheless, in the Yukon study, Smith *et al.* (1988) found that the natural food supply in the control plots appeared to be plentiful in terms of the browse available-to-required ratio. For example, in one of the control plots ('Beaver Pond'), the authors estimated the standing crop of mixed browse (less than 5 mm in basal diameter) per hare at the end of winter: it varied from as low as 165 and 353 (dry) kg/hare when hares were most abundant, to a maximum of 10 225 kg when hares were scarce. Considering that dry weight of woody browse is probably 50–60% of fresh weight, a hare should do well if it eats 165 g of good quality twigs in dry weight. It appears as though, even at peak densities, plenty of food was still left at the end of the winters; 165 kg of twigs is enough to support one hare for another 1000 winter days. Why, then, did the hares in the control plots not maintain weight as much as those in the experimental plots?

The commercial rabbit food is much richer in digestible protein than natural woody browse. Also, the rabbit food was supplied at several feeding stations. In other words, the food was supplemented in a highly nutritious, concentrated form and manner in the experimental plots. Natural foods are not so nutritious and are distributed rather sparsely. Thus, we have to consider how the distribution of food in the habitat influences the foraging effort by the hares, particularly under heavy predation pressure (section 6.7).

Consider, as extreme examples, that there was one hare and 300 g of good quality browse in a 100-ha plot and that there was one hare and 300 g in a 1-ha plot. The food available-to-required ratio, as calculated above, is the same between the two plots. But the hare in the first plot would no doubt starve because the foods are too sparsely distributed in the habitat. Evidently, the ratio is not a well conceived measure. Further research is needed to assess the sufficiency of food supply by taking into account hares' foraging activity in relation to the distribution of foods in the habitat. In the meantime, another factor which could explain the apparent shortage of food is its quality.

6.5.4 Nutritive value and digestibility of foods

In their feeding experiment with captive hares using commercial rabbit food in the Yukon study, Sinclair *et al.* (1982) found that the hares' relative weight changes depended on crude protein concentration in the diet. In particular, the hares lost weight when fed with the diet containing protein less than 11.3% of total dry matter; they gained weight above this critical value (Fig. 6.14). At 10°C the hares consumed slightly less, rather than more, food when the protein concentration was low (7 to 10%) than when it was high (15 to 20%). The authors explain that the lower the quality of food (i.e. lower protein concentration), the longer it takes for the hares to digest. This suggests that the hares are unable to consume a large bulk of poor quality food to secure enough daily intake of protein. Thus, the hares lose weight

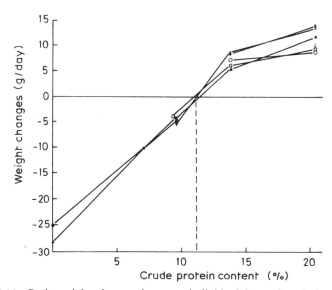

Figure 6.14 Body-weight changes in seven individual hares in relation to crude protein content in their diet. The vertical dashed line: threshold of zero weight gains at 11.3% protein. After Sinclair *et al.* (1982; fig. 3).

even with an apparently plentiful supply of food if the protein concentration is not high enough.

As already mentioned, wild hares usually select twigs which are no thicker than 5 mm. In fact, Sinclair and Smith (1984) and Sinclair *et al.* (1988) found that in the two species of birch (*Betula glandulosa* and *B. pumila*), the crude protein concentration steadily decreased as the diameter of the twigs increased (Fig. 6.15). If fed with twigs thicker than 5 mm, which are low in protein, the hares would not be able to digest enough protein daily to maintain their body-weight. This explains the finding by Pease *et al.* (1979) (section 6.5.1).

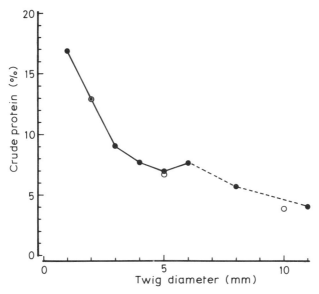

Figure 6.15   Mean crude protein content in twigs of two birch species distal to the clipping point at different diameters. After Sinclair *et al.* (1988; fig. 1) for *Betula pumila* (solid circles) and Sinclair and Smith (1984) for *B. glandulosa* (open circles).

Now, protein content by itself does not adequately indicate the quality of natural foods because their digestibility is also affected by the presence of some secondary plant metabolites (resins and phenols). Sinclair *et al.* (1988) have demonstrated that the digestibility of dry matter and protein in commercial rabbit food, free of secondary plant metabolites, was considerably reduced when mixed with resin and phenols extracted from birch and spruce. The authors further found that dry matter digestibility was twice as high in more preferred willows as in less preferred soapberry and white spruce.

Also, 'juvenile-form' twigs – terminal portions of one- or two-year-old shoots sprouting straight from stumps – in some woody plants (e.g. willows and birch) are lower in digestibility than 'mature-form' twigs (terminal portions of more than three-year-old shoots), even though the nutrient contents are much the same between the two forms (Reichardt *et al.*, 1984; Bryant *et al.*, 1985). [As noted before, all those twigs, terminal portions of both juvenile and mature shoots, can be a current year's growth.] Thus, hares reject twigs on juvenile shoots and eat those on mature shoots.

Furthermore, Bryant *et al.* (1985) found that wild hares' preference for winter twigs (less than 4 mm thick) of *S. alaxensis* increased as the age of the shoots increased (Fig. 6.16). This suggests that the supply of good quality twigs depleted during one winter may not fully recover during the following growing season, even though the total standing crop of palatable woody browse apparently does. Such slow recovery of quality food supply seems to cause a delay in the density dependence of hares' natality as shown below.

Figure 6.16 Effect of shoot age (years after browsing) on the daily consumption of winter-dormant feltleaf willow (*Salix alaxensis*) twigs by wild snowshoe hares in Alaska. After Bryant *et al.* (1985; fig. 2).

6.6 LAGGED DENSITY DEPENDENCE IN NATALITY

The causal chains discussed above imply that mean natality of the hares should depend ultimately on their density at the begining of the preceding winter. So, in Fig. 6.17a, I regressed potential natality (n_t') in the summer of year t on the estimated hare density (y_{t-1}') at the beginning of November of the previous year ($t-1$). [For the estimation of the November hare density (y'), see the appendix at the end of this section.] As it turns out, the relationship is not simple. There is a systematic tendency for n_t' to be higher when the population is increasing (see the solid circles) than when decreasing (open circles), given the same density. I interpret this relationship the following way.

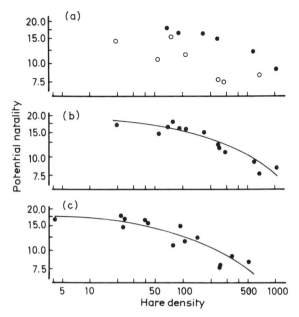

Figure 6.17 Regressions of potential natality (Fig. 6.4a) of the summer of year t on:
(a) the estimated hare density (y'_{t-1}) at beginning of previous November (see
Appendix, section 6.6); (b) density (y'_{t-2}) in November of year $t-2$; and (c) density
(y_{t-1}) in April of year $t-1$ (Fig. 6.2a). The curves in graphs b and c were drawn by
eye.

Recall that heavy depletion of quality woody browse in the winter of a
high hare density may not be fully replenished by the beginning of the
following winter (section 6.5.4). In other words, a heavy grazing by the hares
during the winter beginning year $t-2$ still has an effect on the food supply
for the following winter. So the natality in the summer of year t could still
depend on the hare density (y'_{t-2}) at the beginning of November of year
$t-2$. Then, we should regress n'_t simultaneously on y'_{t-1} and y'_{t-2}. We can
guess the three-dimensional regression surface by looking at its projection
onto the (n'_t, y'_{t-1}) plane (Fig. 6.17a) as well as another onto the (n'_t, y'_{t-2})
plane (Fig. 6.17b). We see that n'_t depends almost entirely on y'_{t-2}. This
means that the surface is curved along the y'_{t-2} axis but it is almost flat
along the y'_{t-1} axis.

Note that the time-lag involved in the above relationship is about one and
a half years, not two full years. However, because the hare density y'_{t-2} in
November is highly correlated with their density y'_{t-1} in the following April
(the coefficient being 0.98), natality n'_t depends practically on density y_{t-1}
(Fig. 6.17c). In other words, the process determining natality is practically
second-order density-dependent. I shall discuss the significance of this
lagged density-dependent natality in the hare population dynamics in
section 6.9.

Appendix
Ignoring their local movement, the total number of hares at the beginning of November (y') of a given year is:

$$y' = y(s + ns'/2) \qquad (6.3)$$

in which y is the adult (both sexes) population on 1 April (Fig. 6.2a); s is adult survival for seven months from 1 April to 1 November, calculated from Keith and Windberg (1978; see below); n is realized natality in Cary and Keith (1979; table 8); and s' is the mean juvenile survival for 125 days from the mean date of birth (25 June) to 1 November. [I calculated s' by interpolation from the mean survival rate for 180 days in Keith and Windberg (1978; table 10).] Keith and Windberg (1978; table 7) provide the mean monthly adult survival rates for the two periods, one from spring to summer and the other from autumn to midwinter. To calculate adult survival s, I used the spring–summer survival rate as the monthly average for the first four months beginning 1 April, and the autumn–winter survival rate as the monthly average for the subsequent three months.

6.7 ANALYSIS OF MORTALITY

6.7.1 Mortality factors revealed by radio telemetry studies

Both in the Rochester and Yukon studies, radio telemetry revealed that predation was the immediate cause of a majority of deaths among wild hares.

In Rochester, the study was conducted twice: first from February 1971 to August 1973 (Brand *et al.*, 1975) and second from December 1981 to April 1982 (Keith *et al.*, 1984). In each period, the hare population in the study area had begun to decline after a peak density. In the earlier attempt, the loss of information (due to transmitter failure, collar-caused death, etc.) was high; only 40 out of 184 collared hares were successfully monitored for the entire 18-month study period.

Table 6.1 shows that, of the 40 monitored hares, 26 (65%) died. In view of the fact that the population was declining during the period, the 65% mortality over the 18-month period no doubt underestimated the actual mortality. Nonetheless, 80% of the 26 known deaths were due to predation.

The second attempt was much more successful, with only 19% loss of information. The 69% mortality over the 4-month study period (Table 6.1) is a reasonable estimate. [It is similar to the 58% mortality (over three months from December 1980 to February 1981) in the Yukon study which is discussed below.] Of 91 known deaths in the second Rochester study, 94.5% was due to predation.

In the Yukon study, the period monitored by radio telemetry extended from June 1978 to May 1982 (Boutin *et al.*, 1986). The study was conducted

Table 6.1 Results of radio telemetry study in Rochester, Alberta, during the two peak-to-decline phases of snowshoe hare cycle. Summarized from Brand *et al.* (1975; table 1) and Keith *et al.* (1984; Table 16).

	Feb. (1971) to Aug. (1973)		*Dec. (1980) to Apr. (1981)*	
Total collared	184		163	
Information loss	144	(78.3%)*	31	(19.0%)
Total monitored	40		132	
Alive at end of study	14	(35.0%)	41	(31.1%)
Dead	26	(65.0%)	91	(68.9%)
Predation	21	(80.8%)	86	(94.5%)
Other	5	(19.2%)†	5	(5.5%)‡

*78 disappearances, one egress, 22 transmitter failures, 16 collar losses, and 27 trapping- and collaring-caused deaths.
†One diseased and four undetermined.
‡All judged starved.

in two groups of plots, i.e. food-supplemented and control plots. In the former group, commercial rabbit food was supplied *ad libitum*, in addition to natural foods, in several feeding stations from September to May. Each year was divided into four three-month seasons (e.g. spring = March to May), and seasonal mean survival, predation and starvation rates were calculated. Figure 6.18 shows the results. The graphs on the right are the results from the food-supplemented plot and those on the left are the pooled results from the control plots.

Figure 6.18 reveals that the consistent decline in survival rate over the study period in the control plots (graph b) is largely due to the increase in predation (graph c). Starvation in winter (connected solid circles in graph d) shows a similar trend. An interesting fact is that starvation during the spring season (open circles connected with dashed lines in graph d) was comparatively high but was virtually unchanged from year to year. Starvation rates in the food-supplemented plot (graph d′) were much lower, though not nil, than in the control plots. Predation (graph c′) was largely responsible for the decline in survival rate (graph b′) in this experimental plot.

Now let us calculate annual survival and mortality, using the seasonal rates in Fig. 6.18. Let s_i be the survival rate during the ith season, with $i = 1$ being spring; also, let m_i be the seasonal predation rate. Then, the annual survival rate is the product $s_1 s_2 s_3 s_4$, and the annual predation rate is given by the sum of weighted seasonal rates,

$$m_1 + s_1 m_2 + s_1 s_2 m_3 + s_1 s_2 s_3 m_4,$$

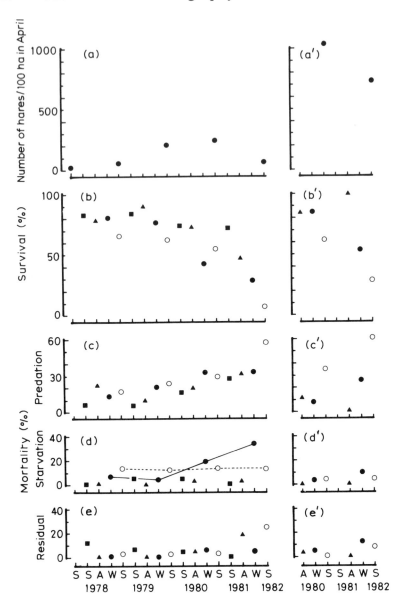

Figure 6.18 Annual and seasonal changes in: (a) number per 100 ha, (b) survival, (c) predation, (d) starvation and (e) residual mortality, of snowshoe hares in control plots, Yukon. Graphs on the right: same as those on the left but from food-supplemented plots. Open circles, squares, triangles and solid circles are seasonal data for spring, summer, autumn and winter, respectively. Numbers in graph a are averages from two control plots 'Silver Creek' and 'Beaver Pond' and those in graph a' from the food-supplemented plot 'Microwave' in Krebs *et al.* (1986; table 7). Graphs b, b', c, c', d and d' are based on Boutin *et al.* (1986; table 2).

and likewise in starvation and residual mortality. Table 6.2 shows the calculated annual rates for the three years from 1979 to 1981. The combination of predation and starvation dominates total mortality.

The results in Table 6.2 are subject to two types of probable error. First, juveniles less than three weeks of age cannot be trapped (Brand et al., 1975). If those young hares were the most vulnerable to predation, their exclusion would underestimate the average rate of predation. Second, radio-collared individuals were slightly more vulnerable to predation than uncollared ones, particularly for the first week or so after collaring (Brand et al., 1975; Boutin et al., 1986). This could somewhat overestimate the rate of predation. Thus, the effects of these types of error tend to cancel each other.

Table 6.2 Annual (March to March) survival, predation and starvation rates (%) among snow-shoe hares revealed by the radio telemetry study in control plots, Yukon. Calculated (see text) from Boutin *et al.* (1986; table 2).

Year	1979	1980	1981
No. hares per 100 ha. in March	52.7	199.2	238.3
Survival	38.4	14.4	5.3
Predation	35.7	54.1	62.9
Starvation	17.8	22.5	20.4
Residuals	8.1	9.0	11.5
Predation and Starvation	53.5	76.6	83.3
% predation among all deaths	58.0	63.2	66.4

Finally, I calculated residual mortality in Fig. 6.18 (bottom graphs) by subtracting the sum of predation and starvation rates from total mortality which equals 100% minus percent survival; see also Table 6.1. [Boutin *et al.* (1986) made no mention of mortality caused by factors other than predation and starvation. Any types of mortality other than predation and starvation (e.g. diseases) as well as estimation errors – particularly the two types of error mentioned above – fall into this category.] The residuals are mostly minor, being comparatively high only during the last autumn and spring seasons in the control plots (Fig. 6.18c).

Clearly, diseases did not appear to play a major role in the cyclic fluctuations in these intensive studies. Thus, the legendary mass die-offs of snowshoe hares (Seton, 1928) allegedly caused by epizootics (MacLulich,

1937) or 'shock diseases' (Green *et al.*, 1939) could not have been an inevitable or universal occurrence at every peak of population cycles, although it could have happened every now and then following an extreme density. However, as Dr J. N. M. Smith (personal communication) points out, a disease like coccidiosis might weaken hares and predispose them to predation or starvation.

6.7.2 The predator complex and its impact on a hare population

Many mammalian and avian predators share a local snowshoe hare population for food. Those predators identified in the Rochester and Yukon studies include: coyote (*Canis latrans*), grey wolf (*C. lupus*), lynx (*Lynx canadensis*), Bobcat (*L. rufus*), red fox (*Vulpes vulpes*), long-tailed weasel (*Mustela frenata*), mink (*M. vison*), great horned owl (*Bubo virginianus*), snowy owl (*Nyctea scandiaca*), hawk owl (*Surnia ulula*), goshawk (*Accipiter gentilis*), northern harrier (*Circus cyaneus*), red-tailed hawk (*Buteo jamaicensis*), rough-legged hawk (*B. lagopus*), golden eagle (*Aquila chrysaetos*) and raven (*Corvus corax*). Of course, they are not equally abundant; many do not stay all year around and they do not all depend on hares to the same extent.

Keith *et al.* (1977) studied hare consumption by several commonly occurring predators near Rochester. Figure 6.19 shows the average hare consumption per individual predator of three non-migratory species in relation to hare abundance. Consumption by a lynx (graph a) rose sharply with hare abundance. In contrast, the increase is slow and S-shaped in the coyote (graph b). The great horned owl (graph c) exhibits an intermediate tendency.

The above relationships can be explained as a result of: (1) a predator can find an abundant prey more readily; (2) a predator spends a greater proportion of its time hunting a more abundant and profitable prey.

The comparatively sharp rise in hare consumption by the lynx (Fig. 6.19a) reflects this predator's dependence on hares as a chief source of food. Indeed, in the Rochester study, snowshoe hares consistently comprised 75 to 90% (in biomass) of the diet of the lynx (Fig. 6.20a). Only when the hares became scarce did the lynx take more alternative prey, which included grouse (Tetraonids) and some small rodents (Nellis *et al.*, 1972; Brand *et al.*, 1976). Also, lynx tend to follow hares (Keith, 1963); they tend to expand rapidly, or even abandon, their home range as hare density declines sharply (Ward and Krebs, 1985).

Coyotes, on the other hand, depend much less on snowshoe hares than do lynx. The dietary repertoire of coyotes is large, ranging from mammals and birds to insects or even to vegetable matter, e.g. berries and grains (Todd *et al.*, 1981). Figure 6.20b shows that the percent occurrence of snowshoe hares in coyotes' scats studied by the above authors was as high as 87% when the hares were abundant but dropped sharply when scarce. The tendency is similar in winter and summer. Coyotes often abandon hare habitat when

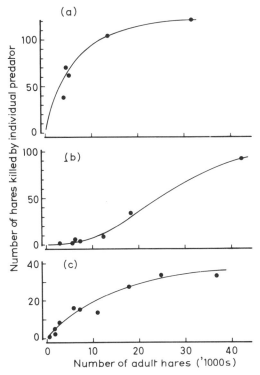

Figure 6.19   Average number of snowshoe hares killed by an individual predator of three non-migratory species: (a) lynx and (b) coyote during the period from December to April, and (c) great horned owl (April to June), in relation to changes in the number of hares in December near Rochester, Alberta. After Keith *et al.* (1977; fig. 4).

hares are scarce and hunt in open habitats where hares do not occur (Todd *et al.*, 1981). The S-shaped response curve in Fig. 6.19b reflects coyotes' foraging behaviour.

Great horned owls also prey on a wide variety of mammalian and avian prey (Adamcik *et al.*, 1978) and their individual consumption curve (Fig. 6.19c) falls between lynx and coyotes. The tendency is evident in the percent biomass of snowshoe hare in the owls' spring diet (Fig. 6.20c; open circles). It is also evident that the owls depend on hares more heavily in winter than in spring. A significant increase in dispersal movement among the owls is associated with local hare scarcity (Adamcik and Keith, 1978).

The impact of predation depends also on the number of predators present. The abundance of the three predator species changed more or less parallel with those in hare abundance (Fig. 6.21). Such numerical changes in the predators depend on their net reproduction (difference between birth and death). Numerical changes in the lynx population depend on its mean pregnancy rate, litter size and the birth-to-winter survival of the kittens

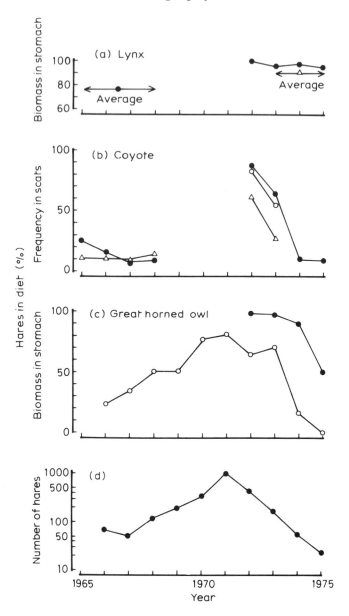

Figure 6.20 Occurrence of hares in the diet of three non-migratory predators near Rochester, Alberta. Graph a: percent hare biomass in lynx stomach (Brand *et al.*, 1976; table 5). Graph b: frequency occurrence of trace of hares in coyotes' scats (Todd *et al.*, 1981; table 4). Graph c: percent hare biomass in stomach contents of great horned owls (Adamcik *et al.*, 1978; tables 4 and 5). Graph d: hare numbers per 100 ha in December (Table 6.3). Solid circles, open circles and triangles are data taken in winter, spring and summer, respectively.

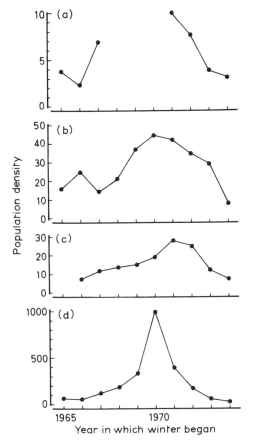

Figure 6.21 Annual changes in winter populations of three non-migratory pred-ators (per 100 ha) near Rochester, Alberta: (a) lynx, (b) coyote and (c) great horned owl (after Keith *et al.*, 1977; fig. 3) in relation to (d) the December snowshoe hare abundance (per 100 ha) (Table 6.3).

(Nellis *et al.*, 1972; Brand *et al.*, 1976; Brand and Keith, 1979). All of these depend directly on hare abundance. Pregnancy rate and litter size in coyotes also depend directly on hare abundance (Todd and Keith, 1983). On the other hand, numerical changes in the great horned owl were due mainly to the proportion of breeding birds; this is directly correlated with the local hare abundance (Rusch *et al.*, 1972; Adamcik *et al.*, 1978). However, mortality among the owls is not clearly correlated with hare density (Adamcik and Keith, 1978).

The overall impact of predation on snowshoe hare by these predators is evaluated in Table 6.3 in terms of estimated numbers and percentages of hares killed. The estimates in each year are for the 5-month period beginning December (Keith *et al.*, 1977: their original table gives percentages only; from them I calculated the numbers).

Table 6.3 Estimated numbers and percentages (in parentheses) of hares killed by the three non-migratory predators in the Rochester area during a 5-month period beginning December; hares present* are estimates for 1 December. Calculated and rearranged from Keith *et al.* (1977; table 1).

Year	Hares present*	Hares killed by			
		Coyotes	Great horned owls	Lynx	Total
1965	62	0.30 (0.5)	— (—)	3.0 (4.8)	— (—)
1966	50	0.20 (0.3)	5.2 (10.4)	1.3 (2.6)	6.7 (13.3)
1967	114	1.60 (1.4)	5.8 (5.1)	4.2 (3.7)	11.6 (10.2)
1968	185	— (—)	10.4 (5.6)	— (—)	— (—)
1969	325	— (—)	10.4 (3.2)	— (—)	— (—)
1970	990	66.30 (6.7)	29.7 (3.0)	15.8 (1.6)	111.9 (11.3)
1971	388	63.60 (16.4)	41.5 (10.7)	15.9 (4.1)	121.1 (31.2)
1972	165	19.60 (11.9)	38.0 (23.0)	12.0 (7.3)	69.6 (42.2)
1973	52	2.30 (4.4)	13.9 (26.8)	3.9 (7.5)	20.1 (38.7)
1974	23	0.05 (0.2)	3.2 (13.7)	3.1 (13.4)	6.4 (27.8)

The number of hares killed by each predator species increased and decreased more or less parallel with the cyclic changes in hare abundance. Although no one predator species singly exerted a profound effect on the hare population, the total percentage of hares taken by all these predators (last column) is substantial and lags in phase about two years behind the hare cycle. This lagged dependence of the overall rate of predation on hare abundance is undoubtedly the direct cause of the hare cycle. Naturally, then, we must consider the action of all major predators as a complex. Similarly, we need to consider a prey complex.

6.7.3 Grouse as part of the prey complex

Many of the mammalian and avian predators of the snowshoe hare also frequently prey on grouse, Tetraonidae (Rusch and Keith, 1971; Rusch *et al.*, 1978). Rusch and Keith (1971) estimated that 12.4% of the ruffed grouse (*Bonasa umbellus*) population in a Rochester study plot were killed during the month of July in 1966 and in 1967 by lynx, coyote, great horned owl, goshawk, red-tailed hawk and broad-winged hawk (*Buteo platypterus*). This is a minimum estimate because the authors did not have information on the impact of several other predators present in the plot. The authors concluded that predation was the predominant source of annual grouse mortality, agreeing with the conclusion by Bump *et al.* (1947) that predation was responsible for perhaps more than 80% of the annual mortality among adult grouse. [At Kluane, grouse are not an important component of the prey complex (J. N. M. Smith, personal communication).]

Interestingly, the tetraonids are a group of prey species known to exhibit a well-marked 10-year cycle (Leopold, 1933) and to fluctuate in close synchrony not only with each other but also with snowshoe hare (Williams, 1954; Rowan, 1954). Figure 6.22 (graphs a to c) shows the relative frequencies of reported peak years for ruffed grouse, ptarmigans (all *Lagopus*), and other grouse (*Pedioecetes phasianellus, Canachites canadensis, Tympanuchus cupido* and *Dendragapus obscurus*) (Keith, 1963: mostly based on Williams, 1954). Figure 6.22d shows the annual hunting-kill statistics in Manitoba (Rusch, 1976). In the same graph, I marked snowshoe hare peak years taken from Fig. 6.1b. We see that all of these prey species cycle in unison over a very wide area. [These statistics demonstrate a periodic fluctuation but do not accurately indicate the abundance of the animals. For instance, an increase in the hunting-kill after 1960 in graph d was mainly due to an increase in sport hunting (Rusch, 1976).]

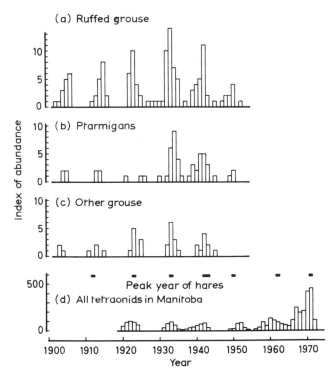

Figure 6.22 Annual population indices of grouse, Tetraonidae. Graphs a, b and c: the relative frequencies of reported peak years in ruffed grouse, ptarmigans, and other grouse, respectively, across Canada (after Keith, 1963; fig. 22). Graph d: number of all tetraonids harvested in Manitoba (after Rusch, 1976; fig. 2) compared with peak years of snowshoe hares (rectangles on the top of the graph) taken from Fig. 6.1b.

6.8 INTERACTION BETWEEN PREDATOR AND PREY COMPLEXES

Figure 6.23 compares the fluctuations in pelt-harvest statistics for lynx, coyote and red fox in Alberta (graph a) and Manitoba (graph b) in relation to those in snowshoe hare and grouse census data from Ontario, Manitoba and Minnesota (graph c). The data are all taken from Keith (1963) except the Ontario hare index which is from MacLulich (1937) (Fig. 6.1b). The unit

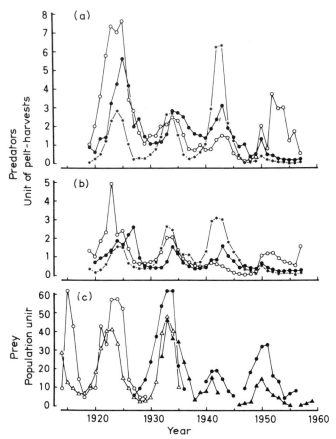

Figure 6.23 Annual population indices of some predators and prey. Graph a: lynx pelts (open circles: unit of 1000 pelts), coyotes (solid circles: unit of 10 000 pelts) and red foxes (asterisks: unit of 5000 pelts) in Alberta. Graph b: same as graph a but in Manitoba with units for lynx, coyotes, and red foxes being 1000, 5000 and 10 000 pelts, respectively. Numerical data for both graphs are given in Keith (1963; tables N, O and P). Graph c: ruffed grouse nests per 50 ha (open circles) in Aweme, Manitoba (Criddle, 1930 in Keith, 1963; table A); ruffed grouse per 100 ha (solid circles) and snowshoe hares per 25 ha (solid triangles) in Cloquet, Minnesota (Marshall, 1954 in Keith, 1963; tables A and B); snowshoe hare index (open triangles) in Oba, Ontario (Fig. 6.1b).

of scale on the vertical axis is conveniently chosen for each species to allow an easy visual comparison of the cyclic patterns between species. For instance, one unit for lynx is 1000 pelts; whereas one unit for coyotes is 10 000 pelts in graph a and 5000 in graph b, etc. The red fox was not a common predator in the Rochester and Yukon studies. However, a recent study in Maine (Major and Sherburne, 1987) has revealed that the hare consitutes a large proportion of the fox's diet.

In Fig. 5.14, section 5.7.2, I have shown that the simple predator–prey interaction model (5.18) can generate a 10-year cycle which simulates the cyclic fluctuation in the lynx fur-return statistics. The model structure is, in principle, comparable to what was revealed in the field studies in Rochester and Yukon; namely, that predation is the dominant, immediate mortality factor among the snowshoe hares and grouse, and that the net reproduction of major predators much depends on the abundance of the prey. The analogy between the model and observed systems suggests that the oscillations in Fig. 6.23 are induced primarily by an interaction between the predator and prey complexes. Only when we look into the detail of their oscillatory patterns does the analogy weaken.

The most noticeable difference between the simulated and observed patterns is the following. In the simulation (Fig. 5.14), the peak of a predator cycle is distinctly lagged (on the average 4.3 years) behind that of the preceding prey cycle; whereas, in the observed relationships in Figs 6.21 and 6.23, allowing for their incompleteness and fuzziness, the phase lag appears much less than that in the model. Particularly in Fig. 6.21, in which the observations were made in a comparatively small area, the lag seems minimal. Movements by the predators, as they quickly respond to local differences in prey density, could have reduced the lag.

To minimize the effect of local movements, we may compare the lynx statistics in Fig. 5.1 g and h with the hare statistics in Fig. 6.1 over the interval between 1850 and 1933. As already mentioned, these snowshoe hare pelts were harvested along the shores of Hudson Bay and, therefore, should be compared with the lynx fur returns from the districts g and h of Fig. 5.2. Wherever one regional series is interrupted or fuzzy, the information from the corresponding section in the other series can be used. These data reveal that the lynx series lags most frequently 2 to 3 years behind the hare series.

Now, the greater phase lag (4.3 years) in the simulation (Fig. 5.14) is in part due to the fact that, both in predator and prey, the cycle is skewed but in opposite ways: the predator cycle is skewed to the right and the prey cycle to the left. However, if we take trough-to-trough differences, the predator series is, on the average, only 1.9 years behind the prey series. A choice between a peak-to-peak lag and a trough-to-trough lag as a measure of the phase lag is arbitrary, of course. Either will be a reasonable measure only when the cycles are symmetric or skewed in the same way. When their cycles are skewed dissimilarly, we may take an average, so to speak.

One such method is to calculate correlations between the two series with the predator series lagging h $(=1, 2, 3, \ldots)$ years behind the prey. The lag h at which correlation is the highest may be taken as such an average phase lag. In the case of the simulation in Fig. 5.14, the correlations for $h = 1, 2, 3, 4$ are, respectively, 0.413, 0.817, 0.847 and 0.567. The high correlations for $h = 2$ and 3 suggest an average lag to be between two and three years, which agrees with the observed lag between the hares and lynx along the shores of Hudson Bay.

Thus, the major difference between the model and observed systems is reduced to one in the degree of asymmetry in the cycles of the species involved. Several conceivable reasons for the discrepancy are the following.

First, the asymmetry is exaggerated in model (5.18) because it is built on the assumption of scramble competition. If contest competition is assumed instead, a cycle could be less skewed. The reader can compare the effects of these two types of competition, using models (4.34) and (4.35), section 4.4.2.

A second and probably stronger reason is as follows. If we consider that x and y in the simulation model represent the predator and prey complexes, then each complex is an unstructured, random mixture of species. Also, the model assumes a closed system: the two complexes are totally dependent on each other. In the field, the system is open: a given species, say a predator, may migrate from habitat to habitat seasonally or irregularly according to the availability of prey. Hence, each complex is not as rigidly tied with the other as in the model. Further, each predator has a repertoire of prey which only partly overlaps that of another predator. Each predator has a preferred prey. But its preference changes from time to time, or from place to place, according to changes in the relative profitability of the prey. The profitability, in turn, changes as the constituents of the complex or their densities change. There must be some ecological rules about prey choice through the food-web structure and community organization which is not incorporated into the model. To reveal the structure and its effect on the population dynamics of the animals involved is a challenging subject for both theoretical and field investigations.

Allowing for its inadequacy, however, model (5.18) explains, in principle, the basic mechanism of the 10-year cycle along the boreal forest zone in simplified terms. In particular, the cycle length in the model is determined solely by the four constants, r_m, r'_m b and c' (sections 4.4.1, 4.4.2 and 5.7.2). The first two are the mean maximum reproductive rates and the latter two represent the overall effects of resource requirement and competition in each of the predator and prey (complexes). It suggests that the observed natural system must happen to have ecological and demographic characteristics, such as food requirement, body mass and natality, which translate themselves into a particular set of parameter values that generates a 10-year cycle.

Conversely, changes in the constituents of the complex system, or a change from one eco-climatic zone to another, may influence the parameter values such that the cycle would change or even disappear. On the other hand, if some changes in the constituents of the complexes make only a small parametric change, the system could be robust enough to resist such perturbations. Thus, in the absence of the lynx or the fox in certain areas, a 10-year hare cycle could still persist. Likewise, changes in the composition of raptor complex from east to west across the continent may not influence the cycle as long as the counterparts on both sides retain similar demographic properties. These are, again, an interesting subject of theoretical and field inquiries.

At one time, there was confusion among some ecologists about the phase lag between the lynx and hare statistics of the Hudson's Bay Company. I take this occasion to clarify it. Leigh (1968) read the fur-return statistics for snowshoe hares and those for the lynx from MacLulich's (1937) figures 3 and 16, respectively. Apparently, Leigh was unaware of the numerical data for hares in MacLulich (1957; table 1) and those for lynx in Elton and Nicholson (1942; table 1, columns 2 and 3). Using Leigh's read-out figures (table III), Gilpin (1973) found an overall tendency of the lynx cycle to be slightly ahead, in phase, of the hare cycle; hence, his quip: 'Do hares eat lynx?'

There are two sources of error in the Leigh–Gilpin analysis, which shifted the lynx cycle a few years backwards from the actual cycle. First, MacLulich mentioned that he plotted each data point for hares against 'the year of the end of biological production' or 'year of the outfit' (section 6.2). However, he somehow plotted each point for lynx in between two successive years, the latter year being the year of the outfit (Elton and Nicholson, 1942). But Leigh mistakenly used the earlier year. Second, the data in MacLulich's graph for lynx (figure 16) was pooled from different districts across Canada, whereas his hare data came from areas along the shores of Hudson Bay (section 6.2). As we can see from Figs 5.1 and 5.2, the 10-year cycles in western regions tend to be slightly ahead in phase of those in eastern regions. Thus, as Williamson (1975) and Finerty (1979) have pointed out, the hare cycle should have been compared with the lynx cycle along the shores of Hudson Bay. As already shown, the lynx cycles along Hudon Bay did lag a few years behind the hare cycle. It turns out, after all, hares did not eat lynx.

6.9 EFFECT OF RESOURCE DEPLETION ON HARE POPULATION DYNAMICS

As we have seen above, the demographic and ecological data from the Rochester and Yukon studies largely agree. In particular, both studies identify predation as a major component of the mechanism underlying the snowshoe hare population cycle. However, they do seem to disagree in their interpretations of some details; this also seems to be true among individual workers in a study. A major source of disagreement is how to interpret the effect of starvation. In this section, I compare their arguments to find a comprehensive interpretation.

6.9.1 Effect of starvation

The Rochester study stresses the importance of starvation among the hares due to resource depletion as the primary cause of the initial decline following a cyclic peak in density (Keith and Windberg, 1978; Keith, 1983; Keith *et al.*, 1984). On the other hand, the Yukon study (in Krebs *et al.*, 1986) seems to place a heavier emphasis on predation as the primary cause of the hare cycle rather than on food shortage. They argued on the basis that under heavy predation, the hare populations declined simultaneously in both food-supplemented and control plots in the Yukon. The truth probably lies in the middle as I shall now argue.

The dynamics of a predator–prey system are determined not only by the effect of predation on both populations (or complexes) but also by the effect of competition intrinsic to each population. [These effects determine the four conditional reproduction curves in Fig. 4.6, section 4.4.1.] Thus, with the predator density fixed, starvation among hares should increase as their density increases because competition intensifies. We can see how the effect of starvation influences the dynamics of the system by changing parameter b' in model (5.18); a larger value indicates more intense competition, and vice versa. Although the model system may still cycle even if the parameter value is changed, it may no longer be a 10-year cycle; smaller b' (less competition) tends to increase cycle length.

Now, in the Yukon study, even though starvation was minimized in food-supplemented plots, the populations cycled much the same way as in control plots. It looks as though starvation in the control plots had little effect on the population cycle. This is not necessarily so, however. We must consider the effect of local movement of the predator complex in a study plot.

Because the hares were more abundant in a food-supplemented experimental plot than in a control plot, more predators must have been attracted to the experimental plot. Such a local movement of predators, exerting more pressure on the hare population in the experimental plot, could offset an otherwise possible increase in cycle length caused by a reduced starvation rate. An important point is, however, that such an effect of the predators' local movement would not show in a population process in a much larger geographical area. If the effect of starvation was minimal everywhere in the large area, fur returns might not have exhibited a 10-year cycle. Thus, the result of the Yukon study must be interpreted with caution.

On the other hand, we cannot literally take the assertion of the Rochester study that food shortage is the primary cause of the initial decline in the hare population. Starvation, at its face value (Tables 6.1 and 6.2 and

Fig. 6.18, section 6.6.1), was of comparatively minor importance compared with predation. The Rochester study contends, however, that starved hares are more susceptible to predation.

Keith *et al.* (1984) compared the physiological state of predator-killed hares with that of those trapped alive. The authors found significant differences between the two groups of individuals in liver and heart weight and stored fat (e.g. bone marrow fat). The predator-killed hares tended to be more malnourished than the trapped ones. This suggests that starved hares were easier targets for predators than healthier ones. However, it is not quite clear whether starvation contributed to a substantial increase in predation or, simply, the starved hares were killed first, while the overall predation rate was not influenced substantially. In other words, it does not necessarily follow that starvation had a greater influence on the initial decline in the hare population than the data indicate.

Even if starved hares did not die, however, their natality would be reduced (section 6.4). Let us consider how this affects hare population dynamics.

6.9.2 Effect of lagged density-dependent natality

In section 6.6, I showed that hares' natality depended on their density with a one-year time-lag. Such a lagged density-dependent process (or second-order process) can generate a cyclic population fluctuation (see sections 2.2.4 and 2.2.5). On this basis, one might suggest that a hare population could cycle even in the absence of density-dependent predation and that this possibility could theoretically support Keith's (1974) theory of hare-vegetation interaction as a primary cause of the 10-year cycle. There is a problem in such an idea, however.

As already shown, the annual changes in natality and mortality due to starvation alone are comparatively minor contributions to the total variation in the net rate of change in hare density. Therefore, even if the natality process did in fact generate a 10-year cycle, it would not necessarily be recognizable as such. If predation were density-independent, a natality-induced cycle could be perturbed too much (because predation was heavy) to be stable.

At this point, it seems most reasonable to consider that the hare cycle is essentially a predator–prey (complex) cycle, and that the hare–vegetation interaction contributes to the remarkable stability of the 10-year cycle.

6.10 INTER-REGIONAL SYNCHRONY OF THE POPULATION CYCLES

Both Figs 6.22 and 6.23 illustrate inter-regional synchrony in population cycles. As I have already discussed in sections 5.3 and 5.5.1, the most likely mechanism to maintain such synchrony is the Moran effect (Moran theorem in sections 2.5 and 5.5.1). Briefly:

> If two regional populations have the same intrinsic, density-dependent structure, they will be correlated under the influence of an extrinsic factor if it is correlated between the regions.

In the snowshoe hare population process, predation constitutes the basic density-dependent structure which governs the cyclic fluctuation in a given locality. Certain climatic conditions during winter are candidates for the density-independent factors that bring independently cycling regional population complexes into synchrony. For example, hares' access to woody browse depends on the depth of snow cover (Keith *et al.*, 1984). During very low temperatures, starved individuals would die more readily than well-fed ones (Pease *et al.*, 1979). Predation rate could also be influenced by temperature. In the Rochester Study, predation during January 1982 rose about three-fold as minimum temperatures dropped from $-25°C$ to $-37°C$, although the precise mechanism causing this correlation remains unclear (Keith *et al.*, 1984).

Keith (1974) considered that certain weather conditions during winter could bring regional hare cycles into synchrony. However, his hypothesis was dependent on the assumption that mild and cold winters occurred in a particular sequence and coincidentally with a particular phase of the hare cycle. Such a particular assumption makes his explanation inflexible and difficult to substantiate. The Moran theorem shows this assumption to be unnecessary.

The Moran theorem is general and universally applicable because it does not depend on any particular assumption about the pattern of weather sequence. It does not matter if mild or severe winters hit a peak or trough or any other phase of the cycle. There just needs to be an extrinsic (density-independent) factor influencing reproduction or survival and correlated between regions. Moreover, the density-independent factor involved need not exert a substantial impact on the reproductive process of the animals concerned. It need not even be highly correlated between the regions (see my simulation in Fig. 2.22, section 2.5). In particular, if we consider the variations in weather conditions across the boreal-forest zone, there may be frequent periods of poor correlations between regions. But the stability of the density-dependent predator–prey cycle could be sufficiently robust to stop regional population complexes from moving out of phase with each other once synchronized.

7 Density effects on the dynamics of a single-species population: Utida's experiments on the azuki bean weevil

'Mathematical investigations independent of experiments are of but small importance ... '

Georgii Frantsevich Gause
(*The Struggle for Existence*, 1934)

7.1 INTRODUCTION

In his well-known laboratory studies on the azuki bean weevil, *Callosobruchus chinensis* (L.), Utida (1941a,b,c,d,e, 1942a,b, 1943a,b, 1947) focused his attention on how the weevil's own density influenced its reproductive rate. Prior to Utida, Pearl's (1927) work on *Drosophila melanogaster*, and Chapman's (1928) and Park's (1932) works on *Tribolium confusum* had shown that, among other factors, the insects' own densities exerted a strong influence on the growth of their populations. The density effect is a crucial intrinsic-regulation mechanism which promotes a logistic population growth.

Utida's work, as we shall see, reveals precise mechanisms of density effects operating at all stages of the weevil's life cycle: on oviposition rate, hatching success and survival of larvae to adults. Even today, Utida's work, I believe, is an unsurpassed model experiment in its thoroughness of design, quality of results and insight. However, his work has not been fully appreciated by ecologists outside Japan because of inadequate circulation of his original papers. Also, it is nearly half a century old. Therefore, I attempt a detailed discussion and reanalysis of the work in the context of the theoretical investigations in Part One of this book.

Utida conducted two types of experiment. In one type, he placed varied numbers of breeding pairs in a set of experimental chambers containing azuki beans (*Vigna angularis*). He measured oviposition rate, hatching success of eggs and survival of larvae to adults. No attempt was made to maintain further generations. I call this a 'discrete experiment'.

Utida found that the oviposition, hatching and survival rates all changed systematically as the breeding density changed. In sections 7.2 to 7.6, I analyse the results of his observation in terms of the two generalized logistic models: one assumes scramble competition and the other contest competition (section 4.3.2). By estimating the models' parameters from data, we can deduce population dynamics.

In the other type of experiment, Utida continually kept several series of weevil populations for more than one year and observed their fluctuations from generation to generation. I call this a 'series experiment'. In section 7.7, I compare the observed population dynamics in the series experiment with the dynamics deduced from the discrete experiment and discuss their differences.

For full comprehension of my arguments in this chapter, I recommend the reader glance over sections 4.2.5 to 4.3.2 for the theoretical basis of logistic process and the two generalized models I use for the data analysis.

7.1.1 Life cycle

The azuki bean weevil, originally from China and now distributed world-wide, is a pest infesting seeds of leguminous plants in storage as well as in the field (Fujii *et al.*, 1990). It can complete its life cycle entirely on beans in storage. Sex ratio is normally 1:1 and mating takes place within a day of adult emergence. A female lays eggs for seven days or longer, depending on temperature and humidity. One egg is laid at a time and is glued onto the surface of a bean. A larva, upon hatching, bores into the bean directly under its own eggshell, which is still firmly attached to the bean. If the eggshell is slightly lifted off the bean, the larva is unable to dig in and dies. Pupation occurs inside the bean just under the testa. The developmental time from the deposition of the egg to adult emergence depends on temperature, decreasing steadily from about 7 weeks at 20°C to 3 weeks at 32°C, but increasing again to 3.5 weeks at 35°C, all under 70–80% relative humidity (Ishikura, 1941; Utida and Nagasawa, 1949).

7.1.2 Experimental setup

A standard experimental chamber that Utida used consists of a pair of petri-dishes measuring 8.5 cm across and 1.8 cm deep. The bottom dish, containing some supersaturated salt solution for humidity control, is covered with a piece of thin cloth which holds a single layer of azuki beans. [Different salt solutions control relative humidity at different specific levels (Zwölfer, 1932).] The other dish covers the beans and keeps moisture inside the chamber. Because Utida was primarily interested in studying density effects, he kept, within a given set of experiments, all physical conditions as constant as he could. The petri-dish chambers were kept in total darkness and at constant temperature.

Except for special purposes, Utida used either 10 or 20 g beans per chamber. On average, 112 beans of standard size with 15.3% moisture content weigh about 20 g.

7.2 REPRODUCTIVE RATE IN DISCRETE EXPERIMENT

Utida prepared several sets (as replicates) of chambers, each containing 20 g beans. He introduced pairs of freshly emerged male and female weevils into each chamber with their number varied in the dual geometric series (1, 2, 4, 8, 16, 24, 32, 48, 64, 96, 128, 192, 256 and 384 pairs).

7.2.1 Observed rate

Table 7.1 shows the relationships between the mean number of progeny per breeding pair and the number of breeding pairs at four temperature–humidity combinations. From the data, I calculated the total number of progeny (y_2) and plotted against the number of parents (y_1 = females + males or $2 \times$ breeding pairs) on a log–log scale (Fig. 7.1). Let us call such a regression a (y_2 on y_1) reproduction curve.

The reproduction curves from the four experiments are alike in pattern. In all experiments, y_2 peaks at about $y_1 = 100$. This pattern – similar to the theoretical curve in Fig. 4.3b (section 4.2.6) or to that in Fig. 4.4b (section

Table 7.1 Mean number of progeny (second-generation adults) per breeding pair of azuki bean weevils in four experimental sets at different temperature-humidity combinations. Data for Experiments I, II and III taken from Utida (1941b; table 2) and IV from Utida (1941a; table 4).

Experiment	I	II	III	IV
Temp. (°C)	30.4	30.4	30.4	24.8
Relative humidity (%)	76	52	32	74
No. breeding pairs/20 g beans	Progeny/pair (no. of replicates)			
1	77.5 (6)	68.5 (6)	50.6 (5)	65.2 (10)
2	68.1 (6)	63.7 (6)	50.8 (6)	63.3 (10)
4	60.1 (5)	57.4 (5)	32.8 (5)	62.5 (10)
8	44.5 (5)	44.8 (5)	33.8 (5)	53.5 (10)
16	31.6 (4)	28.7 (4)	25.9 (3)	39.4 (10)
24	26.8 (4)	25.4 (4)	18.5 (4)	32.9 (10)
32	21.9 (3)	19.5 (3)	16.8 (3)	26.8 (10)
48	15.0 (3)	13.9 (3)	10.7 (2)	18.5 (10)
64	11.1 (3)	9.3 (2)	7.8 (2)	14.3 (10)
96	7.8 (2)	6.3 (2)	3.4 (2)	8.9 (10)
128	5.2 (1)	4.3 (2)	—	5.8 (10)
192	3.2 (3)	—	—	3.8 (10)
256	2.3 (3)	—	—	2.9 (10)
384	0.4 (2)	—	—	1.4 (3)

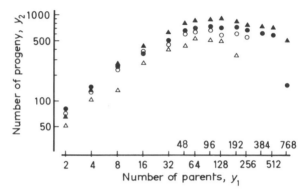

Figure 7.1 Regressions of progeny density (y_2) on parent density (y_1) in Experiments I to IV (solid circles, open circles, open triangles and solid triangles) calculated from Table 7.1.

4.3.2) – is explained by one of the two generalized logistic models, (4.19) or (4.20), section 4.3. I briefly explain the ecological meaning of these models in the context of the Utida experiment.

7.2.2 Logistic nature of reproductive rate

Suppose the weevils are distributed at random over beans in the experimental chamber. Consider an area around a female weevil (e.g. the dashed circle of area s around a point, the weevil, in Fig. 4.2, section 4.2.5) within which competition reduces her reproductive rate. Let the rate decrease by the factor k_i ($0 < k_i < 1$) when the female encounters the ith weevil in the area s (area of interference). Thus, i such encounters reduce the mean potential rate (say, b) by $k_1 k_2 \ldots k_i$.

Let us assume scramble competition, that is, all individuals are equally affected by mutual interference as their density (y_1) increases (section 4.3.2). No individual can breed or survive when the chamber becomes too crowded. An exponential reduction in the mean potential reproductive rate (b) with increasing y_1 represents such a situation. The realized mean reproductive rate (say, $r = y_2/y_1$) is then given by

$$r = b \exp(-c y_1^a). \tag{7.1}$$

Parameter $a > 1$ indicates the effect of interference on each female to intensify as she encounters more individuals, i.e. $k_1 < k_2 < \cdots < k_i$. Conversely, $a < 1$ indicates habituation to interference, i.e. $k_1 > k_2 > \cdots > k_i$. If $k_i = k$ regardless of i, then $a = 1$. Parameter c is a combined effect of s and the k; in particular, $c = s(1 - k)$ if $a = 1$. Model (7.1) is an adaptation of (4.19) (section 4.3.1).

In contest competition, a stronger individual suffers less from mutual interference than a weaker one. As a result, some individuals can breed or survive even when the chamber is overcrowded. Therefore, the realized reproductive rate $(r = y_2/y_1)$ does not decrease as much as in the scramble competition model (7.1). The following model, adapted from (4.20b), describes this situation:

$$r = b/(1 + cy_1)^a. \tag{7.2}$$

Model (7.2) is empirically conceived, not directly deduced from the geometric model of Fig. 4.2. Nonetheless, we can interpret the ecological meaning of parameters a and c as those in (7.1), although their values, estimated from the same data, could differ between the two models. Parameter b should be the same because it is a realized r when the effect of competition is minimal.

A reproduction curve can be a plot of y_2 against y_1 on a log–log scale as in Fig. 7.1, or a plot of reproductive rate $r(= y_2/y_1)$ against y_1 also on a log–log scale. Either way, a reproduction curve in scramble competition model (7.1) declines ever more steeply as y_1 increases regardless of all parameter values: the curve becomes perpendicular in the limit (Fig. 4.3b or 4.5a, solid curve). On the other hand, a reproduction curve in contest competition model (7.2) tends to a constant slope as y_1 increases: the slope of the asymptote is $(1 - a)$ in the (y_2 on y_1) curve and is $-a$ in the (r on y_1) curve (Fig. 4.4 or 4.5a, dashed curve).

In Fig. 7.2, I plotted r against y_1, using the same data used in Fig. 7.1. The left-hand graphs are duplicated on the right. Then, I fitted models (7.1) and (7.2) on the left- and right-hand graphs, respectively. [For the parameter values used, see Table 7.4. For model fitting procedure, see Appendix below.]

The two models fit nearly equally well, a situation similar to Fig. 4.5b. The data do not discriminate between the two different competition processes because the experiments did not extend to a sufficiently high range in the y_1 spectrum. However, if we subdivide the reproductive rate into several stage specific rates we do see a difference in certain of them (section 7.7).

Appendix
Model (7.1) is log transformed to:

$$\log r = \log b - \exp(\log c + a \log y_1). \tag{A7.1}$$

Notice the following: b determines the maximum of $\log r$ when y_1 is minimal; given b, the curvature of the (r on y_1) reproduction curve is determined by a; c translates the curve horizontally along the log-scaled y_1 axis without affecting the curvature. So, first, we roughly determine b by an r realized at a minimal y_1. Set $c \equiv 1$ temporarily. Choose a to draw a curve and translate it horizontally along the y_1 axis to match the data points. Determine c by the distance translated. Repeat the procedure for a better fitting if necessary. Model (7.2) can be fitted similarly.

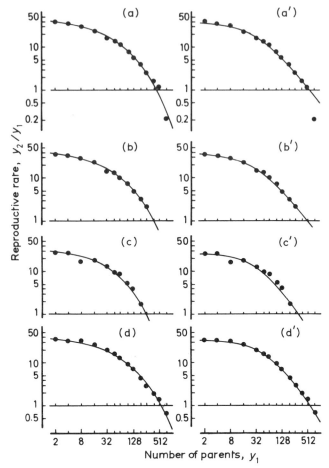

Figure 7.2   Regressions of reproductive rate (y_2/y_1) on parent density (y_1), using the same data as in Fig. 7.1. Each of Experiments I to IV (from top to bottom graphs) is dually shown on the left and right. Models (7.1) and (7.2) are fitted to the left and right graphs, respectively. The estimated parameter values are given in Table 7.4.

7.3 ANALYSIS OF REPRODUCTIVE RATE

In each of the experiments in Table 7.1, a reproductive rate (graphed in Fig. 7.2) can be partitioned into three components: oviposition rate, hatching success of eggs and survival of larvae to adults. Table 7.2 contains information for oviposition and hatching success, and further using Table 7.1, we can compute larval survival.

7.3.1 Density effect on oviposition rate

Figure 7.3 shows the mean number of eggs laid per female decreasing steadily as density of breeding population (y_1) increases. Utida

Table 7.2 Mean numbers of eggs laid and eggs hatched (in parentheses) per breeding pair of azuki bean weevils in four experimental sets of Table 7.1, compiled from Utida (1941c; tables 1, 2, 4 and 5).

Experiment	I	II	III	IV
No. breeding pairs/20 g beans		Eggs laid (hatched)/pair		
1	88.70 (81.00)	80.60 (74.80)	66.20 (60.40)	76.30 (66.20)
2	80.95 (75.65)	74.20 (69.90)	64.45 (58.60)	71.20 (65.70)
4	70.70 (65.30)	64.00 (60.40)	45.75 (39.05)	74.65 (65.75)
8	57.43 (51.43)	56.70 (54.05)	44.60 (39.30)	66.68 (61.70)
16	50.19 (45.84)	44.43 (41.74)	38.83 (33.30)	60.88 (56.16)
24	49.94 (45.42)	49.50 (45.74)	32.30 (25.50)	63.16 (59.65)
32	47.18 (43.67)	49.50 (40.77)	34.06 (26.06)	62.48 (57.21)
48	46.90 (36.58)	42.09 (32.58)	30.76 (16.42)	55.36 (49.61)
64	38.45 (29.03)	42.07 (26.93)	34.19 (14.01)	53.86 (45.18)
96	37.64 (19.45)	34.32 (15.50)	22.66 (5.04)	51.33 (36.02)
128	31.89 (9.64)	40.07 (8.22)	—	44.42 (24.74)
192	28.70 (5.53)	—	—	43.44 (12.78)
256	29.36 (4.14)	32.41 (4.08)	—	40.46 (7.99)
384	16.76 (0.52)	—	—	30.12 (1.93)

(1941d) considered the two most likely causes of the observed relationship: mutual interference among adult weevils and environmental contamination from the accumulation of metabolic wastes. After some experiments and observations, Utida concluded that mutual interference was the major factor.

A typical oviposition procedure is as follows. A female first selects a bean for oviposition and then pauses a while before laying. She smears the bean surface with some gelatinous substance on which she lays an egg. Then, she covers the egg with the jelly. When undisturbed by other individuals, she takes 30 to 60 seconds to complete the entire process. The female stays with the egg for several minutes before she leaves. During that time the egg hardens and is cemented to the bean.

Higher breeding densities promote mutual interference among weevils during site selection and egg laying. This lowers the rate of oviposition. An increase in breeding density also slightly reduces the average longevity of females (Utida, 1941d; table 4), although this is much less important than the direct effect of interference.

The above process of interference should be described either by model (7.1) or by (7.2): here, we interpret r to be the mean number of eggs laid per female, and b the potential oviposition capacity. The two models fit the data equally well (for the parameter values used, see Table 7.4). In both, parameter a is much less than 1, indicating a strong tendency

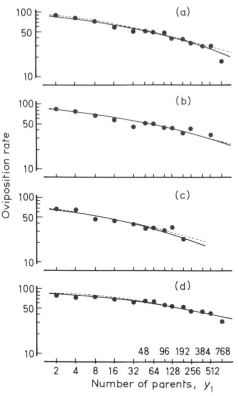

Figure 7.3 Regressions of oviposition rate (eggs/female) on parent density (y_1) in Experiments I to IV (graphs a to d) calculated from Table 7.2. The solid and dashed curves are produced by models (7.1) and (7.2), respectively. The estimated parameter values are given in Table 7.4.

for habituation to interference. The following observations support this interpretation.

A female weevil prefers to oviposit on new beans rather than those with eggs already laid (Yoshida, 1961). [For the chemical basis of this behaviour, see Oshima *et al.* (1973).] As a result, in an uncrowded situation, the deposition of eggs on beans tends to be more evenly distributed than random (Umeya, 1966). However, the egg distribution on beans changes from even to random and eventually to clumped as crowding increases (Utida, 1943a). Evidently, the weevils tend to become habituated to crowding and lay more readily on conditioned beans. Hence, $a < 1$. However, parameter a does not depend on temperature and humidity.

The two models produce curves similar to each other within the observed range in density (y_1). Therefore, we still cannot decide on the type of competition: scramble or contest. Let us move onto the hatching success of eggs.

7.3.2 Density effect on hatching success of eggs

Figure 7.4 shows the hatching success of eggs (data in Table 7.2) sharply decreasing beyond a certain point as the density of their parents increases. Utida (1941d) demonstrated that the main cause of egg mortality was mechanical injuries that eggs received from adult weevils trampling on the beans.

He introduced 128 pairs of weevils into an oviposition chamber and let them lay eggs for one day. He removed the weevils the next day and divided the beans with eggs on them into eight equal groups. Then, he introduced 128 males in each chamber of one group and let them trample on the beans, but only for one day. The next day, he introduced another 128 males in each chamber of a second group and kept them in there for one day, and so on up to the fourth day after oviposition. Thus, each group of eggs was exposed to

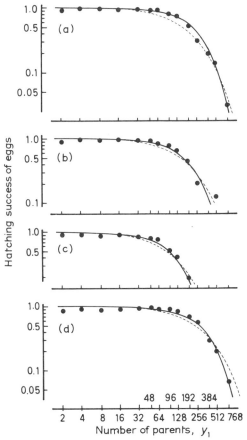

Figure 7.4   Regressions of hatching success of eggs on parent density (y_1) in Experiments I to IV (graphs a to d) calculated from Table 7.2. The solid and dashed curves are produced by models (7.1) and (7.2), respectively. The estimated parameter values are given in Table 7.4.

adult males for a different single day. The remaining four groups were kept free of adult weevils as controls.

Table 7.3 shows a much higher incidence of unhatched eggs in the group exposed to the adult males than in the controls. There is little difference among the controls, which were exposed to adults only during the one-day laying period.

What sort of mechanical damage do adults do to their eggs? Utida (1941d) found that most of the unhatched eggs contained dead, fully developed larvae. Obviously, the males did not damage the developing embryos inside their eggshells. So, why were the larvae dead?

Table 7.3 Influence of adult male weevils on the hatching success of eggs at different ages of eggs. After Utida (1941d; table 10).

Days after oviposition	1	2	3	4
		% unhatched eggs		
Exposed to males	30.4	44.0	45.3	20.6
control	8.7	9.1	8.6	8.1

Although each egg is cemented to a bean with a jelly-like substance, an eggshell can be slightly lifted off the bean when stepped on by an adult weevil. Then, the larva within would be unable to bore into the bean, though it would still develop fully before it died of starvation (Utida, 1972). Utida first thought that a freshly laid egg was soft and would then be most susceptible to mechanical injuries. But he found the contrary up to the third day as seen in Table 7.3. In fact, as the viscosity of the cement becomes more brittle as it hardens, the eggs come off the bean more easily. [The distinctly lower percentage of unhatched eggs in the exposed group on the fourth day resulted, not because the damage was low, but because more than one half of the eggs had successfully hatched by the time the adults were introduced.]

In Fig. 7.4, I fitted models (7.1) and (7.2) in which r is the proportion of eggs hatched and $b = 1$ (i.e. 100% hatching). The solid curves produced by model (7.1) fit the data more satisfactorily than the dashed curves of model (7.2) (for the parameter values used, see Table 7.4). This is probably because the weevils do not avoid eggs on the beans, nor do they have territories to protect their own eggs. They damage eggs at random. The situation is, therefore, more similar to scramble competition than to contest competition. The estimated value $a = 1.3 > 1$ in the scramble model (7.1) indicates that the probability of damage on an egg will increase slightly each time it is stepped on.

A close inspection of the graphs in Fig. 7.4 reveals that observed hatching success tends to fall slightly under fitted curves for $y_1 < 32$. A slightly higher

Table 7.4 Numerical values of parameters a, b and c in models (7.1) and (7.2) fitted (after log transformation) to regressions in Figs 7.2, 7.3, 7.4 and 7.7.

Model Experiment	Figure	(7.1)			(7.2)		
		$\ln b$	$-\ln c$	a	$\ln b$	$-\ln c$	a
I	7.2	3.90	1.700	0.5	3.60	3.70	15.00
II		3.85	1.650	0.5	3.60	3.55	15.00
III		3.50	1.750	0.6	3.25	4.00	15.00
IV		3.75	1.875	0.5	3.50	4.45	15.00
I	7.3	5.00	0.700	0.2	4.57	1.30	0.25
II		4.90	0.800	0.2	4.40	2.00	0.25
III		4.80	0.560	0.2	4.40	0.70	0.25
IV		4.75	1.200	0.2	4.40	3.30	0.25
I	7.4	0.00	7.410	1.3	0.00	8.10	15.00
II		0.00	7.020	1.3	0.00	8.00	15.00
III		0.00	7.355	1.3	0.00	7.50	15.00
IV		0.00	7.540	1.3	0.00	8.50	15.00
I	7.7	0.00	7.600	1.0	0.00	10.58	20.00
II		0.00	7.450	1.0	0.00	10.43	20.00
III		0.00	7.600	1.0	0.00	10.58	20.00
IV		0.00	7.750	1.0	0.00	10.72	20.00

proportion of unfertilized eggs in low densities (Utida, 1941d) probably caused this tendency.

7.3.3 Density effect on survival of larvae

Using the data in Tables 7.1 and 7.2, we can compute larval survival from the time larvae successfully burrow into beans (initial number of larvae, say y_2') to their emergence as adults (y_2). Figure 7.5 shows the relationships between survival rate (y_2/y_2') and parent density (y_1) in the four experimental sets. In every graph, survival is minimal at an intermediate parent density. This minimum survival rate corresponds to the number of breeding pairs that produce a maximum number of hatched larvae (Fig. 7.6). At this maximum number, larval survival is the lowest. Thus, it is primarily the density of larvae, not parents, that determines larval survival, although there is an indication that parents' density has a small influence (see below).

Those larvae which burrow into the same bean compete. Again, the logistic models (7.1) or (7.2) should describe the process. [In the models, r is the larval survival rate; $b = 1$ as the potential (maximum) survival rate; and the number of hatched larvae (y_2') replaces y_1.] Figure 7.7 shows regressions of the survival rate (y_2/y_2') on y_2' fitted with curves produced by the two models. The two curves in each graph are so close that we cannot separate them within the range of observed densities. So, again, we are unable to decide on the type of competition by the observed relationships.

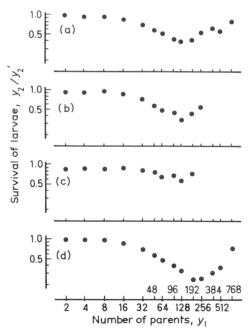

Figure 7.5   Regressions of survival of larvae to adults (y_2/y_2') on parent density (y_1) in Experiments I to IV (graphs a to d) calculated from Tables 7.1 and 7.2.

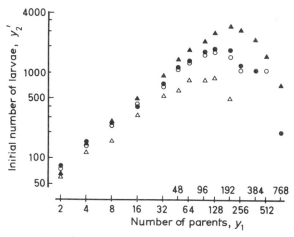

Figure 7.6   Regressions of the initial number of larvae (y_2') on parent density (y_1) in Experiments I to IV (solid circles, open circles, open triangles and solid triangles) calculated from Table 7.2.

Probably a major factor determining the type of competition is synchrony in the time of hatching. If the synchrony is good, all larvae dig into the bean more or less at the same time. This should promote scramble competition. Contest competition should be more likely if hatching spreads over a longer

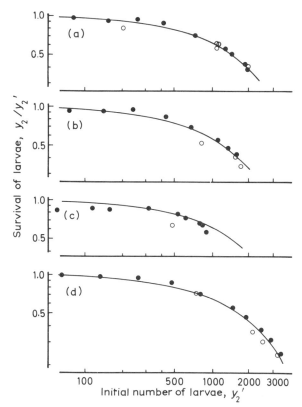

Figure 7.7   Regressions of the survival rate of larvae (y_2/y_2') on the initial number of larvae (y_2') in Experiments I to IV (graphs a to d). The fitted curves are produced by models (7.1) and (7.2): y_2' replaces y_1. The estimated parameter values are given in Table 7.4. The open circles in each graph indicate those larvae produced by parents whose density was above the peak of the regression in Fig. 7.6.

period of time, so that earlier larvae have an advantage over later ones. In the discrete experiment, the breeding stock was uniform in age, and the larvae hatched in a comparatively short interval of time. This situation was likely to have promoted scramble competition. I shall elaborate on this issue in section 7.7.

In Fig. 7.7, solid and open circles distinguish the survival rates of larvae produced by breeding populations below and above the most productive density in Fig. 7.6. For the same initial number of larvae, open circles tend to be very slightly below the fitted curves. In other words, larval survival is not completely independent of parent density. As already mentioned, the distribution of eggs on the beans tends to become more clumped as breeding density becomes higher. This must be the cause of the lower mean larval survival.

7.4 AN INTERPRETATION OF THE LOGISTIC LAW

As larvae grow, their food requirement increases: the area s in the geometric model of competition (Fig. 4.2) will increase. At the same time, their density decreases as some larvae die. Accordingly, the intensity of competition (k) inside a bean will change. In other words, parameter $c = s(1 - k)$ in model (7.1) – when $a = b = 1$ – should change during the larval stage. Why, then, does a curve produced by (7.1) with a constant c fit an observed relationship in Fig. 7.7? I interpret this the following way.

We can assume c to be constant in a sufficiently short interval of time during the larval growth. So, dividing the larval stage into n such intervals with the densities $y_2', y_2'', \ldots, y_2^{(n)}$ and the parameters $c', c'', \ldots, c^{(n)}$, we should have:

$$y_2'' = y_2' \exp(-c' y_2')$$
$$y_2''' = y_2'' \exp(-c'' y_2'')$$
$$\cdots\cdots\cdots$$
$$\cdots\cdots\cdots$$
$$y_2 = y_2^{(n)} \exp[-c^{(n)} y_2^{(n)}]$$

[y_2 is the second-generation adults as before.] Multiplying all from the top to bottom on both sides of these equations and dividing them by $y_2' y_2'' \ldots y_2^{(n)}$, we have:

$$y_2/y_2' = \exp[-c' y_2' - c'' y_2'' - \cdots - c^{(n)} y_2^{(n)}]. \qquad (7.3)$$

Recall that each of the relationships between y_2/y_2' and y_2' in Fig. 7.7 was described by model (7.1) with $a = b = 1$:

$$y_2/y_2' = \exp(-c y_2'). \qquad (7.4)$$

Then, because the exponents on the right-hand side of (7.3) and (7.4) should be equal, we find

$$c = c' + c''(y_2''/y_2') + \cdots + c^{(n)}(y_2^{(n)}/y_2'). \qquad (7.5)$$

Thus, we can interpret the constant parameter c as the sum of the parameters $c^{(j)}$ weighted by the rate of reduction in density $y_2^{(j)}/y_2'$.

Similarly, the relationships in Fig. 7.2, described by model (7.1), are the results of the three distinct processes, oviposition, hatching and larval survival. All of these component processes are, in turn, described by models of the same form as (7.1). Thus, letting r_1, r_2 and r_3 be, respectively, oviposition rate, hatching success and larval survival, all are given by:

$$r_i = b_i \exp(-c_i y^{a_i}), \quad i = 1, 2, \text{ and } 3 \qquad (7.6)$$

in which y on the right is the parent density (y_1) for $i=1$ and 2 and is the initial density of larvae (y_2') for $i=3$. Note that y_2' is the product of female parent density $(y_1/2)$, oviposition rate (r_1) and hatching success (r_2), i.e.

$$y_2' = r_1 r_2 y_1 / 2$$
$$= (y_1/2)b_1 \exp(-c_1 y_1^{a_1} - c_2 y_1^{a_2}). \qquad (7.7)$$

Further, because $y_2/y_1 = r_1 r_2 r_3 / 2$, we have:

$$y_2/y_1 = (b_1/2) \exp(-c_1 y_1^{a_1} - c_2 y_1^{a_2} - c_3 y_2'^{a_3}) \qquad (7.8)$$

and this is equal to the right-hand side of (7.1), i.e. $b \exp(-cy_1^a)$. [Parameters b and $b_1/2$ should be equal theoretically, although their estimations differ: $b < b_1/2$ (Table 7.4). The discrepancy is largely due to the production of a small proportion of unfertilized eggs at low density (section 7.3.2).]

Equating the right-hand side of (7.8) with the same side of (7.1), ignoring the difference between b and $b_1/2$, we see that parameter c is the weighted sum:

$$c = c_1 y_1^{a_1 - a} + c_2 y_1^{a_2 - a} + c_3 y_2'^{a_3} / y_1^a \qquad (7.9)$$

in which y_2' is given by (7.7).

An important point is that there is no mathematical necessity for the right-hand side of (7.9) to be constant; neither does the right-hand side of (7.5) need to be constant. Why are they virtually constant in the data? It is an ecological law under which the logistic model (7.1) holds. Perhaps overall resource requirements by the organism concerned, and the overall effect of competition throughout its entire life cycle, can adequately be represented by the geometric model of Fig. 4.2. If the right-hand side of (7.9) changes with y_1, or the same side of (7.5) changes with y_2', the logistic law should break down.

Model (7.8) is a partitioned model. Incorporating into the model the component processes that the Utida experiment revealed, we understand the mechanism of a logistic process. However, the estimation of a constant in each stage inevitably contains some error, and error tends to accumulate rather than cancel between stages. On the other hand, as long as the logistic law holds, the unpartitioned model (7.1) bypasses all intermediate processes, so that it can predict y_2 directly from y_1 avoiding the problem of error accumulation. Thus, model (7.1), as a recursion model, is a useful device for deducing population dynamics (section 7.7).

7.5 EFFECT OF RESOURCE QUANTITY ON DENSITY EFFECTS

In his experiments, Utida sometimes used 10 g of beans per chamber and 20 g at other times. However, density is a relative measure. It does not distinguish between five pairs of weevils per 10 g and 10 per 20 g. So, it is important to make sure that a density effect is unaffected by the quantity of beans used: for instance, the reproductive rate for five pairs per 10 g is the same as the rate for 10 pairs per 20 g.

Utida (1942a) prepared circular cardboard enclosures with different diameters and placed a single layer of beans in each at 24.5°C and 74% relative humidity (RH). He systematically changed the diameter of an enclosure so that the total weight of beans enclosed changed geometrically from 1.25 to 20 g in five steps. Further, in each weight class, he varied the

Table 7.5 The effects (at 24.5°C and 74% RH) of quantity of beans and density of breeding weevils on: (a) oviposition rate, (b) hatching success and (c) reproductive rate, compiled from Utida (1942a; tables 2, 3 and 4).

Density of weevils: equivalent to breeding pairs/ 20 g beans	*8*	*16*	*32*	*64*	*128*	*256*	*512*
Beans (g)/ chamber	(a) Mean number of eggs/pair						
20.00	73.0	72.3	66.4	64.1	57.2	—	—
10.00	—	65.3	65.6	62.1	58.0	—	—
5.00	—	70.8	58.8	57.7	46.0	—	—
2.50	—	54.0	55.0	53.8	44.4	39.1	—
1.25	—	68.9	55.9	53.9	47.1	44.5	35.2
	(b) Hatching success (%)						
20.00	—	90.0	88.6	84.4	73.6	—	—
10.00	—	94.2	91.2	79.4	64.8	—	—
5.00	—	95.1	93.4	85.8	66.3	—	—
2.50	—	97.0	85.5	89.6	75.0	—	—
1.25	—	94.6	92.5	87.0	74.3	—	—
	(c) Mean number of progeny/pair						
20.00	59.5	40.7	24.9	13.6	6.7	—	—
10.00	—	42.0	26.0	14.1	6.5	—	—
5.00	—	43.6	24.7	14.3	5.8	—	—
2.50	—	40.0	25.1	14.9	7.1	3.3	—
1.25	—	44.2	25.8	14.2	7.4	3.3	1.4

number of weevils equivalent to 8, 16, 32, ..., 512 pairs per 20 g of beans. Table 7.5 summarizes the results.

Oviposition rate in each column (table a) tends to decline slightly as the quantity of beans is reduced, although there is little difference between 10- and 20-g classes. Hatching success (table b) and reproductive rate (table c) show little systematic change within each column. Thus, density effects as measured in 10 and 20 g beans are virtually the same.

7.6 DIFFERENTIAL EFFECTS OF MALE AND FEMALE DENSITIES

In the experiments described so far, Utida used equal numbers of male and female weevils as breeding pairs. The first of Utida's experiments discussed in this section reveals that males and females differentially exert their effects on oviposition and hatching success. Consequently, as the second of his experiment shows, the reproductive rate of a population, for a given total number of females and males, depends on the sex ratio in a certain manner. The results in these experiments are important when we analyse population dynamics in section 7.7.

7.6.1 Experiments with single-sex populations

Utida (1941d) conducted two subsets of experiments, one excluding the effect of males and the other minimizing the effect of females (the effect of females cannot be completely removed as they are laying the eggs).

In the first subset, Utida introduced 128 pairs of newly emerged male and female adults in a mating chamber. Twelve hours later, females only were moved to other chambers, each containing 10 g beans without eggs on them, and were kept at 30°C and 76% RH. The number of females per chamber was varied in the usual geometric series, and they were allowed to lay eggs until death. Thus, no male interfered with oviposition or damaged eggs. Table 7.6 lists the observed oviposition rate and hatching success.

In the second subset, Utida kept equal numbers of males and females together in a laying chamber for three days. Then he divided the beans with eggs on them into two groups, one with no adult weevils at all and the other with males only. In the latter group, the number of males was varied in the dual geometric series per chamber containing 10 g beans with eggs on them. Table 7.7 compares the observed hatching success in the two groups. Clearly, the males continued to damage the eggs after the females had been removed.

Tables 7.6 and 7.7 reveal that, for a given density, males exert a much more detrimental influence on hatching success than do females, allowing that the effect of males cannot be completely separated from the effect of the laying females.

Table 7.6 Effect of female density (at 30°C and 76% RH) on oviposition rate and hatching success when males were excluded after mating. After Utida (1941d; tables 2 and 5).

No. of females/ 10 g beans	Eggs laid/ female	Hatching success (%)
8	46.4	85.4
16	50.0	91.7
32	58.9	89.5
64	52.2	89.3
128	38.8	75.4
256	30.7	59.5

Table 7.7 Effect of male density (at 30°C and 76% RH) on hatching success of eggs when females were removed after three days. In control, both males and females were removed after 3 days. After Utida (1941d; table 6).

| No. males/ 10 g beans | Hatching success (%) | |
	Exposed to males	Control
32	90.34	91.12
48	88.40	91.57
64	77.55	90.57
96	66.27	88.50
128	48.69	80.70
256	14.22	51.54

7.6.2 Experiments with varying sex ratios

In his second experiment, Utida (1947) introduced freshly emerged male and female adults into chambers containing 20 g beans. The chambers were divided into three groups, containing 4, 32 and 128 weevils per chamber with different sex ratios. They were kept without further manipulation until the second-generation adults emerged. Table 7.8 summarizes the results. Utida plotted oviposition and hatching success against sex ratio (proportion of females) (Fig. 7.8).

Utida conducted the above experiment prompted by Zwölfer's (1931) claim that reproduction in some insects depends on the sex ratio by the relationship:

$$y_2 = y_1 [f_1/(f_1 + m_1)] p(1 - E) \tag{7.10}$$

Table 7.8 Effects of male and female densities with varing sex ratios (at 30°C and 75% RH) on oviposition rate, hatching success and survival of larvae to adults. Calculated from Utida (1947; tables 1 and 2).

Total number of weevils/20 g beans	Males	Females	Eggs/female	Hatching success (%)	Survival of larvae (%)
4	3	1	63.0	92.5	91.3
	2	2	61.9	88.7	96.1
	1	3	68.6	89.4	96.7
32	28	4	40.1	89.4	97.1
	24	8	55.0	91.3	90.6
	16	16	52.4	91.4	71.2
	8	24	55.6	92.9	57.8
	4	28	55.0	91.2	52.5
128	112	16	35.2	76.1	90.5
	96	32	37.3	75.4	71.4
	64	64	33.9	81.4	50.0
	32	96	51.5	89.7	23.7
	16	112	45.5	91.6	20.6
	8	120	52.7	87.6	19.8

where y_1 and y_2 are parent and progeny densities; f_1 and m_1 are female and male parent densities; p is the maximum (potential) fecundity; and E ($0 \leqslant E \leqslant 1$) is a 'measure of environmental resistance'. [All notations are mine.]

But how can Zwölfer's model explain the observed relationships in Fig. 7.8? After all, in his formula, breeding density y_1 on the right is cancelled by $f_1 + m_1$ because this sum equals y_1. So, y_2 depends solely on f_1. In other words, sex ratio becomes irrelevant unless the 'measure of environmental resistance' depends on the sex ratio. But we have no idea about what constitutes 'environmental resistance'. In fact, to explain the observed relationships in Fig. 7.8, we do not need the ambiguous notion of environmental resistance; we only need to incorporate differential effects of female and male densities into a logistic process as I now show.

Recall Figs 7.3 and 7.4, section 7.3, illustrating the effect of parent density (y_1) on oviposition and hatching success when the sex ratio was 1:1. Recall also that model (7.6) consistently described these observed relationships. Even when sex ratio is not 1:1, the model should apply as long as the sex ratio is held at an arbitrary, constant value. However, because of the differential density effects, parameters a, b and c would depend on the sex ratio.

Thus, oviposition and hatching success should decrease exponentially as y_1 increases, although their rates of decrease depend on the sex ratio; the mean potential oviposition per individual (males included) also depend on the sex ratio. I idealized the above situation by a family of curves plotted against breeding density (y_1) as well as against sex ratio, $s = f/(f+m)$, in Fig. 7.9.

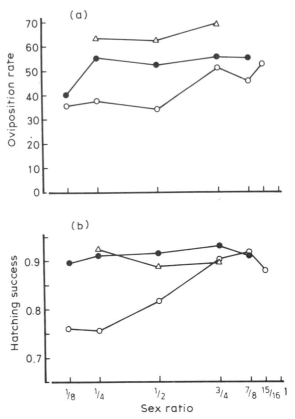

Figure 7.8 Regressions of oviposition rate (graph a) and hatching success of eggs (graph b) on sex ratios (proportions of females) of parent weevils in three density-groups (Table 7.8). The total numbers of parents $(f_1 + m_1)$ are 4 (triangles), 32 (solid circles) and 128 (open circles).

The curve for $s = 1/2$ corresponds to that in Fig. 7.3 or 7.4. For a larger s (more females), a curve in Fig. 7.9 should decrease comparatively more slowly or, conversely, should decrease faster for smaller s. A very large s (small proportion males), however, may reduce mating success and, hence, hatching success (more unfertilized eggs). The observed relationships in Fig. 7.8 are, in fact, the projections of data points (measured at $y_1 = 4$, 32 and 128) onto the (r, s) plane in Fig. 7.9. [Figure 7.9 is an idealization. An experiment designed to quantify the curves by data is needed.]

The differential density effects of both sexes can be incorporated into the unpartitioned logistic model (7.1) by letting the constant parameters a, b and c depend on the sex ratio s:

$$r = y_2/y_1$$
$$= b(s) \exp[-c(s) y_1^{q(s)}]. \tag{7.11}$$

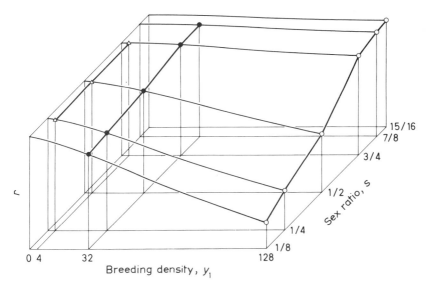

Figure 7.9 An idealized graph of oviposition or hatching success of eggs (*r*) depending on parent density (y_1) with varied sex ratios (*s*).

7.7 POPULATION DYNAMICS

We can deduce population dynamics from model (7.11) with the parameter values estimated from the discrete experiments with a 1:1 sex ratio. The result can be compared with the population dynamics actually observed in the series experiments that Utida (1941e, 1967) conducted over many generations. If the deduced and observed dynamics look alike, the logistic mechanism revealed by the discrete experiments is adequate to explain the population dynamics. As it turns out, however, they do not really look alike. This is because, as I show now, the type of competition differs between the two experiments. Let us look at the results of the series experiment first.

7.7.1 Observed population fluctuation in the series experiment

Starting with 10 g beans and eight pairs of male and female weevils in each chamber, Utida (1941e) continually kept two series of populations for 56 weeks at 30°C and 76% RH. Later Utida (1967) performed two more replicate series. In these experiments, he counted living and dead adults weekly, added 10 g new beans on every third week, and removed the leftover beans. Figure 7.10 shows the weekly counts in the first two series. Because the average length of one generation at 30°C and 76% RH is about three weeks, we see a peak count of living individuals more or less every third week.

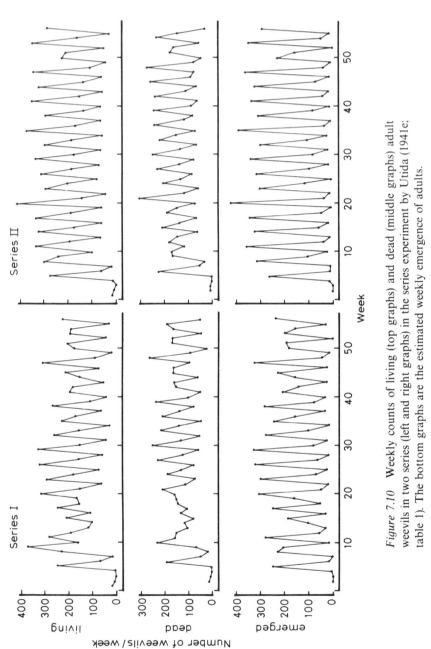

Figure 7.10 Weekly counts of living (top graphs) and dead (middle graphs) adult weevils in two series (left and right graphs) in the series experiment by Utida (1941e; table 1). The bottom graphs are the estimated weekly emergence of adults.

Now the number of living individuals at each count usually contains some survivors from previous counts. So we must somehow estimate the total number of weevils emerging per generation. First, Utida (1941e) computed the total number of newly emerging weevils each week the following way.

Suppose we counted n_{i-1} living individuals on the $i-1$st count, of which m_{i-1} have died by the next count. In the meantime, y_i individuals have emerged between $i-1$st and ith counts, of which m_i have died by the ith count. Then,

$$n_i = (n_{i-1} - m_{i-1}) + (y_i - m_i).$$

Although we cannot distinguish m_i from m_{i-1}, we know their sum. So, transposing, we have the estimate of y_i on the right:

$$y_i = n_i - n_{i-1} + (m_i + m_{i-1}).$$

Figure 7.10 (bottom graphs) show a distinct three-week periodicity in the estimated number emerged (y). Then, Utida took the sum of the ys over three consecutive counts, i.e. $i = (1,2,3), (4,5,6), (7,8,9), \ldots$, as approximate numbers emerging per generation. Figure 7.11 shows these estimated numbers in the four series. All series quickly reached equilibrium levels at about 400 weevils per 10 g beans. Figure 7.11 also shows the sex ratios. Although they are slightly biased in favour of females in series a and b, and males in c and d, the ratios fluctuated, on average, about 1:1.

7.7.2 Population dynamics deduced from the discrete experiment

Replacing y_1 by y_t and y_2 by y_{t+1} in (7.11), and given the initial density y_1, we can recursively generate densities in the subsequent generations. For simulation purposes, it is more convenient to transform (7.11) to logarithms as in (A7.1) in the Appendix, section 7.2.2:

$$\log y_{t+1} = \log y_t + \log b(s_t) - \exp[a(s_t) \log y_t + \log c(s_t)]. \qquad (7.12)$$

Sex ratio (s_t) is now a random variable depending on generation t.

We have already estimated $a(s)$, $\log b(s)$ and $\log c(s)$ for $s = 1/2$ (Table 7.4) from the data in Fig. 7.2, section 7.2.2. We do not have adequate data to estimate them for $s \neq 1/2$. However, based on the fact that s tends to fluctuate at random about a mean of $1/2$, we can approximate the fluctuations in these parameters by setting:

$$a(s_t) = a(1/2) + z_t'''$$
$$\log b(s_t) = \log b(1/2) + z_t$$
$$\log c(s_t) = \log c(1/2) + z_t''$$

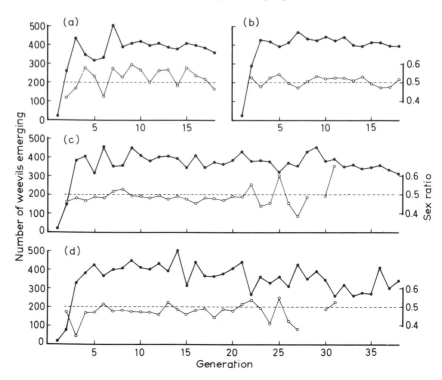

Figure 7.11   The number of adult weevils (per 10 g beans) emerging per 'generation' (i.e. per 3-week interval; see text) in four series (solid circles). Open circles are sex ratios (proportions of females) among the emerging adults. Professor Utida kindly provided me with the unpublished numerical data for graphs c and d.

in which z, z'' and z''' are independent random numbers with zero means. Thus, sex ratios act as vertical, lateral and nonlinear perturbations (section 1.7) of the reproduction curve in Fig. 7.2a. The problem is that we do not have adequate data to determine every perturbation effect. So, assuming that each perturbation effect is comparatively small, a vertical perturbation effect (u, say) can only serve as the first approximation of the total perturbation effect (section 1.7.3). Thus, the final simulation model is:

$$\log y_{t+1} = \log y_t + 3.9 - \exp(0.5 \log y_t - 1.7) + u_t \qquad (7.13)$$

in which u is an independent, normally distributed random number with mean = zero and variance = 0.025. With this variance of u, model (7.13) generates a series $\{y_t\}$ whose variance equals the variance of the observed series in Fig. 7.11. Figure 7.12 shows the first 200 ys generated by (7.13).

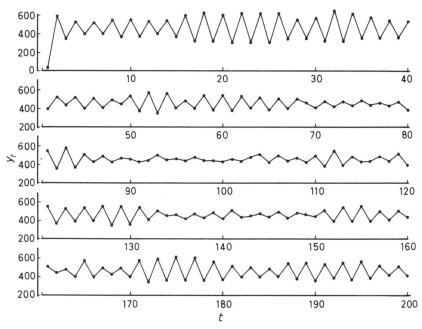

Figure 7.12 A simulation of population dynamics using model (7.13).

7.7.3 Differences between the observed and deduced series

The simulation series (Fig. 7.12) oscillates in a saw-tooth manner far more often than the observed ones (Fig. 7.11). The saw-tooth oscillation is due to the reproduction curve intersecting the horizontal axis at about $-63°$ (Fig. 7.2a). This slope is far steeper than $-45°$, the critical angle below which a series of densities exhibits a saw-tooth oscillation (Fig. 2.3, section 2.2.3).

The simulation series also differs from the observed ones in average density at the plateau. All of the observed series, after having reached the plateau, fluctuate about a mean of 380 weevils. In the simulation, on the other hand, the average number is about 450: the theoretical value is 456 (Fig. 7.2a). This is a substantial difference especially since the amount of beans differed between the two types of experiment: 20 g supplied once at the beginning in the discrete experiment as opposed to 10 g supplied every third week replacing the leftovers from the previous supply. As I have shown, the number of weevils per 20 g beans is virtually equivalent to twice the number per 10 g beans. Thus, after having been adjusted for the amount of beans, the average weevil density is 760 per 20 g beans in the series experiments as compared with only 450 in the simulation based on the discrete experiments. Why does the deduction from the discrete experiments differ so much from the results of the series experiments?

In the discrete experiment, the breeding stock was uniform in age; all females started laying eggs more or less simultaneously. This tended to promote scramble competition. In the series experiments, on the other hand, the emergence of adults tended to be spread over a period. Thus, the weevils in one 'generation' would not begin laying all at once. Consequently, for the same number of breeding weevils per 'generation', there would be less interference among the laying females. Thus, the average number of eggs laid per female per generation could be greater in a series than in a discrete experiment. Furthermore, earlier eggs might escape damage by late-emerging adults. This would promote higher average hatching success and, hence, a larger number of larvae via series than in a discrete experiment.

But, as already shown in Fig. 7.7, a larger initial number of larvae would induce a higher degree of competition among them. This would, in turn, result in a reduced number of adults emerging. However, the age of larvae in a series experiment would not be as uniform as in those in a discrete experiment. Then, older larvae have a competitive advantage over young ones. This should promote contest competition among the larvae in a series experiment. Thus, even though the larval density per 'generation' is apparently higher in contest competition, the reduction in larval survival is not as great as expected in scramble competition. Thus, on the whole, reproduction was higher in the series than in the discrete experiments.

7.7.4 Simulation with a contest competition model

The above argument suggests that model (7.2), based on contest competition, is more appropriate for describing the result of a series experiment. Corresponding to (7.13) based on scramble competition, the recursion formula based on (7.2) is:

$$\log y_{t+1} = \log y_t + \log b - a \log[1 + \exp(\log y_t - \log c)] + u_t. \quad (7.14)$$

The constant parameters estimated from Fig. 7.2a' (given in Table 7.4) are: $a = 1.3$, $\log b = 3.6$ and $\log c = -3.7$. With these parameter values, however, the expected equilibrium density y^* (the intersection between the reproduction curve and the horizontal axis in Fig. 7.2a') is about 600 per 20 g beans. Although much higher than 450 in the simulation in Fig. 7.12, 600 is still short of the observed average of 760 in the series experiments (Fig. 7.11). So the parameter values are not quite the same between the two types of experiments. But which parameter should we adjust to match the equilibrium density in the series experiment?

The equilibrium density y^* in model (7.14) is given by

$$y^* = (b/a - 1)/c. \quad (7.15)$$

[The reader can readily verify this by setting $y_{t+1} = y_t = y^*$ in (7.14).] So we have three options for matching y^* in (7.15) to the observed average. [An average density at a stationary state is approximately equal to the equilibrium density in the nonlinear model (7.14) as long as the variance of perturbation factor u is not too large.]

The first option is to increase parameter b, the potential reproductive rate. Geometrically, this raises the reproduction curve in Fig. 7.2a' vertically so that y^* (the intersect with the horizontal axis) is moved to the right. Ecologically, this option does not make sense because there is no reason to believe that switching from one type of competition to the other influences the potential reproductive rate. By definition, the potential rate is realized when there is no competition.

A second option is to displace the reproduction curve to the right by reducing parameter c. This simulates the reduction in the intensity of competition among the larvae. It makes ecological sense. It is also appropriate in view of the possibility that the total quantity of beans at any moment in the series experiment was not exactly 10 g but more, thus inflating the average density.

The third option is to reduce parameter a to make the slope of the reproduction curve less steep at the intersection with the horizontal axis; so that y^* moves to the right. This simulates the reduction in the competition among the breeding adults in a series experiment. This option also makes ecological sense. At this stage, the second and third options are equally tenable. As a matter of fact, simulations of these two options generate series that are virtually indistinguishable from each other.

Figure 7.13 shows the series of the first 200 points ($y_1 = 32$) generated by model (7.14) with the third option: a is reduced from 1.3 to 1.2. The random term u_t is an independent, normally distributed number (generated by the MINITAB random number package, base 1) about the zero mean with the standard deviation 0.1. With this variance of u_t, the variance of the simulation series $\{y_t\}$ matches that of the observed series in Fig. 7.11. The generated series resembles the observed series much more than does the simulation in Fig. 7.12.

7.7.5 Comparison by sample correlograms

Some readers might wish to make the comparison more quantitative. Then, correlogram analysis (sections 3.2 and 5.4) is in order. There are some technical problems, however. The first is that the observed series (Fig. 7.11) are rather short, particularly series a and b, so that their sample correlograms could not yet have adequately converged (sections 3.2.3 and 3.6). To circumvent that problem, we partition the simulation series into a number of sections with length comparable with the observed series c and d – series a and b are really too short. Then, we calculate a sample correlogram in each short section to compare with those of the observed series.

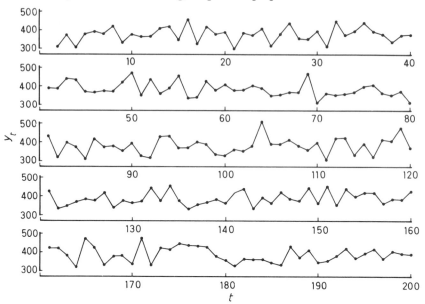

Figure 7.13 A simulation of population dynamics using model (7.14). For the parameter values, see text.

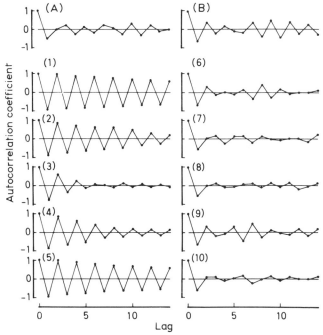

Figure 7.14 Sample correlograms of the differenced population series $\{y_t - y_{t-1}\}$. Graphs A and B: from the observed series c and d in Fig. 7.11. Graphs 1 to 5: from the five 40-point sections of the first simulation in Fig. 7.12. Graphs 6 to 10: from the five 40-point sections of the second simulation in Fig. 7.13.

A second problem is the apparent trend that series d exhibits: its average somehow decreased after the 21st generation for an unknown reason. A sample correlogram of a series with such a trend could be quite different from one when there is no trend. So, we must remove the trend. A simple, effective way of removing a trend is to difference the series $\{y_t\}$, i.e. transform $\{y_t\}$ to $\{y_t - y_{t-1}\}$ (section 3.2.6). After differencing all of the series to be compared, we calculate their correlograms.

In the calculation, I excluded the first two points in the observed series (Fig. 7.11c or d) to use only the stationary section (at the plateau) of the series. Each of the simulation series (Fig. 7.12 and 7.13) is partitioned into five 40-point sections (39 points in the first section). The sample correlograms of the differenced series are compared in Fig. 7.14. Correlograms 6 and 9 from the second simulation (Fig. 7.13) closely resemble correlograms A and B from the observed series.

8 Dynamics of a host–parasitoid interaction system: Utida's experimental study

8.1 INTRODUCTION

Figure 8.1 shows the dynamics of a host–parasitoid interaction system that Utida (1956, 1957a, b) experimentally produced in the laboratory, using the azuki bean weevil, *Callosobruchus chinensis* (L.), as host and the braconid wasp, *Heterospilus prosopidis* (Viereck), as parasitoid. The graphs are routinely cited in ecology text books as a typical example of a host–parasitoid oscillation. In this chapter, I shall delve into the causal mechanism underlying the dynamics of this classic experimental system.

The parasitic wasp, *H. prosopidis*, is native to the southern United States and Mexico. It is not a natural parasitoid of the host, *C. chinensis*, though it readily parasitizes this adopted host in the laboratory. A female wasp attacks by probing a host larva inside the bean just under the testa. It attacks third to fourth instar larvae and pupae, preferentially the late fourth instar to early pupae. As in many parasitic Hymenoptera, females are diploid and males, haploid: fertilized eggs produce females and unfertilized ones, males. An adult female stores sperm after mating and controls the sex of its offspring by fertilizing or not fertilizing the eggs as they pass through the oviduct (Jones, 1982). I have already given the relevant aspects of the biology of the host in section 7.1.1.

Utida kept his populations in the same petri-dish chambers as those used in the previous experiment on the single-species population of the azuki bean weevil (section 7.1.2). Also, there were two types of experiments as in the previous study: series and discrete experiments. In the series experiment, Utida followed a mixed host–parasitoid population in one experimental chamber for many generations; he kept several such series and Fig. 8.1 shows the two longest ones. In the discrete experiment, he introduced different combinations of breeding pairs of the two species into a set of experimental chambers and counted the numbers of emergent offspring.

In section 8.2, I analyse the observed dynamics of Fig. 8.1. I pay particular attention to changes in sex ratio in the wasp population over time as a major source of perturbation of the system dynamics. In section 8.3, I

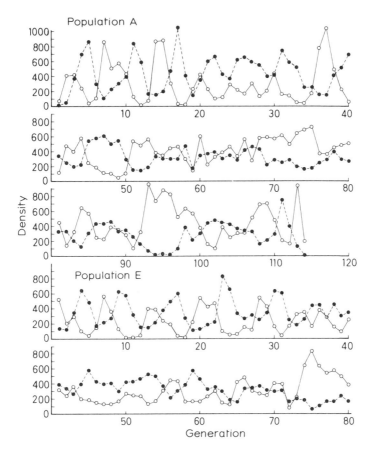

Figure 8.1   Two laboratory series of host and parasitoid populations (A and E) by Utida (1956, 1957a,b). Open circles: the host (*C. chinensis*). Solid circles: the parasitic wasp (*H. prosopidis*). Numerical data by courtesy of Professor Utida.

attempt a simulation of the observed dynamics with two types of model: the simple predator–prey model that I used in Chapter 5 and a more detailed, process model. The first model is a log-linear regression model (section 4.5.3); it can predict a progeny density directly from the breeding density without knowledge of the detail of the entire life stages. With this model, I show an important effect of the sex ratio variation in the wasp population on the system's dynamics.

The log-linear model, however, cannot simulate certain nonlinear aspects of the system process: nonlinear interactions of density-effects between the two species. The second, detailed process model is necessary for analysing the nonlinear aspects of conditional reproduction curves revealed in the discrete experiment (section 8.4). However, it requires knowledge of inter-mediate stages, which are unknown. Therefore, its use is limited to the analysis of qualitative aspects of the nonlinear structure.

The braconid wasp that Utida used completed its life cycle in about half as many days as did the host under a certain temperature–humidity combination. Therefore, to maintain both populations in a series experiment, Utida used two series of host populations out of phase by half a life cycle; the wasp could attack the two series of hosts alternatively. This scheme of experimental manipulation actually occurs in the natural population of the pink salmon, *Onchorhynchus gorbuscha*, in the Pacific coast of northern North America. In this species, two separate lines of breeding stocks return to their fresh-water spawning ground in alternate years.

The peculiar dynamics of this salmon inspired Ricker (1954) with the idea of a multiple equilibrium structure: a structure represented by a sinuous reproduction curve intersecting several times the axis of zero population change, resulting in several equilibrium states (Fig. 2.4, section 2.2.3). A simulation of the salmon dynamics with the process model I develop in section 8.4 shows, however, that the salmon dynamics is more readily explained by the Utida scheme rather than by the Ricker scheme.

8.2. ANALYSIS OF SYSTEM DYNAMICS REVEALED IN THE SERIES EXPERIMENT

8.2.1 Experimental setup

Utida used the same setup as his series experiment in Chapter 7, except a parasitoid population was added. The environment was kept constant at 30°C and 75% relative humidity (RH).

Because developmental time is much longer in the weevil than in the wasp, it was not possible to maintain a two-species system over many generations unless the emergence of adult wasps was synchronized, by some means, with the development of susceptible hosts. Utida devised an ingenious method, utilizing the fact that the length of life cycle in each species depends on temperature and humidity (Utida and Nagasawa, 1949). As it happens, at 30°C and 75% RH, the host completes one generation cycle in about 20 days, while the parasitoid requires just half as many days. So Utida used two series of host populations with the phase shift of one half cycle, i.e. 10 days, so that a single series of wasp populations could attack the two series of hosts alternately.

Figure 8.2 shows the detail of the scheme. An initial breeding stock of weevils (y_1) was introduced into a petri-dish chamber containing 10 g beans to lay eggs. After 10 days, the host larvae (y_3'') produced by y_1 had become ready to be parasitized. At this moment, an initial breeding stock of wasps (x_1) as well as a second breeding stock of weevils (y_2) were introduced together with a further 10 g beans. The wasps (x_1) then attacked the host larvae (y_3''). After another 10 days, both the second-generation weevils (y_3) and wasps (x_2) emerged as adults. By this time, the host larvae (y_4'') produced by the second stock of adult weevils (y_2) had become ready to

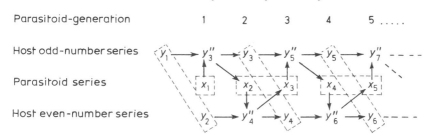

Parasitoid-generation   1   2   3   4   5 . . . . .

Host odd-number series

Parasitoid series

Host even-number series

Figure 8.2   Experimental scheme of the host–parasitoid population interaction by Utida (personal communication). The number of host adults, susceptible hosts and parasitoid adults are denoted by y, y'' and x, respectively; y_1, y_2 and x_1, in particular, are twice the numbers of initial breeding pairs.

be parasitized by the second-generation wasps (x_2). In the meantime, the adult weevils (y_3) laid eggs, and their offspring (y_5') became susceptible to the third-generation wasps (x_3) which had come out of the host larvae (y_4''); and so on. Thus, the system continued to maintain itself without being manipulated but with 10 g new beans supplied, and the leftovers removed, every 10 days.

8.2.2 Result of the series experiment

Since it is comprised of two series, the host population has two correspond-ing generations with one half cycle of phase shift. Utida combined those in one (enclosed by a diagonal dashed rectangle in Fig. 8.2) to make one 'host generation'. Similarly, he combined two successive generations in the wasp series (enclosed by a horizontal dashed rectangle in Fig. 8.2) as one 'generation' comparable to one 'host generation'. Thus, the first, second, third, . . . points in the wasp series in Fig. 8.1 are, respectively, the sums (x_1), ($x_2 + x_3$), ($x_4 + x_5$), . . . in Fig. 8.2. Likewise, the sums ($y_1 + y_2$), ($y_3 + y_4$), ($y_5 + y_6$), . . . in Fig. 8.2 are the first, second, third, . . . points in the host series in Fig. 8.1.

Figure 8.3 shows the sequence of population densities in the original setup: uncombined, original data points are plotted against wasp gener-ations. The solid and open circles separate the host population into odd-number series ($y_1, y_3, y_5, \ldots$) and even-number series ($y_2, y_4, y_6, \ldots$). Correspondingly, the same marks distinguish the wasp generations (though they form a single continuous series) that attacked the hosts in the odd- and even-number generations.

The system oscillated for a while as we expect from a simple host-parasitoid interaction. But, then, the pattern of fluctuations markedly changed around the 110th wasp generation: the system oscillation rapidly damped down thereafter. The tendency is particularly noticeable in the host series. Curiously, this change in the pattern of fluctuation coincided with similar changes in sex ratio in the wasp population. This is not a mere coincidence as I now show.

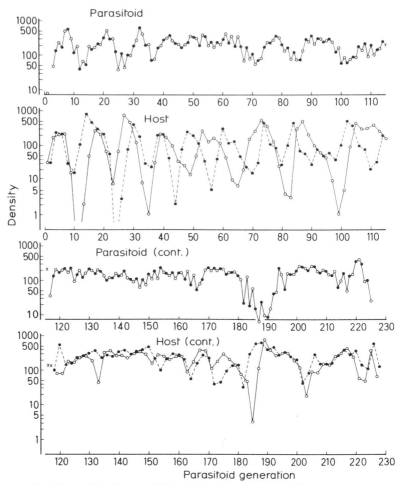

Figure 8.3 The original, unpublished data from the series experiment by Utida (Population A in Fig. 8.1) under the scheme of Fig. 8.2. Open and solid circles indicate the odd- and even-number wasp generations (Fig. 8.2). Numerical data by courtesy of Professor Utida.

8.2.3 Causal connection between the wasp's sex ratio and the system dynamics

Figure 8.4 shows generation-to-generation changes in sex ratio (the proportion of females) in the wasp series. The proportion was, on average, 36.8% (standard deviation = 13.76) until the 110th generation but, then, rose to 48.7% (s.d. = 10.42). [For convenience in a later argument, the series of sex ratios for the wasp generations attacking odd- and even-number host series are separated in the graph, even though they are one continuous series.] On the other hand, there is little change in the average sex ratio in the host population (Fig. 8.5): 46.7% before and 46.9% after the 110th

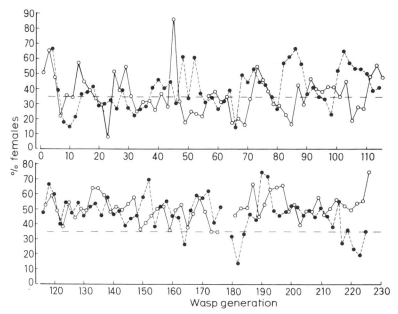

Figure 8.4 Sex-ratio (% females) changes in the wasp population in the series experiment of Fig. 8.3. Open and solid circles indicate the generations parasitizing the odd- and even-number host series, respectively. A dashed horizontal line is the mean percentage up to the 110th wasp generation.

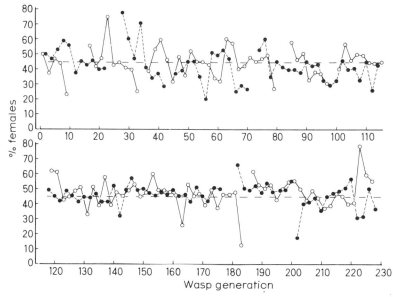

Figure 8.5 Same as Fig. 8.4 but in the odd-number (open circles) and even-number (solid circles) host series plotted against wasp generation (Fig. 8.2).

generation, although the standard deviation was somewhat reduced from 16.35 to 10.08.

Carefully examining the original numerical data (courtesy of Professor Utida), I found that an increase in the female proportion after the 110th generation was solely due to a decrease in the average (absolute) number of males while the average number of females changed little (Table 8.1). This resulted in a decrease in the average number of wasps per generation from about 210 to 145. In other words, the average reproductive output (the rate of parasitism) of a female wasp, ignoring the sex distinction among the offspring, somehow decreased after the 110th generation. It must be, intuitively, this change in the rate of parasitism which changed the pattern of population dynamics; and the parasitism was negatively correlated with the wasp's sex ratio. As I show in a simulation (section 8.3.4), such changes in the rate of parasitism do change the pattern of population dynamics as in Fig. 8.3.

Table 8.1 Average numbers (standard deviations) of male and female wasps (*H. prosopidis*) per generation before and after the 110th generation in the series experiment in Fig. 8.3. Calculated from the original data provided by Professor Utida.

	Males	Females	Total
Before	137.1 (96.0)	72.0 (39.8)	209.0 (103.9)
After	75.5 (42.6)	71.0 (37.0)	146.5 (56.4)

But what caused the change in the average reproductive output of the wasp in terms of the change in the production of male progeny? We need to know the cause in order to incorporate its effect into an appropriate model to describe the system dynamics.

8.2.4 Cause of sex-ratio changes in the wasp population

Females of some parasitic wasps are known to control the sex of their offspring by fertilizing or not fertilizing the eggs as they pass through the oviduct (Chewyreuv, 1913): a fertilized egg produces a female, and an unfertilized one a male. It has been demonstrated in a variety of species that a female wasp tends to lay a fertilized egg on a large host and an unfertilized egg on a small one (Arthur and Wylie, 1959; Aubert, 1961; Assem, 1971; Sandlan, 1979; for earlier literature, see the bibliographies of these papers). It so happens that Jones (1982) has demonstrated that this is the case with *H. prosopidis* parasitizing *C. chinensis*. Figure 8.6a shows Jones' result. Jones used host larvae of different ages to vary their size, although he did not describe the relationship between age and body size. Sandlan's results with several ichneumonid wasps (Fig. 8.6b) show a direct

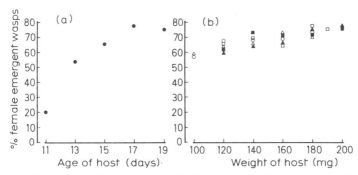

Figure 8.6 Correlations between sex ratios in some parasitic wasps and size of the hosts from which the wasps issued. Graph a: *H. prosopidis* parasitizing *C. chinensis* (adapted from Jones, 1982; fig. 1). Graph b: *Coccygomimus turionellae* parasitizing pupae of five lepidopterous species: *Pieris rapae* (solid squares); *Trichoplusia ni* (open squares); *Autographa californica* (open triangles); *Colias eurytheme* (open circles); *Galleria mellonella* (solid triangles). Adapted from Sandlan (1979; fig. 1).

correlation between host body-weight and the proportion of female wasps emerging.

Now, in *C. chinensis*, the body size of a larva susceptible to parasitism is determined not only by its age but also by the density of its cohort. Figure 8.7 demonstrates that the higher the initial density of larvae, the lower the average weight of the emergent adults. Furthermore, as mentioned in section 7.3.3, the highest number of weevil larvae hatch when their parents' density is 128 per 20 g beans at 30°C and 76% RH (Fig. 7.6; solid circles). As a result, the mean body-weight of the offspring is minimal when the density of their parents is 128 per 20 g (Fig. 8.8).

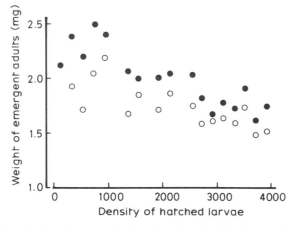

Figure 8.7 Correlations between the density of hatched *C. chinensis* larvae (per 20 g beans) and their body-weight as emergent adults (after Utida, 1941a; fig. 5). Open circles for males and solid circles, females.

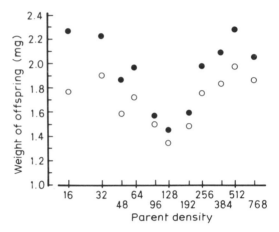

Figure 8.8   The body-weight of the second-generation *C. chinensis* in relation to the density of their parents (per 20 g beans) computed from the relationships in Figs 8.7 and 7.6 (section 7.3.3). Open circles for males and solid circles, females.

Then, from a positive correlation between the body-weight of host larvae and the sex ratio of the wasps issuing from them (Fig. 8.6), we would expect that the proportion of females among the wasps in the series experiment should be minimal at the host-parent density which produces the highest number of (therefore, on average, the lightest) larvae. In other words, if we replace body-weight axis in Fig. 8.8 by the proportion of female wasps, we would expect a similar relationship: the proportion of females would be minimal when their parent density is 128 per chamber.

On the above expectation, I regressed the sex ratio of wasps in one (wasp) generation against host density in the previous (wasp) generation (Fig. 8.9). [I took the regressions separately for the three sections of the sex-ratio series in Fig. 8.3 beginning at the 2nd, 111th and 176th wasp generations.] Contrary to the above expectation, we see no such relationship. The sex ratio among the emergent wasps in the series cannot be predicted by host density as we expected from the results in Figs 7.6, 8.6, 8.7 and 8.8.

But why did our prediction fail? The crucial point is that these results, from which we made the prediction, were all from discrete experiments. The series experiment differs from the discrete experiment in one important aspect: age uniformity. Looking into the difference will provide insight into the process underlying the observed population dynamics.

8.2.5 Effect of age uniformity and life-stage synchrony

In a discrete experiment, host larvae were similar in age because of simultaneous breeding by their parents of a uniform age. In the series experiment, there must have been a wide variation in age among the individuals within one generation (section 7.7.3) in both species. Generations

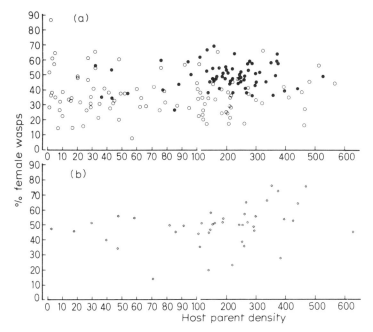

Figure 8.9   The sex ratio (% females) of the wasps issuing from the host at a given wasp generation (t) in relation to the density of host parents (at generation $t-1$) in the series experiment of Fig. 8.3. The whole series is divided into three sections, i.e. $t=2$ to 110 (open circles, graph a), $t=111$ to 175 (solid circles, graph a), and $t=176$ to 226 (asterisks, graph b). Notice a scale change in the host density at 100.

were not ideally discrete so that the timing of the emergence of adult wasps in relation to the maturity of host larvae was, probably, not maintained uniformly in all generations. Besides, the distribution of developmental stages among host larvae depends on their own density (Utida, 1959). Under these circumstances, the average age and size of host larvae at the time they are parasitized would be practically unpredictable by the density of their parents alone. Thus, the sex ratio of the wasp would become virtually independent of the host density.

However, sex ratios in the wasp population (Fig. 8.4) do not fluctuate at random. First of all, the average sex ratio changed around the 110th generation, the main motive of the present investigation. Moreover, it looks as though the sex ratios in Fig. 8.4 tend to cycle in a complex way: it appears that a high frequency cycle is superimposed upon a longer cycle. The shift in the average ratio around the 110th generation, I believe, is due to a long-term oscillation.

The existence of a superimposed high frequency cycle is easy to demonstrate by a correlogram analysis. Figure 8.10 reveals that sex ratios in the wasp tend to oscillate with the average cycle of nine consecutive points (graph a): odd-number series peaking at lag 10 and the even-number series

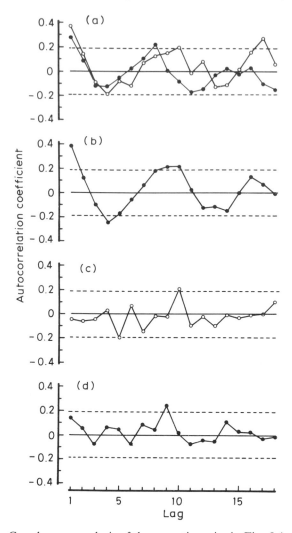

Figure 8.10 Correlogram analysis of the sex-ratio series in Figs 8.4 and 8.5. Graph a: the odd- and even-number wasp series (open and solid circles, respectively). Graph b: the series of differences between the odd- and even-number wasp series. Graphs c and d: the odd- and even-number host series. A pair of dashed lines is the Bartlett band, an approximate 95% confidence interval of the zero correlation.

at lag 8. [Because the averages of both series changed around the 110th wasp generation, I equalized them between the two sections and computed a correlogram for one whole section in each series.] Although the auto-correlation coefficients at peaks and troughs are not high, barely exceeding the Bartlett band (an approximate 95% confidence interval for zero correlation; section 3.2.3), the correlograms clearly exhibit systematic oscillations.

Further, in graph b, I calculated the correlogram of the series of differences between the odd- and even-number series, i.e. between the sex ratios at the $2t - 1$st and $2t$th host generations. Its periodic oscillation with a peak at about lag 9 indicates that the odd- and the even-number series oscillate with the phase shift of one half-cycle: one series tends to be above the other for a few host generations and vice versa for the next few generations. Consequently, when we remove the tentative distinction between the odd- and even-number series (used for the purpose of the correlogram analysis), the sex-ratio series in the wasp (as a continuous series) oscillates with two- and nine-generation cycles. These high frequency cycles are, I believe, further superimposed upon the longer cycle which caused the shift in the average ratio around the 110th generation.

In contrast, there is no systematic tendency in the series of sex ratios in the weevil population. Both odd- and even-number series are virtually at random (graphs c and d).

But what causes the oscillations in the sex-ratio series in the wasp population? We need to know in order to incorporate its effect into a model.

8.2.6 Cause of cyclic sex-ratio changes in the wasp population

If the critical life stages (development of susceptible hosts and emergence of adult wasps) are not completely synchronized, there must be a phase shift. Then, the age composition among the susceptible hosts, when a majority of wasps emerge, would change over generations. If the age compostion shifts periodically over generations, it would cause a periodic oscillation in the wasp sex ratios.

For example, if a majority of wasps emerge comparatively early in relation to the average life stage of the hosts, then an average size of susceptible hosts would be comparatively small, and, hence, the wasps would produce more male than female offspring. In the meantime, the density of susceptible hosts should be, on average, higher in an earlier, than later, developmental stage because of their mortality; thus, more of smaller susceptible hosts are available to the wasps. This would result in a comparatively higher reproductive output of the wasps by producing more male offspring (hence, higher parasitism) for their density.

In the following, I shall incorporate the above hypothesis into a simulation model to see if it reproduces the observed pattern of population dynamics.

8.3 SIMULATION OF POPULATION DYNAMICS

8.3.1 Procedure of simulation

The predator–prey model (5.17) used to simulate the lynx cycle, section 5.7.2, serves as a basic model of the density-dependent structure in the present experimental system. Because y_t hosts produce y_{t+2} progeny in the experiment (Fig. 8.2), the reproductive rate of the host (r'_t) in each series is y_{t+2}/y_t rather than y_{t+1}/y_t.

We can fit the model to data in three steps: (1) select a set of four parameters and assign them values with which the model reproduces a similar cycle length in the first section (up to the 110th generation) of the observed series in Fig. 8.3 (for the number of constants required, see section 4.5.4); (2) transform the densities x and y in such a way that the equilibrium density and the amplitude of a simulated cycle match those of the observed; (3) subject the density-dependent structure to perturbation due to sex-ratio changes in the wasp population.

8.3.2 System model

In step (1), I use the four-parameter basic (canonical) form:

$$x_{t+1}/x_t = a_1[1 - \exp(-b_1 y_t)]\exp(-x_t) \tag{8.1a}$$

$$y_{t+2}/y_t = a_2\exp[-(y_t + b_2 x_t)]. \tag{8.1b}$$

Writing the logarithm of a lower case-letter in the corresponding capital letter, e.g. log $x \equiv X$, (8.1) is log transformed to:

$$X_{t+1} - X_t = A_1 + \log\{1 - \exp[-\exp(Y_t + B_1)]\} - \exp(X_t) \tag{8.2a}$$

$$Y_{t+2} - Y_t = A_2 - \exp(Y_t) - \exp(X_t + B_2) \tag{8.2b}$$

In step (2), X and Y are transformed to X' and Y' by $X' \equiv m_1 + n_1 X$ and $Y' \equiv m_2 + n_2 Y$; and by inversion:

$$X \equiv (X' - m_1)/n_1 \tag{8.3a}$$

$$Y \equiv (Y' - m_2)/n_2. \tag{8.3b}$$

Parameters m_1 and m_2 adjust the equilibria, and n_1 and n_2 the variances, of the log-transformed series $\{X_t\}$ and $\{Y_t\}$ of (8.2). Substituting the right-hand sides of (8.3) for X and Y in (8.2), and rewriting X' and Y' as X and Y, we have:

$$X_{t+1} = X_t + n_1 A_1 + n_1 \log\{1 - \exp[-\exp(Y_t - m_2)/n_2 - B_1]\} - \exp[(X_t - m_1)/n_1] \tag{8.4a}$$

$$Y_{t+2} = Y_t + n_2 A_2 - \exp[(Y_t - m_2)/n_2] - \exp[(X_t - m_1)/n_1 - B_2]. \tag{8.4b}$$

Figure 8.11 shows the deterministic series generated by (8.4) with the constant values:

$$(A_1, B_1, m_1, n_1) = (1, 1.25, 6.85, 0.5)$$
$$(A_2, B_2, m_2, n_2) = (1.95, -3.2, 3.0, 2.0).$$

The deterministic series in Fig. 8.11 are not directly compared with the observed series under the influence of random perturbation. Therefore, for the determination and validation of these constant values, we can use conditional reproduction curves (similar to those in Fig. 5.12 for the lynx).

In the present experimental system, we have data for both the host and parasitoid populations. So there are four conditional reproductive rates, $r(x|y)$ and $r(y|x)$ for the wasp and $r'(y|x)$ and $r'(x|y)$ for the host (Fig. 4.6, section 4.4.1). The regression of a rate, e.g. $r(x|y)$ on x, given a range of y, gives a conditional reproduction curve. Figure 8.12 is an example set of many similar graphs. If we choose n intervals of density for each species to fit n pairs of conditional reproduction curves, we get $4n$ such graphs altogether. We choose a parameter set so that the estimated curves fit the data points satisfactorily in all graphs. Note that not every pair of conditional curves fit equally well, e.g. graph d. A selected set of parameter values

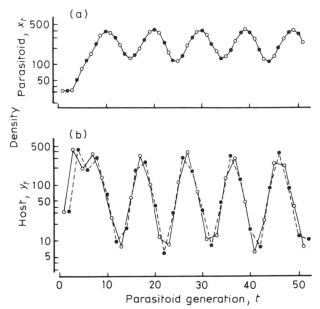

Figure 8.11 Deterministic series generated by the regression model system (8.4): wasp series (graph a) and host series (graph b). Open and solid circles indicate the odd- and even-number (wasp) generations as in Fig. 8.3.

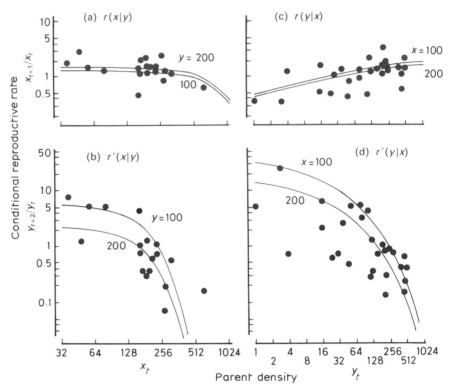

Figure 8.12 An example set of conditional reproductive rates realized in the series experiment of Fig. 8.3 for the interval of fixed densities (100, 200). The fitted curves are calculated with the system model (8.1)/(8.2).

must, of course, generate the dynamics similar to the observed. Many iterations may be necessary before arriving at a satisfactory result.

8.3.3 Simulation of sex-ratio changes in the wasp population

Let us now simulate the observed sex-ratio changes in the wasp population (Fig. 8.4) as a major source of perturbation of the density-dependent structure (8.4a). The simulation need not describe the actual mechanism of sex-ratio changes as discussed in section 8.2. A simple deterministic periodic function with random deviation should serve the purpose.

Let us use a square wave function (as the deterministic component) for each (odd- or even-number) series of observed sex ratios (Fig. 8.4), taking one value (say, *a*) for the first five generations and switching to another value (*b*) in the following four generations, making the total periodicity equal to the observed average period of nine generations. Let one series repeat the sequence and let the other series repeat the same but in the reversed order, i.e. first 4*b*s followed by 5*a*s. So the two series oscillate with a nine-

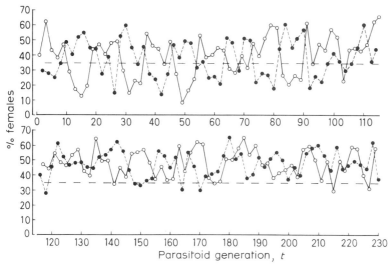

Figure 8.13   Simulation of the observed sex-ratio changes in the wasp series in Fig. 8.4. A dashed horizontal line is the mean ratio in Fig. 8.4 as a reference line for a visual comparison between the observed and simulated series.

generation period with the phase shift of one half-cycle as was observed (section 8.2.5). Putting the two series together, the combined series alternates between a and b, repeating the sequence $(a, b, a, b, a, b, a, b, a)$. To simulate the observed series in Fig. 8.4 in which the average ratio changed around the 110th generation, the a and b values are changed after that generation with a transient period of ten generations. The actual values used in the simulation are:

$$a = 0.8 \text{ and } b = 0.6 \text{ for } t \leqslant 99,$$
$$a = 0.775 \text{ and } b = 0.85 \text{ for } t \geqslant 110;$$

the transient series between $t = 100$ and 109 being:

$$(0.7, 0.8, 0.7, 0.8, 0.7, 0.8, 0.75, 0.8, 0.75, 0.8).$$

A random deviation from each a and b in the above deterministic periodic functions is independent, normally distributed, with the mean -0.356 and variance 0.00316. With these values, the simulated sex-ratio series in Fig. 8.13 have approximately the same averages, variances and periodicity as the observed series in Fig. 8.4.

8.3.4 Simulation of the population dynamics

The final step of simulation (step 3 of section 8.3.1) is to subject the density-dependent structure (8.4) to perturbation. The perturbation effect must be qualified in two aspects: it should damp down the system oscillation as Fig. 8.3, section 8.2.2, and should lower the equilibrium after the 110th generation.

Recall now the three kinds of perturbation effects: vertical, lateral and nonlinear (section 1.7). A vertical perturbation acts on parameters A_i, $i=1$ and 2 in (8.4a) and (8.4b), respectively; a lateral perturbation acts on m_i; and a nonlinear one acts on n_i. Because of the definitions (8.3), parameter n_i can damp an oscillation without affecting the system equilibrium, while parameter m_i does the opposite. Therefore, the lateral and nonlinear perturbations singly do not qualify as candidates. On the other hand, a vertical perturbation effect can damp an oscillation and can lower the equilibrium density at the same time (Fig. 2.20, section 2.4.5). So this type of perturbation is certainly qualified.

An $n_i A_i$ in (8.4) is a potential (log) reproductive rate (say, R_m or R'_m for $i=1$ or 2) which is perturbed by sex-ratio changes. So we consider the situations:

$$R_m + u_t \tag{8.5a}$$

$$R'_m + v_t. \tag{8.5b}$$

The quantities (8.5) must satisfy three stipulations: (1) (8.5a) must be negatively correlated with the simulated sex-ratio series in Fig. 8.13; (2) (8.5b) is correlated with sex-ratio changes in the host which are a series of independent random numbers (Figs 2.5, 2.10c and d, section 2.2.5); (3) both (8.5a) and (8.5b) must be such that the means and variances of the simulated log population series $\{X_t\}$ and $\{Y_t\}$ are the same as those in the observed series in Fig. 8.3.

A simple way to satisfy the above conditions is to consider R_m to be a periodic function, $R_m(t)$, which repeats the sequence $(a, b, a, b, a, b, a, b, a)$ as in section 8.3.3 but the values of a and b are:

$$a=0.4 \text{ and } b=0.8 \text{ for } t\leqslant 99,$$
$$a=0.45 \text{ and } b=0.3 \text{ for } t\geqslant 110;$$

and the transient series between $t=100$ and 109 is:

$$(0.6, 0.4, 0.6, 0.4, 0.6, 0.4, 0.5, 0.4, 0.5, 0.4).$$

The random numbers u_t and v_t are independent, normally distributed with mean $=0$ and variance $=0.01266$.

Figure 8.14 shows a result of the simulation which compares with the observed series in Fig. 8.3. However, the model does not simulate one aspect of the observed series: the two host series tend to be separated from one

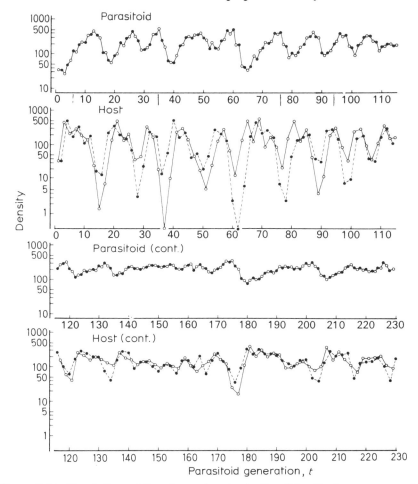

Figure 8.14   Simulation of the observed series in Fig. 8.3 with the system model (8.1)/(8.2), using the parameter values given in the text.

another. In the simulation, they tend to fluctuate in unison. The separation between the two host series in the experiment occurs because the density-dependent structure of the system is actually nonlinear. Model (8.4) has a log-linear structure (section 4.5.3) – after the log transformation, the functions of X are independent of those of Y – and is unable to reproduce this aspect of nonlinear dynamics.

However, it so happens that if we combine two successive generations in Fig. 8.14 after the Utida scheme in Fig. 8.2, we can bring out the aspects of dynamics unaffected by the experimental manipulation of using two host series. Figure 8.15 is the result directly comparable with Fig. 8.1.

Some qualitative aspects of the nonlinear structure can be explained by a detailed process model (partitioned stage-process model). The conditional

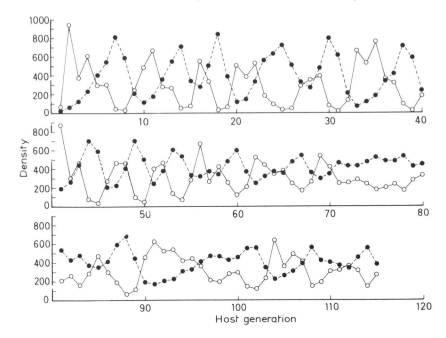

Figure 8.15 Same as Fig. 8.14, but the densities in the adjacent odd- and even-number wasp generations are combined in both species after the scheme of Fig. 8.2 and plotted in the linear scale on the vertical axis to compare directly with the observed series of Fig. 8.1.

reproduction curves determined in the discrete experiment provide a basis for formulating the partitioned model as I now discuss.

8.4 ANALYSIS OF CONDITIONAL REPRODUCTION CURVES

8.4.1 Experimental setup and results of the discrete experiment

Utida (1944, 1948) conducted two sets of discrete experiments. In the 1944 set, he introduced 16 breeding pairs of hosts into each petri-dish chamber containing 10 g beans. After about 14 days, when the second-generation host larvae had become late 4th instar and ready to be parasitized, he introduced breeding pairs of wasps at various densities in the dual geometric series (as in the previous experiment, section 7.2). Table 8.2 shows the numbers of second-generation host and parasitoid adults that emerged.

In his 1948 set, Utida varied the number of host breeding pairs while holding wasp density at one pair per chamber with 10 g beans. He repeated the same experiment with one pair of wasps per 20 g beans. Table 8.3 shows the results. [Because the quantity of beans used differs from experiment to

Table 8.2 Results of the 1944 discrete experiment at 30°C and 75% RH with 10 g beans per chamber. The host (*C. chinensis*) breeding density was held at 32 (16 pairs) per chamber, while the wasp (*H. prosopidis*) breeding density was varied in the dual geometric series. The symbols mm and ff are males and females, respectively. After Utida (1944; table 4).

	Numbers of insects per chamber					
Parasitoid parents (*mm/ff = 1*)		*Host offspring*			*Parasitoid offspring*	
	mm	*ff*	*total*	*mm*	*ff*	*total*
0	185.3	203.0	388.3	0.0	0.0	0.0
2	183.4	188.7	372.1	7.7	22.9	30.6
4	171.8	159.0	330.8	12.8	43.5	56.3
8	135.3	155.3	290.6	21.3	69.8	91.1
16	108.3	98.8	207.1	45.8	114.5	160.3
32	72.0	59.2	131.2	68.8	151.6	220.4
48	64.5	36.5	101.0	76.5	139.5	216.0
64	24.7	17.0	41.7	120.0	124.0	244.0
96	19.7	20.0	39.7	86.7	164.3	251.0
128	28.0	26.7	54.7	113.3	138.3	251.6
192	25.5	20.5	46.0	95.0	126.5	221.5

experiment, I standardize the densities of the insects as numbers per 20 g beans: numbers per 10 g will be doubled (section 7.5).]

From these tables, we can calculate conditional reproductive rates (section 4.4.1):

$$x_2/x_1 = r(x_1|y_1) \quad \text{(wasp)}$$
$$y_2/y_1 = r'(x_1|y_1) \quad \text{(host)}$$

for the fixed host density $y_1 = 32$ pairs, and

$$x_2/x_1 = r(y_1|x_1)$$
$$y_2/y_1 = r'(y_1|x_1)$$

for the fixed wasp density $x_1 = 1$ or 2 pairs. [Note that y_2 in these discrete experiments is the progeny of y_1 and should not be confused with the initial breeding stock (y_2) of the even-number host series in the series experiment (Fig. 8.2).] These conditional rates are then plotted against the varied breeding density (y_1 or x_1) in Fig. 8.16. [A curve fitted in each graph is calculated with the model which I formulate in section 8.4.2. The dashed curve in graph d is a replicate of Fig. 7.2a, section 7.2.2: a reproduction curve of the host without parasitoids.]

Comparing these curves with their idealized counterparts in Fig. 4.6 (section 4.4.1), we notice the following. [Allow for the difference in scaling

Table 8.3 Results of the two 1948 discrete experiments at 30°C and 75% RH. The quantity of beans per chamber: 10 g in table (a) and 20 g in table (b). In both experiments, the wasp (*H. prosopidis*) breeding density was held at two (one pair) per chamber, while the host (*C. chinensis*) breeding density was varied in the dual geometric series. [The numbers of insects in (a) are doubled to make them comparable with those per 20 g beans in (b): $x_1 = 4$ in (a) and 2 in (b).] The notations *mm* and *ff* are males and females, respectively. After Utida (1948; tables 1 and 2).

Host parents $(mm/ff = 1)$						

| Host parents $(mm/ff = 1)$ | Host offspring | | | Parasitoid offspring | | |
	mm	ff	total	mm	ff	total
(a)						
1	4.6	1.0	5.6	9.0	17.0	26.0
2	7.6	4.4	12.0	17.0	41.6	58.6
4	54.0	43.6	97.6	18.4	51.6	70.0
8	115.6	118.8	234.4	32.4	60.0	92.4
16	*	*	338.0	14.4	59.6	74.0
32	*	*	549.6	15.2	57.6	72.8
64	*	*	734.0	16.4	41.2	57.6
96	*	*	840.4	12.4	50.0	62.4
128	*	*	904.0	18.4	44.8	63.2
192	*	*	967.2	22.0	23.6	45.6
256	*	*	868.4	14.4	24.0	38.4
384	*	*	800.0	8.0	36.4	44.4
512	*	*	738.4	9.2	33.2	42.4
(b)						
2	5.5	2.0	7.5	12.8	15.7	28.5
4	54.8	41.0	95.8	10.9	24.8	35.7
8	73.7	63.5	137.2	6.5	28.0	34.5
16	181.0	166.8	347.8	18.8	25.8	44.6
32	281.8	270.5	552.3	17.3	32.0	49.3
48	314.5	318.0	632.5	10.0	24.0	34.0
64	397.5	387.3	784.8	6.8	29.0	35.8
96	427.0	455.5	882.5	5.5	24.3	29.8
128	412.0	422.3	834.3	6.8	22.5	29.3
192	374.8	389.5	764.3	8.3	22.8	31.1
256	361.0	358.5	719.5	12.0	25.0	37.0

Numbers of insects per chamber

the axes between the two figures: linear scale in Fig. 4.6 and log scale in Fig. 8.16.] The observed curves in Fig. 8.16a and b are much like their counterparts, Fig. 4.6a and b. Those in Fig. 8.16c and d differ from those in Fig. 4.6c and d.

The pattern in Fig. 8.16d (solid curve) is comparable to Fig. 4.8b (dashed curve), section 4.5.3, based on Bulmer's (1976) contention: the parasitism may substantially reduce the reproductive rate of a host (compared to its

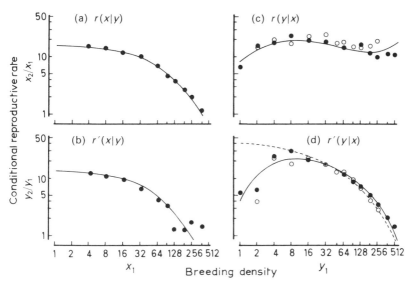

Figure 8.16 A set of conditional reproductive rates determined in the discrete experiment. Graphs a and b: the conditional rates $r(x|y)$ of wasp and $r'(x|y)$ of host, for the host breeding density fixed at $y_1 = 64$ per 20 g beans, plotted against the wasp breeding density x_1. Graphs c and d: the conditional rates $r(y|x)$ and $r'(y|x)$, for the wasp breeding densities fixed at $x_1 = 2$ (open circles) and $x_1 = 4$ (solid circles), plotted against the host breeding density y_1. The fitted curves are computed from the stage-process model system to be formulated in section 8.4.2. Dashed curve in graph d is a replicate of Fig. 7.2a without parasitoids.

rate in the absence of parasitoids) towards the lower end of the host density. The effect of parasitism, however, quickly dissipated as the host density increased because the wasp density x_1 was held low in this experiment.

The observed conditional rate $r(y|x)$ in Fig. 8.16c is rather unexpected. Instead of a monotonic increase to a plateau like Fig. 4.6c, the observed trend is concave about $y_1 = 128$. The stage-process model which I formulate now can explain this pattern.

8.4.2 Formulation of stage-process models

Figure 8.17 shows the sequence of major developmental stages. In the host sequence, $y_1/2$ breeding pairs produce y_2'' larvae and pupae that are susceptible to parasitism. Of these y_2'' susceptible hosts, y_2''' are parasitized by $x_1/2$ breeding pairs of wasps. The unparasitized hosts ($y_2'' - y_2'''$) would be further reduced because of continued competition among themselves before y_2 survivors emerge as adult weevils.

As for the wasp sequence, the number of hosts parasitized (y_2''') is the potential number of second-generation wasps because usually only one wasp can mature inside a single host. Utida (1944) noticed, however, that some of

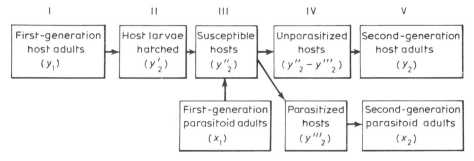

Figure 8.17   Sequence of major developmental stages in the host–parasitoid system.

the parasitized hosts died without producing mature parasitoids. Thus, the number of emergent wasps (x_2) tends to be less than y_2''.

Although little information is available in stage III and onwards, we can formulate a model from the two basic principles: the logistic and random search processes. As shown in Chapter 7, the logistic law governs the rate of change in density between stages in the host without parasitoids. Thus, the host density is reduced exponentially from stage to stage as in (4.32a), section 4.4.1. The wasp, *H. prosopidis*, lays eggs indiscriminately on unparasitized and parasitized hosts. So the eggs are likely to be distributed over hosts in the Poisson fashion, and the rate of parasitism is described by (4.32b).

Both processes involve the exponential function which, for simplicity, can be written:

$$\alpha \exp(-\beta\theta^\gamma) = E(\theta;\ \alpha,\ \beta,\ \gamma). \tag{8.6}$$

The argument θ is the initial density at a given stage in Fig. 8.17 (with modifications at certain stages), and α, β, and γ are non-negative constant parameters.

Stages I to III in host
These stages are free of parasitoids and should repeat the previous Experiment I (section 7.3) up to the susceptible stage. Because the sex ratio is 1:1 in the breeding population (y_1), the number of breeding females is $y_1/2$. Thus,

$$y_2'' = (y_1/2) \times (\text{oviposition rate}) \times (\text{hatching success}) \times (\text{survival to}$$
$$\text{susceptible stage})$$
$$= (y_1/2)E(y_1; k_1, k_2, k_3)E(y_1; 1, k_4, k_5)E(y_2'; 1, k_6, 1) \tag{8.7}$$

The values of k_1 to k_5 have been determined in Experiment I, section 7.3. k_6 is unknown but should be less than $c = \exp(-7.6)$ for the whole larva-to-pupa period (see $-\ln c$ for Experiment I, Fig. 7.7, in Table 7.4).

Parasitism
Suppose x_1 wasps (including males) lay n eggs over y_2'' hosts at random. So the proportion parasitized (y_2'''/y_2'') is 1 minus the zero-term of the Poisson distribution with the mean n/y_2''. Thus,

$$y_2''' = y_2''[1 - E(n/y_2''; 1, 1, 1)].\tag{8.8}$$

Now the average oviposition by a wasp (n/x_1) depends on the number of susceptible hosts (y_2'') as well as on the wasp density (x_1). In particular: (1) given y_2'', n/x_1 decreases as x_1 increases as a logistic process due to competition among the wasps (Fig. 4.6a); (2) given x_1, n/x_1 increases with y_2'' but levels off sooner or later, the familiar functional response of a parasitoid (Fig. 4.6b). Thus, in combination:

$$n/x_1 = E(x_1; k_7, k_8, k_9)[1 - E(y_2''; 1, k_{10}, 1)].\tag{8.9}$$

Substitution of n from (8.9) in (8.8) gives y_2'''. All k constants in (8.9) but k_7 are unknown: k_7 is one-half of the maximum oviposition rate of a female wasp which is about 50 eggs (Dr Shimada, Tokyo University, personal communication).

Stages IV to V in host
After parasitism, the unparasitized hosts ($y_2'' - y_2'''$) further decrease as they compete with each other until y_2 survivors emerge as second-generation adults. The logistic process should continue but with one modification: argument θ in (8.6) is not $y_2'' - y_2'''$ but greater. The reason is the following. Usually, a host larva stops feeding as soon as it is parasitized (Dr Shimada, personal communication). Even so, it has already consumed a considerable part of the bean. Then, an unparasitized larva that happened to be in the vicinity may not find enough food until pupation. In other words, the effect of the parasitized larva would be carried over to still influence the subsequent survival of the unparasitized larva nearby. As a result, the effective density should be $y_2'' - k_{11}y_2''' \, (0 < k_{11} \leqslant 1)$: $k_{11} = 1$ means no carry-over effect. Thus,

$$y_2 = (y_2'' - y_2''')E[(y_2'' - k_{11}y_2'''); 1, k_{12}, k_{13}].\tag{8.10}$$

All k constants are unknown in (8.10).

Stages IV to V in parasitoid
Although multiple parasitism is usual in this wasp (a single host receiving many eggs), only one larva matures within one host. So the total number of host parastitized (y_2''') is the potential number of second-generation wasps, i.e. x_2. However, a proportion of parasitized hosts usually die without producing wasps (Utida, 1944).

Two conceivable processes are: (1) multiple egg laying by the wasps fatally injures some hosts; (2) an unparasitized host may damage a parasitized host nearby. In the first situation, the ratio x_2/y_2''' – the proportion of x_2 emergent wasps from y_2''' parasitized hosts – should decrease exponentially (assuming random stubbing) as the ratio x_1/y_2'' increases. In the second situation, the ratio x_2/y_2''' decreases also exponentially (assuming random encounters) as the effective density of competing host larvae (which is greater than $y_2' - y_2'''$ for the reason mentioned above) increases. Thus, combining the two situations:

$$x_2/y_2''' = E[(x_1/y_2'); k_{14}, k_{15}, k_{16}]E[(y_2' - k_{17}y_2'''); 1, k_{18}, k_{19}]. \quad (8.11)$$

All k constants in (8.11) are unknown but $0 < k_{17} < 1$ as k_{11} in (8.10).

The assumption of process (8.11) is crucial to the explanation of the concave trend in Fig. 8.16c. As already shown in Chapter 7, the number of second-generation weevils (both hatched larvae and emergent adults without parasitism) peaks when their parent density is 128 per chamber (solid circles, Figs 7.6 and 7.1). Therefore, by interpolation, the number of susceptible hosts (y_2') is likely to peak at $y_1 = 128$. If the functional response of the wasp is the usual monotonically increasing function of host density (as Fig. 4.6b), we would expect that the number of parasitized hosts (y_2'') should peak at $y_1 = 128$. If, also, all of y_2'' survive and produce mature wasps, then we would expect that the rate $r(y|x)$ in Fig. 8.16c should also peak at $y_1 = 128$. The observed trend shows just the opposite: it troughs at $y_1 = 128$! So we must conclude that a substantial proportion of y_2'' in fact die as Utida (1944) remarked.

Figure 7.5a, section 7.3.3, shows that survival of weevil larvae is the lowest at $y_1 = 128$ because it produces the highest number of larvae at which larval competition is most intense. If parasitized larvae are very susceptible to damage by unparasitized ones, the number of emergent wasps (x_2) would be reduced substantially. This can explain the concave conditional curve of Fig. 8.16c.

8.4.3 Model fitting

As noted in section 8.4.2, parameters k_1 to k_5 were already determined in Experiment I in section 7.3. These are given in Table 7.4: $k_1 = b, k_2 = c, k_3 = a$ (for Fig. 7.3); $k_4 = c$ and $k_5 = a$ (for Fig. 7.4). Some other ks are unknown, but we know their restrictions as already noted: k_6 should be positive and less than $\exp(-7.6)$; k_7 should be in the order of 25; both k_{11} and k_{17} are positive but not greater than 1. The remaining 10 ks are completely unknown positive constants.

After many trials, starting with a rough, initial guess, I obtained the set of values in Table 8.4 (I) which produced the curves fitted to the data points in Fig. 8.16. Whether or not the ecological mechanism the model assumes is correct needs to be tested.

Table 8.4 Values of parameters k_1 to k_{19} in the stage-process models (8.7) to (8.11) used in: (I) Figs 8.16 and 8.18; (II) Figs 8.19 and 8.20; (III) Figs 8.21b and 8.23; (IV) Figs 8.22 and 8.24.

	I	II	III	IV
k_1	74.2066	74.2066	25.0000	80.0000
k_2	0.4966	0.4966	0.4966	0.4966
k_3	0.2000	0.1800	0.1800	0.1800
k_4	0.0006	0.0006	0.0006	0.0006
k_5	1.3000	1.2000	1.2000	1.2000
k_6	0.0001	0.0002	0.0001	0.0001
k_7	60.0000	50.0000	25.0000	30.0000
k_8	0.0800	0.0800	0.1500	0.1500
k_9	0.5000	0.2000	0.2000	0.2500
k_{10}	0.4500	0.0020	0.0009	0.0009
k_{11}	0.5000	0.5000	0.0000	0.0000
k_{12}	0.0010	0.0010	0.0010	0.0010
k_{13}	0.8000	0.8000	0.8000	0.8000
k_{14}	2.7183	1.8221	0.5000	0.5800
k_{15}	2.2500	1.3000	0.0000	0.0000
k_{16}	0.0008	0.0008	—	—
k_{17}	0.9500	0.4000	0.5000	0.5000
k_{18}	0.5000	0.5000	0.0498	0.0498
k_{19}	1.0000	0.1000	0.4000	0.4000

8.4.4 Deduction of population dynamics

The partitioned model formulated above defines a pair of reproduction surfaces from which we can deduce the system dynamics. In order to make the deduction comparable with the experimental setup in the series experiment (a single wasp series attacking two host series alternately), we only need to set: $x_1 \equiv x_t, x_2 \equiv x_{t+1}, y_1 \equiv y_t$, and $y_2 \equiv y_{t+2}$; cf. the unpartitioned model (8.1). In this section, I only deal with the deterministic behaviour of the model process.

With the set of parameter values estimated from the result of the discrete experiment (Table 8.4, I), and starting with the initial set of densities ($x_1 = 8, y_1 = y_2 = 32$) as the series experiment, I found the model system did not persist. The model wasp quickly wiped out the odd-number host series and itself, in turn, perished. In Fig. 8.18, using these parameter values, I calculated a set of conditional reproduction curves (like the set in Fig. 8.12) with the fixed interval of densities (100, 200) in both species. We see that the calculated curves fit the data points rather poorly. A major reason for the poor fit is, probably, the following.

In his discrete experiment, Utida fixed the wasp breeding density at one or two pairs per chamber (section 8.4.1). However, the wasp densities in the series experiment (Fig. 8.3) were mostly within the range of 50 to 500 with the average

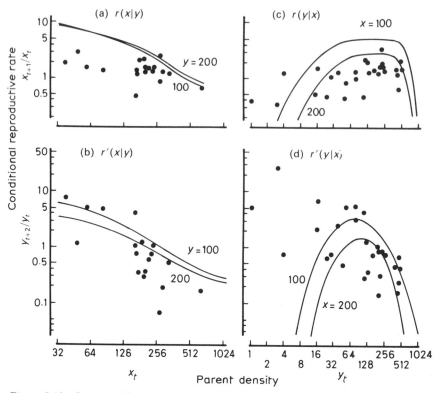

Figure 8.18 Same as Fig. 8.12, but the curves are calculated from the stage-process models (section 8.4.2) with the set of parameter values in Table 8.4(I).

of 210 (before the 110th generation) which declined to 150 after that generation (Table 8.1). This means that the conditional reproductive rates determined in the discrete experiment, i.e. $r(y|x)$ and $r'(y|x)$ (for $x = 2$ or 4) in Fig. 8.16c and d, were way outside the observed range of densities in the series experiment. An extrapolation from a single pair of conditional reproduction curves at an extreme fringe area could considerably distort a central part of the reproduction surface. A discrete experiment should be designed to estimate a reproduction surface centred about the equilibrium density.

We see a serious disagreement between the model and data in $r'(y|x)$ (Fig. 8.18d). The calculated conditional curves bend sharply downwards towards a lower range of host density. In that range the model host could not survive because of extremely high parasitism. The observed reproductive rates of the wasp in the series experiment (data points in Fig. 8.18a and c) are, on the whole, much lower than the calculated curves indicate. Probably, the process model (8.9) tends to overestimate (k_{10} too large) the actual parasitism at low host densities in the series experiment.

If we reduce the k_{10} value in (8.9), the model system begins to persist. With some more minor changes in other parameters (Table 8.4, II), we can reduce the deviation of calculated curves from the data points (Fig. 8.19). At the same time, a pattern of dynamics similar to that in the early part of the series experiment begins to emerge (Fig. 8.20); in particular, the oscillations in the two host series are now well separated as in the observed series in Fig. 8.3.

Because of so many unknown parameters in the model, a quantitatively accurate simulation is difficult at this stage. More experiments are necessary for a further improvement. However, a qualitative aspect of the dynamics that this model system generates happens to provide insight into the complex dynamics of a Pacific salmon.

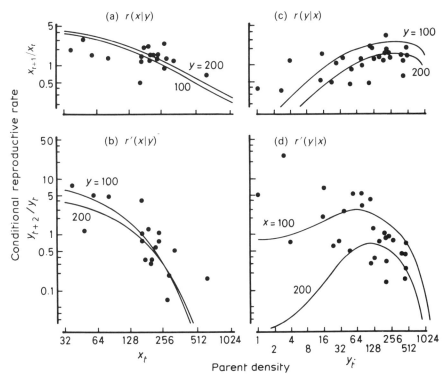

Figure 8.19 Same as Fig. 8.18, but the curves are calculated with the set of parameter values in Table 8.4(II).

8.5 ANALOGOUS FISH POPULATION DYNAMICS

The use of the dual breeding stocks in the host population in Utida's experiment is an experimental manipulation. As it turns out, however, an almost analogous situation actually occurs in nature.

The pink salmon, *Onchorhynchus gorbuscha* (Walbaum), one of the several Pacific salmon occurring along the west coast of Canada and the adjacent coasts of the United States, has a rigid life cycle of two years (Neave, 1953). The

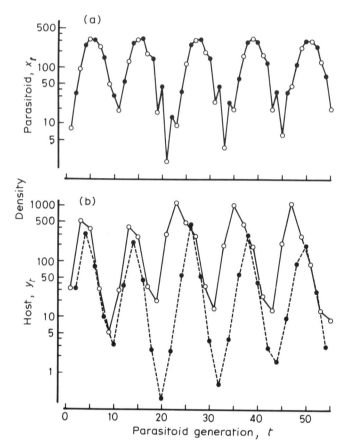

Figure 8.20 Deterministic series of wasp (graph a) and host (graph b) generated by the process model system of section 8.4.2, with the same set of parameter values as Fig. 8.19. Open and solid circles indicate the odd- and even-number wasp generations as in Fig. 8.3.

adults return to fresh water from the sea in the late summer to spawn. The fry, soon after their emergence from the gravel, migrate to the sea in the spring. After having spent two summers in the sea, the mature fish return to their natal river and, as is well known, all die after spawning.

Because of this uniformly maintained two-year life cycle, the fish returning to the same river in odd- and even-years constitute two completely separate breeding stocks (so-called 'lines'). If the two lines are alternately preyed upon by the common predator complex, the situation becomes analogous to Utida's experiment.

Figure 8.21a shows the population fluctuations of the two lines of the salmon: the data from 1918 to 1950 in graph a represent estimated numbers of the fish caught annually in the Skeena River area of British Columbia (Neave, 1953); those from 1951 to 1975 are the annual runs of the salmon (total catches

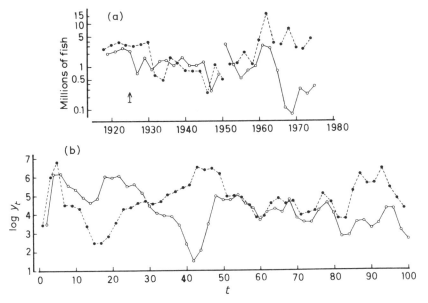

Figure 8.21 Graph a: Population fluctuations in the pink salmon, *Onchorhynchus gorbuscha*, in British Columbia: open and solid circles indicate the odd- and even-year lines. Before 1950: annual catches in the Skeena River area (after Neave, 1953; fig. 4). After 1951: total run up the Atnarko River system (after Peterman, 1977; fig. 2). The vertical arrow indicates the magnitude of a 60% increase in the Skeena series to make them comparable with the Atnarko series (see text). Graph b: a simulation of the observed dynamics in graph a, using the process model system of section 8.4.2 with the parameter set in Table 8.4(III). Open and solid circles are the realized values of log y_t for odd and even t.

plus escapements) at the Atnarko River system about 300 km southeast of the Skeena (Peterman, 1977).

Neave (1953) quotes the statistics between 1934 and 1949 which show that, on average, a steady 60% of total run was caught annually by commercial fishing. Therefore, if the data points in graph a for the Skeena River area are raised by the length of the vertical arrow shown in the graph, they would be comparable to total run up the Atnarko. After the adjustment, the two sections of the graph seem more or less to merge into each other. This is not necessarily a coincidence in view of the Moran effect (sections 2.5 and 5.5.1) which could have synchronized the fluctuations in the two local populations only a few hundred kilometres apart; also the two populations could have similar dynamics if they share a common feeding ground in the sea.

Ricker (1954; p. 602) describes the peculiarity of this species' dynamics:

The species is peculiar in that there are a few areas where one of the two 'lines' is completely absent in a series of streams. In other areas one line exists at a level much below the other, and the two lines will have maintained approximately the same relative position for decades, though not necessarily for the whole period of record. In still other areas there is no consistent difference between the lines in respect to abundance. Furthermore there have been cases where one line has dropped suddenly from a high to a low level of abundance and has remained at that level for a considerable period; in one or two instances such a reduced stock has bounced back up to its former level.

To explain the above behaviour of the pink salmon populations, Ricker (1954) proposed a multiple-equilibrium process, represented by a sinuous reproduction curve which intersects several times the axis of zero population change, i.e. equilibrium points (Fig. 2.4, section 2.2.3). The idea has been elaborated by Peterman (1977). It is hypothesized that, as the environment fluctuates, the equilibrium state of the process may shift from low to high, and vice versa, generating a complicated pattern of dynamics.

I do not believe that Ricker's multiple-equilibrium structure applies here. Rather, the peculiar pink salmon dynamics are more readily explained by the Utida scheme of 'multi-line' system.

Figure 8.21b shows a simulation with the system model of section 8.3.2, in which the log reproductive rates of the 'fish' (y) and 'predator' (x) populations are subjected to vertical perturbations; only the 'fish' population series (i.e. the two $\{y_t\}$ series) are shown in the graph. The second half of the generated series (b) is much like the observed series (a). Figure 8.22 is another example of simulations with the same model (different parameter set). It resembles some aspects of the observed dynamics that Ricker described: typically, one line dropping suddenly from a high to a low level but bouncing back after a period.

The behaviour of the simulated populations is easy to understand if we look at their deterministic behaviour before subjected to random perturbation. Figure 8.23, the deterministic version of Fig. 8.21b, shows the two lines of populations exhibiting limit cycles at different equilibrium levels. Which line takes the upper or lower level depends on the initial density of the line relative

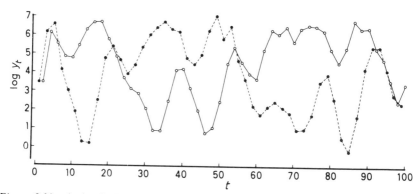

Figure 8.22   A simulation similar to Fig. 8.21b but with the parameter set given in Table 8.4 (IV).

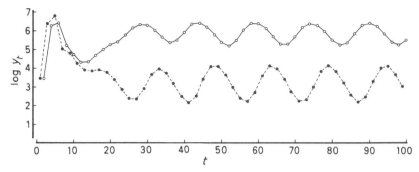

Figure 8.23   Deterministic version of Fig. 8.21b.

to the other. In the simulation, the two lines have the same initial densities, and the odd-number line took the lower equilibrium. When subjected to random perturbation (Fig. 8.21b), the two lines may switch over between the two levels; or they may even stay together at an intermediate level for a while before they are separated again.

Figure 8.24, the deterministic version of Fig. 8.22, exhibits a deterministically irregular behaviour (section 2.2.5). It behaves like one subjected to random perturbation: one series switches between two levels or both series fluctuate together without a consistent difference at an intermediate level. If the model is subjected not only to vertical perturbation but also to lateral and nonlinear perturbations, almost all aspects of the salmon population dynamics that Ricker described would be reproduced in simulations with my system model of the Utida scheme. [In these simulations, the selection of a parameter set was arbitrary; the model does not even have to be exactly as formulated in order to show the principles involved.]

We may say that the Utida scheme is an example of a multiple-equilibrium system in that the two lines of prey (host) populations have different equilibrium levels. But its structure is altogether different from the Ricker scheme.

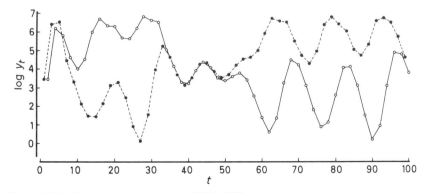

Figure 8.24   Deterministic version of Fig. 8.22.

9 Dynamics of the spruce budworm outbreak processes

'Dedicated to those budworms who have fought and died in the battle against science.' Signed *Supreme Budworm* (an obituary posted on a wall in the Green River Field Laboratory, July 1957)

9.1 INTRODUCTION

Despite the obituary, the spruce budworm (*Choristoneura fumiferana* [Clem.], Lepidoptera: Tortricidae) is alive. It occurs along the North American boreal-forest zone, although tending to concentrate more towards the eastern part. Its chief foods are the needles of balsam fir, *Abies balsamea* (L.), and several spruce, *Picea*, species.

Local budworm populations fluctuate between extreme levels. At high densities, a single, average-sized branch sampled from a mature fir or spruce tree may carry a few hundred larvae. Severe defoliation caused by such high budworm densities for several successive years seriously retards the growth or even kills the tree. Severe infestations often cover vast areas over the eastern part of its geographic distribution. In contrast, when scarce, sampling a large number of branches on many trees may yield only a few larvae.

Finding a way to minimize timber loss from outbreaks of this notorious pest has been a major mandate of the Canadian Forest Service for nearly half a century. In the summer of 1945, a field laboratory was established on the Green River watershed in northwestern New Brunswick. A major achievement of the Green River Project, led by Dr R. F. Morris, was the famed life-table study carried out during the period of province-wide outbreaks in the 1950s.

The project has yielded a number of influential papers culminating in the monograph edited by Morris (1963). During the 1960s, however, the budworm populations in the province decreased and stayed so low that the continuation of a full life-table study was impractical. And, only a few years after I joined it, the project was officially terminated in 1972, despite an early sign of population increase which eventually developed into another widespread outbreak during the 1980s. Nevertheless, I continued my analysis and re-examination of the vast amount of data left by retired senior members of the project.

I re-examined the data because I was not fully convinced by the previous analyses and interpretations by the earlier investigators published in the 1963 monograph. Briefly, their idea was that the budworm–forest ecosystem has a multiple-equilibrium structure; and an epidemic originates in an epicentre where the budworm population is released from an endemic state by climatic factors favouring a population increase; then, egg-carrying moths disperse from the epicentre to infest surrounding areas to cause widespread outbreaks.

I published the results of my re-analysis in 1984. However, I was frustrated by the fact that the Green River data contained a large amount of unidentified mortality among the larvae when the populations were declining. Because of this I was not satisfied with my argument. Towards the end of the 1970s, I proposed to resurrect a life-table study in order to find what exactly causes the decline of a population; what maintains the population at a low level for a period of time; and how the population starts increasing again. In the early 1980s, parallel studies were started in New Brunswick and Ontario, and later in Quebec, by a new group of research scientists from the Canadian Forest Service establishments in these provinces. A large amount of information has been accumulated since then, although most of it is yet to be published.

The present chapter is largely based on my analysis of the Green River data that I published in 1984. However, I update my interpretation of the data and my theory of outbreak processes, incorporating, as much as permitted, new information from our ongoing project in Canadian Forest Service.

In the first few sections, I give a brief description of the life cycle (section 9.2); define notation and terminology to be used throughout the chapter (section 9.3); and describe the main source of data to be analysed (section 9.4). In section 9.5, I describe the cyclic recurrence of budworm outbreaks in the past two centuries. To find the mechanism underlying the recurrence process is the ultimate goal of this chapter. Sections 9.6 to 9.9 include the analysis of the Green River life-table data and the identification of factors that determine survivals at major developmental stages and recruitment rate. In the final section, I first review some early theories of budworm outbreak processes, then, I propose my own view by synthesizing the results of the foregoing analysis.

9.2 LIFE CYCLE

Figure 9.1 illustrates an average pattern of numerical changes in a spruce budworm population in the course of its annual life cycle.

In most areas of New Brunswick, moths emerge towards the end of June and early July. Females lay eggs in masses on the needles of balsam fir or several spruce species. Each egg mass contains an average of about 20 eggs. Females raised under normal feeding conditions during their larval stage lay from 100 to 300 eggs, with an average of 200. Heavy defoliation caused by high larval density can reduce fecundity to one-half or less.

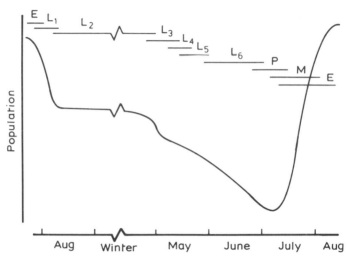

Figure 9.1 An average pattern of numerical changes in a spruce budworm population in the course of its annual life cycle. The notation for each developmental stage on top of the graph is given in Table 9.1.

The eggs hatch in about 10 days. A major source of egg mortality is the parasitic wasp, *Trichogramma minutum*. Some predatory insects, e.g. pentatomids, and birds feed on budworm eggs.

After hatching, the first-instar larvae do not feed but immediately seek overwintering sites under bark, behind lichens or in old staminate flower bracts. Many larvae disperse to another tree. Some migrate even further by ballooning with silk. Often, a substantial proportion (50% or more) of the dispersing larvae disappear permanently even in the midst of a vast, continuous forest stand. After having found suitable sites, the larvae spin hibernacula within which they moult to the second instar. Two universally occurring parasitic wasps, *Apanteles fumiferanae* (braconid) and *Glypta fumiferanae* (ichneumonid), attack the larvae during dispersal and within hibernacula. The offspring of these parasitoids overwinter inside the host larvae. Mortality among the overwintering larvae, parasitized or not, is usually small (about 10%) and does not vary much from year to year.

When sufficient heat units have accumulated, in late April or early May, to an average of 46 degree days above 5.6°C, the larvae emerge from their hibernacula and actively move within a tree or disperse by silk between trees. After the dispersal, they mine into one- or two-year-old needles (or seed and pollen cones, if available) of fir and spruce to feed. A substantial proportion, e.g. 70%, of larvae could be lost during their active movement and dispersal.

During the third to sixth instars, larvae feed on the current-year shoots of host trees. If larval density is high and current-year shoots become depleted, 'back feeding' (feeding on older foliage) occurs. This often results in reduced size and fecundity of adults.

The two common wasps, *A. fumiferanae* and *G. fumiferanae*, that had parasitized the first- or second-instar larvae in the late summer of the previous year, mature and begin to come out of the host larvae when the larvae are fourth to fifth instar. The parasitoids immediately spin cocoons by their hosts. The growth of the parasitized larvae is retarded compared to that of healthy ones. At the end of sixth instar, pupation takes place on the foliage towards mid-June. A complex of parasitoids, comprising some braconid and ichneumonid wasps and several tachinid flies, attacks the fourth- to sixth-instar larvae and pupae.

Many of the parasitic wasps are in turn attacked by several (chalcid, braconid and ichneumonid) hyperparasitic wasps. The tachinid flies which drop to pupate in the forest litter are often heavily utilized by invertebrate predators and small mammals.

Avian and invertebrate predators also utilize budworm larvae and pupae. Several pathogens, e.g. viruses, protozoan parasites (microsporidia), pathogenic fungi, bacteria and yeasts, are always present and could cause heavy mortality among older larvae and pupae.

Mortality during these larval and pupal stages varies substantially from year to year and is the major factor determining the dynamics of a local budworm population.

About 10 days after pupation, moths eclose and lay eggs to complete their life cycle. Both male and female moths can fly long distances. Females usually emigrate after laying part of their egg complement at their natal forest and lay the rest at new sites. An emigration flight (a spectacular mass exodus when moth density is high) occurs in the evening when meteorological conditions are suitable. The moths emigrate to new sites which are often 50 to 100 km downwind (Greenbank *et al.*, 1980), but which can be as far as 450 km (the distance between the east coast of New Brunswick and the west coast of Newfoundland).

9.3 NOTATION AND TERMINOLOGY

As described in the preceding section, there are five ecologically distinct stages, i.e. egg, early-instar (overwintering) larva, late-instar (feeding) larva, pupa and moth stages.

Table 9.1 lists the code and notation to be used to denote these five developmental stages, the population density at the beginning of each stage, and the rate of change in density between two successive stages. Stage and generation are denoted by the subscripts s and t, respectively. Thus, y_{st} is the density at the beginning of stage s at generation t, and h_{st} is the rate of change in density during stage s in generation t. The rate is defined as the ratio of the density at the beginning of stage s and that of stage $s+1$, i.e.

$$h_{st} = y_{s+1,t}/y_{st}.$$ (9.1)

Table 9.1 Code and notation for densities at five stages and the rate of change in density between successive stages.

Stage	Year	Generation	Initial density	Rate of change
1. Egg (E)	$t-1$	t	y_{1t}	
2. Early-instar larva (L_1, L_2)	$t-1/t$	t	y_{2t}	h_{1t}
				h_{2t}
3. Late-instar larva (L_3 to L_6)	t	t	y_{3t}	
				h_{3t}
4. Pupa (P)	t	t	y_{4t}	
				h_{4t}
5. Moth (M)	t	t	y_{5t}	
				h_{5t}
6. Egg (E)	t	$t+1$	$y_{1,t+1}$	

Note: L_1 and L_2 = non-feeding larvae; L_3 to L_6 = feeding larvae.

The subscript t may be called 'generation year' as distinguished from the calendar year t. As already mentioned, generation t in the budworm life cycle extends from the late summer of year $t-1$ to the same time in year t. In most graphs in the present chapter, population parameters will be plotted against generation year rather than against calendar year.

The rates of change in density, h_1, h_3 and h_4, are survival rates during the stages concerned. The rate h_2 is an apparent survival rate because a considerable exchange in individual larvae between trees or stands often occurs during the two dispersal periods. The rate of change $h_{5t}(=y_{1,t+1}/y_{5t})$ is the ratio of all eggs laid to all moths emerged in the study plot – or the egg-to-moth (E/M) ratio. The total number of eggs in a given area is the sum of those laid by local moths before emigrating and those gained from immigrants. Therefore, the E/M ratio is an apparent per-capita recruitment of a new-generation cohort for the local population.

In the following analysis, I shall mostly use the (base 10) log transformation of the above parameters. Thus, using upper-case letters to denote the logarithms of the corresponding lower-case letters, equation (9.1) is transformed to

$$H_{st} = Y_{s+1,t} - Y_{st}. \tag{9.2}$$

The log recruitment rate (E/M ratio) H_5 is usually positive (or $h_5 > 1$), despite the physical possibility of a negative value of H_5 (or $h_5 < 1$) resulting from an extremely high rate of emigration. The log rates Hs in all other stages are normally negative (or $h < 1$), i.e. net loss. H_2 can be positive if net gain of larvae occurs at the time of second-instar dispersal in the spring, but this occurs only rarely (Miller, 1958).

To denote the inter-generation rate of change (the rate of change in density at the beginning of stage s in generation t to the same stage in generation $t+1$), I shall use the notation r_{st} (or $R_{st} \equiv \log r_{st}$). Thus, for example,

$$R_{1t} = Y_{1,t+1} - Y_{1t}$$
$$= H_{1t} + H_{2t} + H_{3t} + H_{4t} + H_{5t} \tag{9.3}$$

is the log egg-to-egg inter-generation rate of change. As already mentioned, H_1 to H_4 are the log survival rates. Their sum is the log (overall) survival rate during the whole span of generation t. In this sense, the sum can be called the log intra-generation survival rate (or simply, generation survival), denoted by H_{gt}. Then, (9.3) is reduced to

$$R_{1t} = H_{gt} + H_{5t}. \tag{9.4}$$

In the following analysis, I shall also use R_{3t}, the log inter-generation rate of change in density measured at the time when the majority of larvae were third to fourth instars, i.e.

$$R_{3t} = Y_{3,t+1} - Y_{3t}$$
$$= H_{3t} + H_{4t} + H_{5t} + H_{1,t+1} + H_{2,t+1}. \tag{9.5}$$

In earlier chapters, I have called the rate r 'net reproductive rate' (or simply 'reproductive rate') because the above expression is more economical than the precise but lengthy expression 'inter-generation rate of change in density'. In some publications from the Green River Project, the same parameter was often referred to as the 'index of population trend' denoted by the letter I, since Balch and Bird (1944) coined the term. This is unfortunate terminology, however, because the year-to-year rate of change by itself cannot indicate population trend in the usual sense. Usually, trend is a more or less consistent tendency of a series of events over a comparatively long period of time. Therefore, to reveal a trend, observations must extend many more than two generations.

A tendency for a population to increase or decrease over a comparatively short period of time (e.g. not much more than 10 years) may be called a short-term trend. However, further observations might reveal that the system is merely oscillating without exhibiting any long-term trend. An example is the lynx 10-year cycle from Mackenzie River District (Fig. 5.1b, section 5.2). Then, we may call it an oscillatory trend. In the spruce budworm, as we shall see, the length of an oscillation is very long (30–40 years). So it may be permissible to say that the population exhibits an increasing trend for many years before it declines.

Altogether, the use of the term depends on which aspect of comparatively long-term population changes is to be emphasized. Note, however, that a consistent increase or decrease for as long as seven consecutive points occurs quite frequently in a trend-free series of purely random numbers (Fig. 2.14a, section 2.4.3). Therefore, if an observed series as short as seven points exhibits a consistent increasing or decreasing tendency, we must be careful about using the term 'trend'.

9.4 SOURCE OF INFORMATION

Life-table studies of the spruce budworm in the Green River Project were carried out at localities free from aerial spraying of pesticides. [Pesticide spray in New Brunswick began in 1952.] A number of permanent plots of 8 to 10 ha were selected in relatively homogeneous sections of forests predominantly balsam fir mixed with white spruce (*Picea glauca*), and foliated branches were sampled periodically from selected trees of both species. The data that I use in the following analysis are those taken from the fir trees: they are more complete than those from the spruce trees.

The construction of life tables was based on sampling at three stages in the life cycle of budworm: (a) soon after all eggs had hatched; (b) when the majority of larvae were in the third and fourth instars; and (c) at the time of 60–80% moth eclosion. At stage (a), most of the eggs laid (both hatched and unhatched) were still retained in the foliage. Also, at stage (c), most pupal cases after moth eclosion were still attached to the foliage. Therefore, the above sampling scheme enabled the investigators to determine the total numbers of: (1) eggs laid, (2) first-instar larvae hatched, (3) larvae with the majority being third to fourth instars, (4) pupae and (5) emergent moths.

There were some 20 study plots altogether in the unsprayed area. However, many plots were sampled only for a few years, not long enough for the analysis of temporal changes in budworm density. Only one plot, G4, yielded 12 years of uninterrupted sampling data between 1947 and 1958. These years covered a major part of the outbreak in New Brunswick during the 1950s (Fig. 9.2, section 9.5). The set of data from this plot is the main source of information in the present analysis. However, in this plot the budworm density never reached a level high enough to cause heavy defoliation and tree mortality. Three other plots provide supplementary data, i.e. G5 between 1949 and 1957, and K1 and K2 between 1952 and 1958. Plots K1 and K2, although covering only the peak to declining phases of the outbreak, were situated in stands where the budworm density reached an extremely high level, causing heavy timber losses. Despite the fact that peak density differed between these plots, the dynamic pattern of (log-transformed) population changes was much the same in all plots.

Another supplementary source of information about budworm abundance is the egg-mass sampling data gathered for the purpose of monitoring budworm infestation levels in the province's programme of aerial spraying of insecticide which began in 1952.

After 1959, budworm populations in the Green River area fell so low that a full life-table study became impractical. Only one larval stage (when the majority were third to fourth instars) continued to be sampled at plots G4 and K1 until the termination of the Project in 1972. During the 1970s, a study of long-range moth dispersal was carried out using aircraft and radar (Greenbank *et al.*, 1980).

Populations in most parts of New Brunswick had begun to increase again around 1968 and eventually reached an outbreak level in the late 1970s and early 1980s. This latest widespread outbreak began to subside in the mid to late 1980s.

After my re-analysis of the life-table data from the Green River Project (Royama, 1984), the need for further intensive and systematic studies of budworm population dynamics prompted the resumption of the project. It began in the early 1980s in the Canadian Forest Service establishments in New Brunswick, Quebec and Ontario. The budworm populations in these provinces were already high then, so we missed an opportunity to study an increase phase of these outbreaks. Because most quantitative results from the resumed project have not yet been published I cannot discuss them here. However, I shall update, as much as possible, my theory of budworm outbreak processes I published earlier (Royama, 1984).

9.5 PATTERN OF POPULATION FLUCTUATION

Figure 9.2a shows annual changes in spruce budworm density near the Green River field station. [In the Green River Project, budworm density was expressed as the number of insects per square metre (approximately equal to the original unit of $10\,ft^2$) of 'foliage surface area' calculated by multiplying the length of a foliated sample branch by its midpoint width.] Solid circles represent the log densities (Y_{3t}), when the majority of larvae were third to fourth instars, observed at plot G4 up to 1958; after 1959, these are average densities in plots G4 and K1. Open circles are the log egg densities (Y_{1t}): the first series is those from plot G4 and the second series is based on the egg-mass monitor data for the provincial aerial spraying programme.

The cyclic pattern of population change in Fig. 9.2a was not local but observed widely over eastern Canada and the adjacent areas of the United States during more or less the same time period. Figure 9.2b and c show examples of such population changes observed in balsam fir stands near Black Sturgeon Lake and in Chapleau, Ontario. Although differing in phase and length, the cyclic pattern of the Ontario populations is much the same as that of the Green River population.

As we focus our attention on comparatively more restricted areas, such as the whole of New Brunswick (Fig. 9.3), we see that, despite some regional

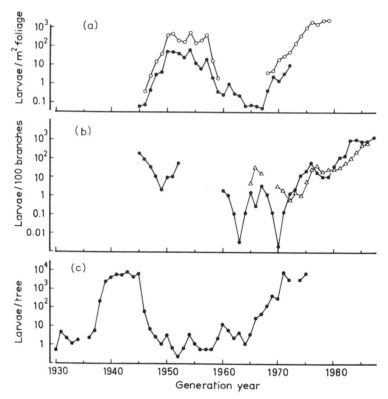

Figure 9.2 Annual changes in spruce budworm density recorded at three different locations in eastern Canada: the Green River area (graph a); Black Sturgeon Lake area (graph b); and near Chapleau (graph c). Solid and open circles in graph a represent larval and egg densities, respectively. Open triangles in graph b are male moths caught in pheromone-baited traps (after Sanders, 1988). Graphs b and c are based on the data deposited in Canadian Forest Service–Ontario Region, Sault Ste. Marie, Ontario.

differences in amplitude, most local populations tended to oscillate more or less in unison. Such synchrony in the population cycles (particularly in peaks) is, of course, recognized as a province-wide outbreak of this insect.

Even during the low-density period between the latest two outbreaks, a few scattered patches of comparatively high density areas always remained on the budworm infestation maps of eastern Canada (Brown, 1970; Kettela, 1983). These were probably the areas where the troughs of population oscillations stayed comparatively high, e.g. central regions in New Brunswick in the 1960s (Fig. 9.3).

Earlier budworm outbreaks in New Brunswick and Quebec have been traced in recent and archival records or by examining radial growth patterns of old trees, especially white spruce (Tothill, 1922; Swain and Craighead, 1924; Blais, 1958, 1962, 1965; Greenbank, 1963a). These trees experienced

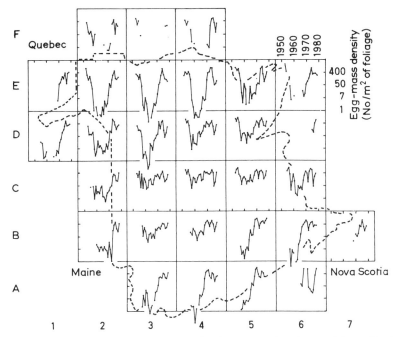

Figure 9.3 Annual changes in egg-mass density across New Brunswick (outlined by broken lines) since 1952. In each block, density is plotted against calendar year as scaled in block E6. After Royama (1984).

and survived several years of severe defoliation that could be caused by nothing but high-density spruce budworms.

Open arrows in Fig. 9.4 indicate the approximate years of the beginning of outbreaks since 1770 in New Brunswick. Solid arrows indicate the same in the Laurentide Park region of Quebec since 1710. Based on these records of outbreaks, the oscillatory curve I drew in Fig. 9.4 crudely restores the

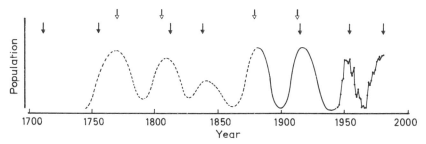

Figure 9.4 Spruce budworm outbreak cycles in New Brunswick and Quebec in the past two centuries, restored from: radial growth-ring analysis (broken line); archival records (solid line); and sampling data in Fig. 9.2a after 1945. Open and solid arrows on the top indicate approximate years of beginning of outbreaks in New Brunswick and Quebec, respectively. After Royama (1984).

pattern of budworm population changes during the past two centuries in these provinces. It shows that a majority of outbreaks occurred at an interval of 25 to 45 years.

The exceptions were the interval between 1806 and 1878 in New Brunswick and another between 1838 and 1914 in Quebec. Both of these were about twice as long as most intervals. Interestingly, however, the growth rings in trees from Quebec indicated light defoliation for a few years around 1838. Whereas, during the same time period, there was little sign of defoliation in the sample trees from New Brunswick. Conversely, there were light but clear signs of defoliation in trees from New Brunswick around 1879, when there were few in the Quebec samples. Perhaps the populations in both provinces peaked more or less at the same time as in other outbreaks, but the population in one province did not reach a level high enough to affect the tree growth rings. The following facts tend to indicate the likelihood of such hidden peaks.

Other outbreaks in the two adjacent provinces tended to occur fairly close together in time. In recent years, most regional populations in New Brunswick have oscillated in unison as in Fig. 9.3. During the 1950s, all local populations in the Green River area reached their peaks more or less simultaneously. A peak density was extremely high in most localities but, in a few places, it was not high enough to cause severe defoliation. The population already shown in Fig. 9.2a was, in fact, one such low density case. Thus, placing a small peak at about 1840 in Fig. 9.4 restores the oscillatory tendency of the budworm population in the two provinces with the average interval of about 35 years, i.e. eight peaks (including the hidden one about 1840) over 270 years.

9.6 MAJOR COMPONENTS OF POPULATION FLUCTUATION

In Fig. 9.2a, I have shown the annual changes in budworm density in and near the Green River area. We can look at the same data in terms of the inter-generation rate of change in density. Instead of plotting the log density at stage s, i.e. Y_{st}, against t as in Fig. 9.2a, we plot the difference $Y_{s,t+1} - Y_{st}$ ($= R_{st}$, the log rate of change) against t (Fig. 9.5). The solid circles represent the series $\{R_{3t}\}$ which corresponds to $\{Y_{3t}\}$ (series of log $L_{3/4}$ densities) in Fig. 9.2a. The open circles represent $\{R_{1t}\}$, corresponding to the second series of log egg density in Fig. 9.2a.

Figure 9.5 clearly reveals that the annual changes in R_t are made up of two major components: a principal oscillation (or low-frequency cycle enhanced with the sinusoidal curve) and high-frequency, erratic fluctuations about the principal oscillation. [The principal oscillation corresponds to the population oscillation in Fig. 9.2a with a shift in phase: the peak and the trough in the principal oscillation in Fig. 9.5 correspond to the steepest ascent and descent, respectively, in the population oscillation in Fig. 9.2a.] It

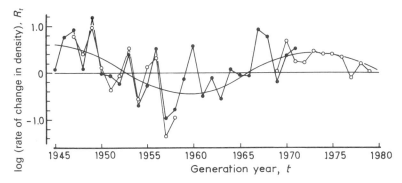

Figure 9.5   Equivalent to Fig. 9.2a, but the log rate of change in density from generation t to $t+1$ is plotted against t. Solid and open circles are the rates of change in larval and egg densities, R_{3t} and R_{1t}, respectively. The sinusoidal curve is drawn by eye. After Royama (1984).

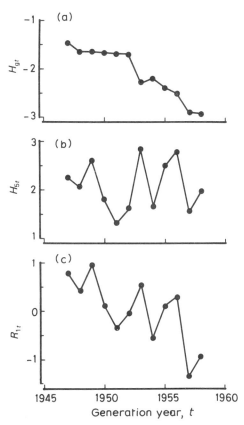

Figure 9.6   The annual variations in the log generation survival rate (H_{gt}), recruitment rate (H_{5t}) and their sum (R_{1t}) observed at plot G4 in the Green River area. After Royama (1984).

is this principal oscillation that periodically brings the budworm population to an epidemic level.

Now recall the relationship (9.4) in which the log rate of change R_{1t} is partitioned into the two components H_{gt} (log generation survival rate) and H_{5t} (log recruitment rate or E/M ratio). Annual changes in these parameters are shown in Fig. 9.6. During these years, H_{gt} exhibited a consistent declining trend, while H_{5t} showed no trend but fluctuated frequently and erratically. Thus, noting that the annual changes in R_{1t} were nearly identical with those in R_{3t} (Fig. 9.5), we see that H_{gt} was the source of the principal oscillation in $\{R_{3t}\}$ in Fig. 9.5, and H_{5t} was responsible for the secondary, high-frequency fluctuations about the principal oscillation.

I now analyse generation survival further into several components.

9.7 ANALYSIS OF GENERATION SURVIVAL RATE

The generation survival rate (h_{gt}) is partitioned into survival rates during the four stages of generation t: h_{1t}, egg survival; h_{2t}, survival of early-instar (overwintering L_1 and L_2) larvae; h_{3t}, survival of late-instar (feeding L_3 to L_6) larvae; and h_{4t}, survival of pupae (Table 9.1). Thus, after the log transformation ($H \equiv \log h$):

$$H_{gt} = H_{1t} + H_{2t} + H_{3t} + H_{4t}. \tag{9.6}$$

In each stage, distinct factors cause mortality. But, first, let us look at these four stage survival rates as depicted in the Green River life-table studies (Fig. 9.7). [When there is no risk of confusion, I may call a log survival rate, H, simply 'survival rate'. Also, I may drop the generation subscript t.]

The variability in generation survival (H_g) in Fig. 9.6 is mostly attributable to those in H_2 and H_3 in Fig. 9.7. Curiously, H_3 tended to be negatively correlated with H_2. As a result, the fluctuation in H_3 tended to cancel the fluctuation in H_2, and the sum $H_2 + H_3$ revealed the downward trend of H_g in Fig. 9.6.

I believe that this curious negative correlation between H_2 and H_3 resulted from the timing of sample collections. Since the end of one stage in a life table constitutes the beginning of the next stage, the calculated survival rate in one stage could be negatively correlated with that in the other (section 3.3.2). In the present case, the division between the early- and late-instar stages was set by the date on which samples were collected. This date was about the time the majority of lavae were third and fourth instar. However, the date of sample collection is somewhat arbitrary because these instars overlap on any one day. Thus, an early date of collection, relative to a given actual developmental stage, tends to overestimate H_2 and underestimate H_3, and vice versa if the collection date was relatively late (Fig. 3.18, section 3.3.2).

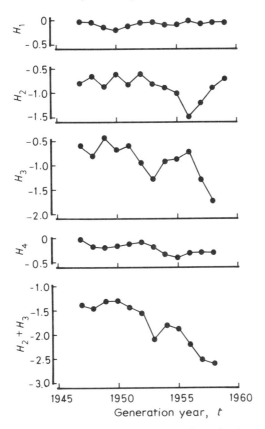

Figure 9.7 Annual changes in log survival rates at four developmental stages: egg (H_1), early-instar larva (H_2), late-instar larva (H_3) and pupa (H_4), observed at plot G4 in the Green River area. After Royama (1984).

When H_2 was adjusted by interpolation, supposing that each year's samples were taken exactly on the mid-date between the peaks of third and fourth instars, no clear tendency appeared (Royama, 1984). Mortality in L_1 and L_2 occurs mostly during their dispersal. Perhaps such loss of larvae is a random event with no trend. Our recent studies in New Brunswick (Eveleigh and Royama), Quebec (Régnière and Perry), and Ontario (Nealis, Sanders and Lysyk) – all unpublished – indicate that H_3 is a major component which determines a trend in H_g.

Egg survival (H_1, Fig. 9.7) did not vary much from year to year and contributed little to the variability in H_g. The variability in pupal survival (H_4) was just as small as that in H_1, but it exhibited a declining trend which contributed, to a small extent, to the declining trend in H_g. Our recent studies reveal, however, that pupal mortality contributes to a much greater extent to the reduction in generation survival in a declining population than the Green River data indicated. There is every reason to believe that the sampling schedule used in the earlier project tended to substantially under-estimate pupal mortality (section 9.3).

Thus, we can conclude that the sum $(H_3 + H_4)$, i.e. survival of feeding larvae and pupae, determined the declining trend in generation survival (H_g) and, hence , must be the main source of the principal population oscillation as depicted in Fig. 9.5 (smoothed curve). I now analyse mortality factors which determined H_3 and H_4.

9.8 ANALYSIS OF MORTALITY FACTORS AMONG FEEDING LARVAE AND PUPAE

Some have believed that weather and food supply play major roles in budworm outbreak processes, but, as I shall argue shortly, there is no hard evidence to support the belief.

On the other hand, feeding larvae (L_3–L_6) and pupae are killed by many insect parasitoids, predators and pathogens. So I argue that these mortality factors must be the primary cause of the cyclic generation survival extrapolated in Fig. 9.5. In the light of the theoretical investigations in Part One, it makes sense that the complex of these natural enemies acts as a second-order density-dependent factor to produce a population cycle (section 2.3).

However, individually, none of the budworm's natural enemies is an efficient control agent. Most of them are slow to react (if they do) to changes in host density. Only as a complex, these agents can produce a long cycle with a high amplitude shown in Figs 9.2 and 9.5. The complexity of the budworm–natural enemy system cannot be described by the simple pred-ator(complex)–prey(complex) model used in Chapter 5, although the prin-ciple is the same. I shall discuss the web of interactions within the natural enemy complex and their role in budworm outbreak processes in section 9.10. In this section, I describe the ecology of natural enemies with reference to their action and efficacy as individual natural control agents.

9.8.1 Parasitoids

Many hymenopterous and dipterous parasitoids attack budworm larvae and pupae at different developmental stages. Table 9.2 lists those species which occur regularly (some common and others not so common) in our samples in New Brunswick. Some are univoltine and others multivoltine species. Univoltine species can complete their life cycles on the spruce budworm alone. They require no alternate host species to maintain their populations. Many of them, however, have repertoires of alternative hosts. In multivoltine species, the overwintering generation requires an alternate, non-budworm host.

The two most common univoltine parasitic wasps, *Apanteles fumiferanae* and *Glypta fumiferanae*, are ubiquitous across the geographical range of budworm distribution. They attack the first- and/or second-instar host larvae

Table 9.2 Parasitoids of spruce budworm occurring regularly in southern New Brunswick (Eveleigh, personal communication).

	Host stages		Voltine	
	Attacked	Killed	Uni	Multi
Hymenoptera:				
Trichogramma minutum (chalcid)	Egg	Egg		x
Apanteles fumiferanae (braconid)	$L_1 - L_2$	$L_4 - L_5$	x	
Glypta fumiferanae (ichneumonid)	L_2	L_5	x	
Meteorus trachynotus (br.)	$L_5 - L_6$	L_6		x
Ephialtes ontario (ich.)	P	P		x
Itoplectus conquistor (ich.)	P	P		x
Phaeogenes hariolus (ich.)	P	P	x	
Diptera (tachinid)*:				
Winthemia fumiferanae				
Lypha setifacies				
Phryxe pecosensis				
Eumea caesar				
Pseudosarcophaga affinis				
Actia interrupta				

*All tachinids attack L_6; may come out of pupae; are probably univoltine.

in the late summer. The second-generation wasps overwinter inside the hosts and, in the following summer, consume and kill the hosts when the latter are at their fourth and fifth instars. A parasitized host, however, develops more slowly than a healthy one. After having consumed the hosts, the wasp larvae come out and immediately spin cocoons by the hosts. After a few weeks, second-generation adult wasps emerge and attack the early-instar host larvae to complete the life cycle.

The rate of parasitism by these wasps can be determined accurately by collecting host larvae just before their spring emergence and rearing them in the laboratory. Figure 9.8 shows parasitism by *A. fumiferanae* (the commonest parasitoid in New Brunswick) observed at plot G4 during the Green River study. This wasp largely depends on the spruce budworm in the Green River area. So, as we expect from the host–parasitoid system of Chapter 8, this wasp should be able to respond reproductively to an increase in the host population to control it.

Contrary to such an expectation, a peak parasitism by this wasp never becomes higher than 30% even though the host density stayed comparatively high for as long as eight years after 1950. Parasitism by *G. fumiferanae* is even lower than that. Eveleigh (unpublished recent study) found that, in New Brunswick, many parasitoids were attacked by several different chalcid, braconid and ichneumonid hyperparasitic wasps. Because of this, these primary parasitic wasps are unable to multiply fast enough to control the budworm.

Several parasitic flies (tachinids) attack late-instar (mainly L_6) host larvae, and many of the second-generation flies come out (as maggots) after the

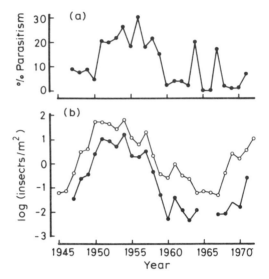

Figure 9.8 Annual changes in parasitism by the braconid wasp, *A. fumiferanae*, observed at plot G4 in the Green River area. Graph a: percent parasitism. Graph b: estimated number of parasitizing wasp larvae (solid circles) compared with the number of $L_{3/4}$ budworm larvae from Fig. 9.2a (open circles).

hosts have pupated. The maggots drop to the ground, pupate under the forest litter and most of them overwinter as pupae. There, they are heavily preyed upon by invertebrate and mammalian predators (Eveleigh, unpublished recent study). Because of this, these flies cannot multiply fast enough to control budworm.

Most parasitic wasps, that attack budworm late-instar (mainly L_5 to L_6) larvae and pupae (Table 9.2), come out of the hosts as adults in the autumn, indicating that they are multivoltine; their second generation requires an alternate host in which to overwinter. Recent studies of *Meteorus trachynotus* have identified *Choristoneura rosaceana* as an alternate host (Maltais *et al.*, 1989; our own observation in New Brunswick). [An exception is *Phaeogenes hariolus* which is reported to overwinter as adult (Miller, 1963b) and probably does not require an alternate host.] The efficacy of a multivoltine parasitoid to control the spruce budworm depends on the abundance of the alternate host species. Whenever the population of alternate hosts decreases well below the budworm's as often happens, the parasitoid is unable to multiply to control the budworm.

Altogether, none of the budworm parasitoids acts as a consistently efficient natural control agent. This, no doubt, is a major reason for the regular occurrence of budworm outbreaks (section 9.10.3).

Table 9.3 shows the annual changes in total parasitism estimated in the Green River Project between 1945 and 1958. The rate of parasitism

Table 9.3 Total percent parasitism in feeding larvae (Royama, 1984; based on Miller,
1963b; table 34.1 and unpublished data).

Year	G4	G5	K1	K2
1945	13	..	..	..
1946	10	8	..	..
1947	18	16	..	..
1948	6	6	..	..
1949	19	21	..	..
1950	12	17	..	..
1951	41	19	..	..
1952	47	39	8	16
1953	36	37	20	29
1954	52	52	18	19
1955	33	36	25	26
1956	38	..	19	18
1957	25	..	16	..
1958	48	41	35	34

increased steadily as generation survival (H_g in Fig. 9.6a) decreased during
the same period. On its face value, however, the rate of increase in
parasitism accounts only for a comparatively small part of the much faster
decline in the generation survival rate. Probably, this is not quite true. There
is a reason to believe that annual total parasitism in the Green River study
was substantially underestimated; for a similar reason raised by Van
Driesche (1983), the sampling schedule was not adequate as I now explain.

Consider that a parasitoid species starts parasitizing the host species in
the field on day 1. Every subsequent day, some additional hosts are parasitized. Thus, the proportion parasitized in samples increases until: (1) all
adult parasitoids have finished parasitizing; or (2) some second-generation
parasitoids already start coming out of the hosts, even if some adults are still
parasitizing. In the first case, provided that differences in mortality between
parasitized and unparasitized hosts (if any) are not large, the proportion
parasitized remains the same until the second-generation parasitoids come
out of the hosts. In the second case, the proportion parasitized in samples
may start to decrease before the end of parasitization.

Figure 9.9 illustrates the above processes. Let us define the following
quantities: $f(t) =$ the number of hosts parasitized in the field on day t;
$g(t) =$ the cumulative number of hosts parasitized by day t; and $h(t) =$ the
cumulative number of second-generation parasitoids which have issued from
hosts by day t. Thus, $f(t) = g(t) - h(t)$. The curves $g(t)$ and $h(t)$ probably
increase in a sigmoid manner to a plateau, while $f(t)$ is bell shaped (Fig. 9.9).
The quantity we wish to know is the total number of hosts killed by the
parasitoid. This is given by the height of the $h(t)$ plateau. Then, letting t_0

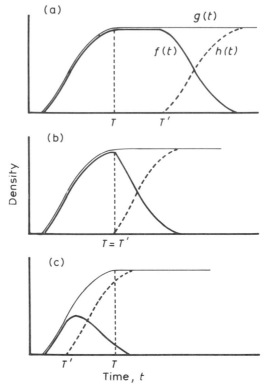

Figure 9.9 Idealized daily parasitism curve $f(t)$ (thick curve); cumulative parasitism curve $g(t)$ (thin curve); and cumulative parasitoid emergence curve $h(t)$ (dashed curve). T: end of parasitism. T': start of parasitoids issuing from hosts.

be a day before the first parasitoid issues from a host, the proportion of the initial host population, i.e. $y(t_0)$, killed by parasitism by day t is given by $h(t)/y(t_0)$. The curve $g(t)$ can be used to calculate total parasitism, although it is often difficult to measure in the field.

In the Green River study, $f(t)$ was the only quantity that was measured, and this is not always adequate to calculate total parasitism. Figure 9.9 illustrates three possible situations depending on day T (on which curve g reaches a plateau) relative to day T' (on which curve h starts rising, i.e. when the earliest parasitoids start issuing from hosts). Graph a idealizes the parasitization by *A.* and *G. fumiferanae*: T is the late summer of one year and T' the early summer of the following year. Here, the proportion found parasitized, $f(t)/y(t_0)$, in a sample taken in the interval $T < t < T'$, is a reliable estimate of total parasitism.

In graph b, days T and T' are so close that curve f has a pronounced peak. This happens with most parasitoids that attack late-instar budworm larvae and pupae (Table 9.2). In the Green River Project, the peak of $f(t)$

divided by the host density $y(t_0)$ was used as an estimate of the total parasitism. As we found recently (Eveleigh *et al.*, 1990), the peak is often so sharp that, unless we take daily samples, we would probably miss it. Parasitism in samples taken even a day or two before or after the peak could considerably underestimate total parasitism.

Moreover, in the Green River Project, all tachinids but *Actia interrupta* were lumped as a complex. As a result, the $f(t)$ curve of the tachinid complex became like one low flat hill, instead of a collection of small peaks, each representing a separate species. Of course, the sum of those individual peaks is an estimate of total parasitism by the complex. A considerable under-estimation must have resulted in the Green River estimate which used a peak of the lumped curve.

If we can measure curve $h(t)$ directly, this would be best. Some parasitoids, like *M. trachynotus*, spin a cocoon next to the host from which it issued. By counting these cocoons, we can estimate the $h(t)$ curve directly. Most tachinid maggots drop to the ground to pupate. By collecting the dropping maggots with a funnel trap, we can determine the $h(t)$ curve directly.

Graph c (Fig. 9.9) illustrates that some second-generation parasitoids start issuing from hosts well before the first-generation adults have finished laying eggs. Then, it is impossible to estimate total parasitism by the $f(t)$ curve, that is, by sampling and rearing host larvae. As far as I am aware, few budworm parasitoids have such a life history. However, some microbes pathogenic to budworm might act like graph c. That would create a serious problem in evaluating the effect of the microbes (section 9.8.3).

9.8.2 Predation

Major predators of the budworm are small insectivorous birds, particularly warblers of the family Parulidae, and a complex of invertebrates, predomi-nantly spiders. When budworm density is high, many birds consume larvae and pupae. About a dozen bird species contained budworms in substantial proportions (in volume) of their stomach contents (Mitchell, 1952). Most of the budworms taken by birds were fully grown larvae and pupae (Dowden *et al.*, 1953). Mook (1963) reported a similar result: 90% of the larvae found in sample gizzards from many birds were L_6 and pupae. Evidently, these birds rejected young and small larvae as unprofitable (Royama, 1970).

The Green River Project used two methods to estimate the impact of bird predation on budworm populations. One was based on: the stomach contents of sample birds; the rate at which a bird digests a mature budworm larva or pupa (determined using caged birds); the number of hours and days the bird spends searching for budworm larvae; and the total number of birds in the forest stand. The other method was to enclose a branch with a coarse-meshed cage to exclude birds and large invertebrate predators. The difference in the mortality between the caged and uncaged groups was ascribed to predation.

Figure 9.10 shows percent predation by the bay-breasted warbler (*Dendroica castanea*) in relation to budworm density, estimated by Mook (1963) using the first method. A peak rate of consumption was about 1.8% of the budworm (L_6) population. The bay-breasted warbler made up about 10% of the total insectivorous bird population during his study. Its stomach always contained a large proportion of budworms. Thus, even assuming that other birds ate budworms at a similar rate, the budworm consumption by all birds in the Green River stand should not be much more than 15% at a peak rate when L_6 density was about 2 per m² of foliage surface. The rate of consumption fell steeply on both sides of this density.

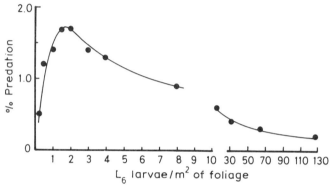

Figure 9.10   Estimated rate of predation on L_6 budworm larvae by the bay-breasted warbler in the Green River area. After Mook (1963).

Miller and Renault (1981) used the exclosure method to determine the effect of bird predation for 5 years during the low density period beginning 1959 in a Green River fir stand. The average density of budworm larvae (L_3) in the study plot was 0.07 per m² of foliage surface in 1961. It decreased further to 0.02 in 1963. The authors did not measure budworm density for the two earlier years, although it was probably in the magnitude of about 0.5/m² in 1959. The authors 'planted' 500 extra L_2 larvae at the rate of one larva per tree so as to minimize the influence on the natural density of budworms, an important consideration in carrying out this type of field experiment. Their result shows little difference in the rate of disappearance between the (coarse-mesh) caged and uncaged groups (Fig. 9.11). Thus, the effect of bird predation on low density budworm populations was minimal. This conclusion agrees with Mook's (*loc. cit.*) curve in Fig. 9.10.

The above results make ecological sense for two reasons: (1) bird populations cannot grow in steps with budworm population; (2) more birds occur in a stand where budworm density is high than where it is low.

First, bird populations cannot totally depend on budworm abundance. Simply, the budworm, no matter how numerous it becomes, is available to the birds only for four to six weeks during the breeding season each year. After the budworm season, the birds have to find enough food from other

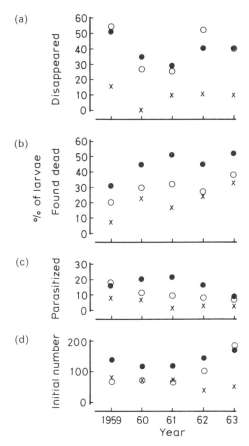

Figure 9.11   Results of the cage experiments by Miller and Renault (1981). The percent of larvae: (a) disappeared; (b) found dead; (c) parasitized. (d) initial number of L_3 larvae. Solid circles: open branches; open circles: coarse-mesh cages; and crosses: fine-mesh cages. After Royama (1984).

sources until their autumn migration to Central and South America where no spruce budworm occur. Thus, by the beginning of the next breeding season, the bird population would show little, if any, influence of the previous year's budworm abundance.

Some birds have been reported to be more abundant during budworm outbreaks than at other times (Kendeigh, 1947; Morris *et al.*, 1958). In the Green River area, there were twice as many birds during the budworm outbreak period in the 1950s compared to the post outbreak period in the 1960s (Gage and Miller, 1978). This must happen, I believe, as a result of local habitat selection by birds according to budworm abundance. During an outbreak period, birds are attracted to stands where budworm density is comparatively high. During a post-outbreak period, these stands should lose much of their attractiveness. So, in a sufficiently large area, annual changes in bird population should be more or less independent of changes in budworm abundance.

The consumption of budworms by individual birds will increase with budworms, but their consumption is limited. So the proportion of budworms eaten will decrease as a budworm population increases. Conversely, birds either move out of a stand of low budworm density or switch to a more profitable prey. Thus, predation by birds also decreases towards the lower end of a budworm density spectrum as we indeed see in Fig. 9.10.

A similar argument probably applies to omnivorous invertebrate predators that utilize budworms at certain developmental stages of both species. Perhaps few predator species depend entirely on the budworm throughout their life cycle. If so, few can respond reproductively to an increase in budworm abundance over the years. In fact, Renault and Miller (1972) have shown that some spiders (e.g. of genus *Dictyna*) consume many young (mainly L_2) budworm larvae. Perhaps many (no data) of the dispersing L_1 and L_2 budworms are caught by these spiders. However, the species composition and abundance of spiders in the Green River area stayed remarkably constant during the 1950s through to the 1960s, despite enormous changes in budworm abundance (Loughton *et al.*, 1963; Renault and Miller, 1972).

Dowden *et al.* (1953) and Crawford and Jennings (1989) reported a much higher rate of predation by birds than the Green River study indicated. However, I believe these authors overestimated an actual rate.

Dowden *et al.* planted budworms on selected experimental trees about 3m tall in a stand of mostly 12m-tall balsam firs. They caged some of the experimental trees to exclude birds. They found that budworm mortality (from L_6 to pupae) was 20% higher on an uncaged tree than on a caged tree in 1950 and it was 40% higher in 1951. However, I roughly estimated from their tables 3 and 4 (Dowden *et al.*, 1953) that the budworm density on the experimental trees was about 1.5 times as high as the natural density in 1950 and seven times as high in 1951. Thus, these experimental trees, particularly those in 1951, attracted birds like a bird feeder, hence, giving a much higher predation rate compared to Mook's estimation.

Crawford and Jennings (1989) studied bird predation in mixed spruce-fir stands in New Hampshire and Maine in the 1982 and 1983 seasons. They used Mook's method (see above). They estimated that 84% of L_4 larvae were preyed upon when their density was low (100 000 L_4/ha; or, according to my guess, roughly 20 L_4/tree in their study areas) and 22% when budworm density was moderate (550 000 L_4/ha). However, for the following reason, I believe these are overestimations.

The estimated 84% predation was the average of six data points in the lower end of the density spectrum in their figure 6. Two of these estimates were about 120% and 145%! Moreover, the authors fitted a concave quadratic polynomial curve to the data points; its minimum fell about a moderate budworm density. Such a regression model can grossly inflate the estimated consumption rate towards the lower end of the budworm density spectrum. In fact, this regression model estimated that more larvae in absolute number were eaten when they were scarce than when they were moderately abundant! Thus, while I appreciate the tremendous technical difficulties in estimating the rate of predation in the field, I cannot rely on their estimations.

9.8.3 Diseases

Many types of microbes are pathogenic to the spruce budworm. These include: the protozoan parasites *Nosema* (microsporidia); viruses (granulosis, nuclear and cytoplasmic polyhedroses, and entomopox viruses); pathogenic fungi (e.g. *Paecilomyces*, *Verticillium*, *Hirsutella*, *Entomophaga* and *Beauveria*); pathogenic bacteria (e.g. *Bacillus thuringiensis*); and yeast. The evaluation of budworm mortality caused by these pathogens is the same, in principle, as that of parasitism (Fig. 9.9), but it is technically much more difficult.

In the Green River study, live budworm larvae were collected at weekly intervals and were reared individually for one week in small vials containing fresh shoots of balsam fir as food. At the end of the week, the larvae were replaced by a new collection from the field which was reared for another week. The mortality that occurred during each weekly rearing was summed over the season, and the sum was used as an estimate of the total seasonal mortality. The dead specimens were examined microscopically to determine the proportion infected by a given pathogen.

There are several problems in the Green River method. The weekly collection and rearing must rely on the assumption that no diseases infect a larva and kill it within one week. If this happens, an underestimation would result. But we know little about the action of some microbes. A more serious problem was a large proportion of unaccountable deaths during the rearing classified as 'unknown-death'. [I called this the 'fifth agent' syndrome in my 1984 paper.]

Several causes of such unaccountable deaths in the Green River study are conceivable. In our recent study (Eveleigh *et al.* 1989), we have found that the most troublesome microbes are bacteria, as they are found in almost every specimen, and it is difficult to determine if they are pathogenic or saprophytic. Also, granulosis and nuclear polyhedrosis viruses are difficult to detect microscopically. A test with molecular probes (Perry *et al.*, 1989a,b) is being developed to identify certain pathogens more accurately and efficiently. But such technology was not available in the Green River study.

Another possible cause of 'unknown' deaths in the Green River study was the failure to attribute some deaths to certain parasitoids. A budworm larva, parasitized by *Meteorus trachynotus*, will die before pupation, but it usually lives for a week or longer after the parasitoid larva has completed feeding and left the host. A weekly collection must have contained some doomed budworm larvae which the parasitoids had already left. These would have died during the rearing of 'unknown' causes. The parasitic wasps, *Mesopolobus verditer* and *Psychophagus tortricis*, could be, as Dowden and Carolin (1950) suspect, hyper-parasitoids of several ichneumonid wasps which parasitize budworm pupae (Table 9.2). They come out of budworm pupae after a prolonged period of time (Eveleigh in recent study). If not reared through this long period, such pupae (looking sick without a sign of infection by a pathogen) could have been classified as dead from unknown causes.

In our recent study in New Brunswick (Eveleigh *et al.*, 1990), we estimate disease-caused deaths in the fields by sampling dead budworms hanging on branches and those falling into drop trays. The results (to be published) indicate that most specimens carry evidence of being killed by some pathogens. There is only an insignificant proportion of deaths that is unaccountable.

How pathogens are spread among a host population is not well documented except for *Nosema*. We always find a proportion of hosts infected by any of those microbes mentioned. Only under certain (unknown) circumstances some of them can become epizootic. An example is *Entomophaga aulicae* which killed a large number of budworm larvae but only in the 1984 season in Ontario (Perry and Régnière, 1986). During the Green River study, high mortality occurred sporadically and locally among the late-instar budworm larvae. A conspicuous dip in H_3 in 1953 in Fig. 9.7 is an example. [Similar dips in H_3 were observed in plot G5 in 1947 and 1951, and in plot K1 in 1952 (figs 5 and 6 in Royama, 1984).] Such a high non-persisting mortality was probably due to an epizootic disease caused by a pathogenic fungus.

The life cycle of *Nosema fumiferanae* is reasonably well known. The protozoan parasite infects the midgut of a host and must overwinter within the host. Its spores are ovarially transmitted from female moth to its offspring (Thomson, 1958). Feeding larvae may be infected during summer when the larvae ingest spore-contaminated needles. Thomson (1960) and Wilson (1973, 1977) documented a steady increase in the incidence of *Nosema* infection over several years during the 1950s and 1970s in Uxbridge, Ontario. Although the host populations were not quantified in these studies, budworm populations in that general area were increasing.

Because the virulence of *Nosema* is comparatively low, many seasons probably pass before the parasites build up in a host population to effectively influence its annual rate of increase (Régnière, 1984).

9.8.4 Food supply

If budworms kill a large proportion of trees in a stand, as happened in some Green River study plots, budworms cannot survive: their population per unit area of the stand should naturally decrease. However, I argue now that tree mortality is not a primary cause of the collapse of a budworm outbreak.

The budworm population in the Green River area declined simultaneously in every plot regardless of the degree of defoliation and tree mortality. As we see in Fig. 9.3, the budworm populations in all parts of New Brunswick declined during the late 1950s more or less simultaneously. They declined regardless of the degree of defoliation; and, incidentally, no matter whether the areas were sprayed or unsprayed with insecticide. Exactly the same has occurred in the recent population decline towards the end of the 1980s in the southern half of the province, including two of our study plots, one of which was heavily defoliated and the other little defoliated.

After the latest budworm outbreak in Cape Breton Highlands, Nova Scotia, a high proportion of balsam firs died. However, a decline in budworm population preceded the occurrence of major tree mortality (Piene, 1989). Moreover, many white spruce stands in the lowland areas survived heavy defoliation. [The white spruce is as susceptible to budworm as the balsam fir but does not die as easily (MacLean, 1980; Blais, 1981; Ostaff and MacLean, 1989).] Nonetheless, budworm populations in the lowland spruce forests declined at the same time as in the highland fir forests.

Evidently, heavy defoliation or tree mortality was neither a primary nor a universal cause of budworm population decline.

At a peak budworm density, current-year foliage (the prime source of food for the budworm) is often completely consumed. Then, the larvae are forced to feed on needles from previous years' growth. This so-called 'back feeding' would result in reduced size and fecundity of moths. This does not necessarily cause mortality. Blais' (1952) experiment has shown that survival of larvae, younger than the 6th instar, was reduced if fed on old foliage, but, once having reached the 6th instar, their survival was little affected by the age of the foliage on which they fed. Since 90% of the total food consumption by a larva occurs after it has reached the 6th instar (Miller, 1977), back feeding would usually have little influence on the survival of the larvae in the field. After a 100% defoliation of current-year foliage, many trees can produce enough new foliage the following year to feed budworm larvae again. Only after 4 or 5 years of repeated defoliation do some trees, particularly matured fir trees, begin to die. But many immature firs and white spruce survive and provide food for budworms before they decline.

The chemical and physical characteristics of foliages change after defoliation; often nitrogen content increases dramatically and new needles are unusually elongated (Piene, 1980). However, there is no adverse effect on the field survival of budworms which feed on them. The Green River data show that the late-instar larvae survived even better in the heavily damaged plots K1 and K2 than in the little damaged plots G4 and G5 (section 9.10.2).

9.8.5 Weather influences

An earlier theory of climatic control of insect populations (Wellington *et al.*, 1950) contended that a series of dry, warm summers would favour the development of the feeding larvae and would allow their population to increase; a series of wet, cool summers would have the opposite effect.

Prompted by this theory, Greenbank (1963a) took 5-year moving averages of the mean precipitation and the mean daily range of temperature in June and July (averages of nine weather stations from all New Brunswick). He chose June and July because these two months cover most of the larval feeding period in northern New Brunswick. Daily range of temperature is the difference between the daily maximum and minimum, supposed to indicate the clearness of the day.

I reproduced Greenbank's graphs in Fig. 9.12 to compare with budworm population changes in Fig. 9.2a. We see an apparent correlation: an on-average dry, warm period in the mid 1940s preceded the period of population increase; during an intermediate condition between 1950 and 1955 the population was at its peak; and the population declined during an on-average wet, cool period between 1956 and 1960.

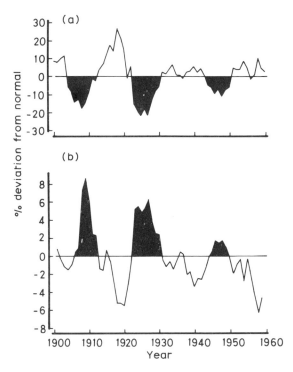

Figure 9.12   Five-year moving averages (each average plotted against the fifth year) of precipitation (graph a) and mean daily range of temperature (graph b) during June and July from 1900 to 1960 in New Brunswick. After Greenbank (1963a).

Figure 9.12 also shows two previous intervals of similar weather pattern. The first one appears to have preceded the well-known outbreak period of 1912–20. However, no budworm outbreak occurred during the second interval. Greenbank explained that an outbreak could not occur during this interval because the forests had not fully recovered from the devastating damage caused by the previous outbreak, which indeed killed 80% of firs but only 25% of spruces in the province (Morrison, 1938).

As I discussed in section 3.4, a moving-average transformation of weather series tends to create a cyclic pattern. Its cycle length depends on the number of consecutive data points to be averaged. Thus, one can create a

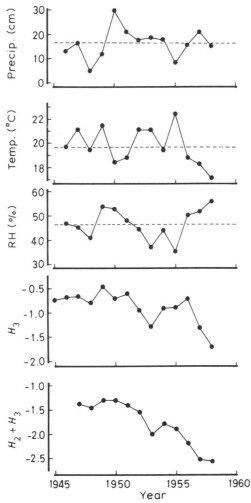

Figure 9.13 Annual changes in some climatic records and survival of larvae observed at a Green River study plot. Dashed lines are averages. The graphs H_3 and $H_2 + H_3$ are taken from Fig. 9.7.

cyclic weather pattern that is correlated with a budworm outbreak cycle. A more direct comparison of a climatic record with budworm survival does not substantiate the climatic theory as I now argue.

Figure 9.13 shows annual changes in total precipitation, mean daily maximum temperature, and mean daily minimum relative humidity, over the period of 1 June to 15 July (a main period of larval development: Morris, 1963a) recorded at the Green River field station between 1945 and 1958. These are compared with log larval survival H_3. We see no clear sign of influence from these weather factors on H_3.

As already mentioned in section 9.7, the estimation of H_3 contains some error because of an arbitrary division between the early- and late-instar stages. The sum $H_2 + H_3$ reduces the error. The resultant, consistent declining trend in the sum (shown also in Fig. 9.13) is even more difficult to explain by much more erratic fluctuation in the local weather records. [No climatic record is available during the early instar period (autumn to spring when the station was closed) to compare with H_2. However, climate tends to fluctuate erratically.]

After all, the larvae in the Green River study area could not possibly have ignored the detail of the local weather conditions, while responding to the smoothed, moving average series of the provincial mean weather conditions.

Thus, it is doubtful that climatic factors dictate the long-term oscillatory trend of budworm populations through influencing generation survival. However, as we shall see in section 9.9, weather does influence dispersal of egg-carrying moths and, hence, recruitment rate (or the E/M ratio).

9.8.6 Roles of natural enemies in budworm dynamics

After having eliminated improbable mortality factors, I conclude that the complex of natural enemies is the major factor which governs the oscillatory trend in generation survival. I now discuss possible roles played by different types of natural enemies.

We can classify natural enemies according to their mode of action depending on their life cycles:

1 Those which can complete their life cycles only on the spruce budworm. No alternate host is required. These include: univoltine parasitoids (*A. fumiferanae*, *G. fumiferanae*, and many tachinids; see Table 9.2) and protozoan parasites (e.g. *Nosema fumiferanae*).
2 Those which require alternate hosts or prey to complete their life cycles, including: multivoltine parasitoids (e.g. *T. minutum*, *M. trachynotus*, most ichneumonids attacking budworm pupae) and many predators, particularly birds. [Probably many hyperparasitoids and predators of the primary natural enemies of budworm (call them secondary natural enemies) belong to this category.]

3 Those pathogenic microbes (viruses, bacteria, fungi) which can stay latent in the habitat and propagate in an opportunistic manner. Because their life cycles are mostly unknown, it is difficult to discuss their roles.

Those in category (1) can respond reproductively to changes in budworm density. These are second-order density-dependent factors (section 2.3.2) which can generate a population oscillation.

Those in category (2) cannot multiply as budworm increases unless their alternate hosts increase with the budworm; this does not actually happen. Thus, these natural enemies can respond to changes in a budworm population only functionally: each individual kills more budworms within limits as their number increases. Potentially, these are first-order density-dependent factors which will not generate an oscillation of the system. They may modify the length and amplitude of the oscillation generated primarily by factors in the first category.

Alternate hosts or prey of the natural enemies – both primary and secondary in category (2) – may act as density-independent factors on the budworm because their densities depend primarily on species other than budworm. In particular, they may act as nonlinear density-independent factors (section 1.7.3): changes in their densities influence the curvature of budworm reproduction surface (the plot of budworm reproductive rate against the budworm's own density as well as against the density of its primary natural enemies). I shall synthesize these roles of the natural enemies in the final section. Let us now look at the recruitment process.

9.9 ANALYSIS OF RECRUITMENT RATE
(Egg/Moth ratio)

The per capita recruitment of a new cohort in a local population is the ratio of the number of eggs to the number of locally emerged moths (both sexes) – or 'E/M ratio' for short – on the unit foliage surface area.

After having laid a portion of their egg complement at their natal site, female spruce budworms often fly long distances to deposit the rest of their eggs at new sites (section 9.2). Therefore, at a given site, some of the eggs could be deposited by the locally emerged females before they migrated and others could be carried in by immigrants from elsewhere. Consequently, the number of eggs deposited at the given site tends to deviate, in either way and often considerably, from the number expected from the mean fecundity of local females.

9.9.1 Annual changes in E/M ratio

Figure 9.14 shows the annual changes in E/M ratios at the four study plots in the Green River area. The dashed line in each graph is one-half of the mean potential fecundity of a local female (because the budworm sex ratio is about 1:1). [Potential fecundity was estimated by the weight of female pupae

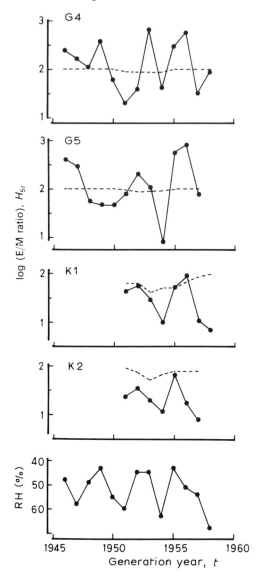

Figure 9.14 Annual changes in the log E/M ratio in four study plots (G4, G5, K1 and K2) in the Green River area and the fluctuation in inverted mean daily minimum relative humidity during estimated moth period. After Royama (1984).

(Miller, 1963a).] An E/M ratio above the dashed line indicates a net gain of eggs from immigrant moths; one below the line indicates net loss of eggs due to mortality and/or to emigration of locally emerged moths, outweighing gain from immigration. Because only 60–80% of the egg complement might be laid before a female dies (Thomas *et al.*, 1980), a 60% deposition of potential eggs at the site does not necessarily indicate emigration of the local moths. However, many points in Fig. 9.14 are well below that level, which do indicate emigration.

We immediately notice two things in Fig. 9.14: (1) E/M ratios fluctuate more or less the same way in all four plots; (2) E/M ratios in relation to the potential fecundity tend to be lower in the severely defoliated K plots than in the lightly defoliated G plots. Climatic conditions and defoliation levels influence moth dispersal activity and, hence, E/M ratio.

9.9.2 Climatic influence on E/M ratio

Long-range dispersal (exodus) flights usually begin early in the evening, often around 17:30 h (Atlantic Daylight Saving Time) in the Fredericton area (my own observation), peaking between 20:00 h and 22:00 h (Greenbank *et al.*, 1980), and continue until midnight or even later (A. W. Thomas, Forestry Canada, personal communication). [An exodus flight is a decisive take-off into the air above tree canopy, clearly different from 'buzz-around' activity around the tree crown seen during most of the day.]

A study by radar and aircraft by Greenbank *et al.* (1980) found that temperature and wind conditions above canopy level were major factors that determined the exodus flight. Exodus tends to peak and finish earlier on cold nights than on warm nights (little activity is observed below 14°C). Cold nights tend to reduce the incidence of exodus, and a cold air mass or rain-storm cell over a given area will force the moths on the wing to land if they happen to be flying over the area. Thus, in a given area in a year when cold nights prevail, the rate of emigration would be low but the probability of immigration could be high, tending to result in a comparatively high E/M ratio in the area, and vice versa.

Greenbank (1963b) showed that an average E/M ratio over the Green River area was inversely correlated with the mean daily minimum relative humidity of a moth period. Indeed, the graph of E/M ratios in Fig. 9.14 appear to be correlated with the inverted graph of mean daily minimum RH during an estimated moth period. [I estimated an annual moth period by accumulated degree-days (Royama, 1984).]

Although the data series are short, I believe the correlation is genuine. It does make sense because, in northern New Brunswick, a low daily minimum relative humidity, which is usually registered early in the afternoon, tends to be followed by a cold, crisp night in the summertime, and vice versa. [A summer night can be cold in New Brunswick.] In other words, dispersal activities in the evening must be correlated with the daily minimum RH.

9.9.3 Effect of defoliation on E/M ratio

Figure 9.14 shows that E/M ratios tend to be less in the severely defoliated K plots than in the little defoliated G plots. Evidently, heavy defoliation promotes moth exodus and, together with reduced fecundity of the local moths, lowers the E/M ratio. In other words, E/M ratio should be density-dependent. As it turns out, however, the relationship is not simple.

There are two ways of looking at the dependence of E/M ratio on population density: spatial and temporal relationships. To see the spatial relationship, we assemble pairs of E/M ratio and density data collected in a given year from many study plots and regress the ratio on density. To reveal the temporal relationship, we assemble pairs of data collected at a given plot over a number of years and regress the ratio on density. These two relationships found in the Green River study apparently differ from each other as I now discuss.

Spatial relationships
Three data sets are available for the analysis of the spatial relationship: namely, those from 13 plots in 1954, ten plots in 1955, and six plots in 1956 (Fig. 9.15). Both 1955 and 1956 (solid and open circles) were the years of

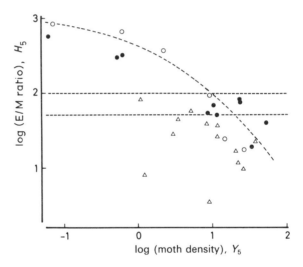

Figure 9.15   The dependence of log E/M ratio (H_5) on log moth density (Y_5) in several plots in the Green River area in years 1954 (triangles), 1955 (solid circles) and 1956 (open circles). The dashed curve is drawn by eye through the 1955 and 1956 data points. The upper dashed line indicates the full potential E/M ratio and the lower one, one half of the potential ratio. After Royama (1984).

large influx of immigrants into the Green River area. In these years, we see a clear inverse relationship. Low-population plots must have received a large number of immigrants because six data points are well above the upper dashed line of 100 eggs/moth, a full potential E/M ratio without moth dispersal. In high-population plots, net emigration must have occurred because many points are below the lower dashed line, a 50% realization of the full potential E/M ratio. [In the heavily defoliated K plots, average fecundity of a local female was reduced to about 140 eggs in 1955 and 1956 (a 30% reduction from the mean full fecundity of 200). Also, moth mortality could result in a further 30% reduction in oviposition by a local female. Therefore, a 50% reduction in E/M ratio could possibly happen even without emigration. A further reduction does indicate net emigration.]

In 1954, population was comparatively high in every plot, and net emigration is apparent in most plots. This was the year when the weather favoured emigration (see a high mean daily minimum relative humidity in Fig. 9.14). Thus, for a given population density, most data points are below those in 1955 and 1956.

No doubt, severely defoliated plots promoted moth emigration and discouraged immigrants, whereas undefoliated plots attracted the immigrants.

Temporal relationships

Figure 9.16 shows E/M ratio regressed on population density over several years at each of the four study plots, G4, G5, K1 and K2. The relationships in the G plots (Fig. 9.16a) are not as clear as those in 1955 and 1956 in Fig. 9.15. Those in the K plots (Fig. 9.16b) are completely obscured. A comparison with the time-series data from each plot shows that, over the period under observation, the time series of E/M ratios (Fig. 9.14) exhibited no trend, whereas budworm density showed a cyclic trend, during the same period (Fig. 9.17a,b). Why is the density effect on E/M ratio apparent in the spatial data set but not in the temporal one?

Explanations

We can explain this apparent paradox if we assume: (1) deposition of eggs (by both local moths and immigrants) at a given forest stand depends primarily on the level of defoliation, not directly on the local population density; (2) local moths tend to emigrate whenever weather permits regardless of density; (3) immigrants deposit their eggs at a little defoliated stand but lay less at a more defoliated stand. [Immigrants do not necessarily lay eggs where they landed; they might leave the site next evening.]

By hypothesis (1), we expect that, in the K plots where defoliation was already heavy when the data were collected, the average E/M ratios were low compared to those in the little defoliated G plots (Fig. 9.14). It also makes sense, by hypotheses (2) and (3), that the E/M ratios in all plots fluctuated similarly from year to year about the different average levels; we expect this to happen if the trend in emigration and immigration was similar (because of a similar climate) over the Green River area.

Good correlations between E/M ratios and population densities in 1955 and 1956 (Fig. 9.15) makes sense by hypothesis (1). In these years, the populations in all plots were still at a plateau or had just begun to decline. Thus, in these years, the level of defoliation in each plot was correlated with population density.

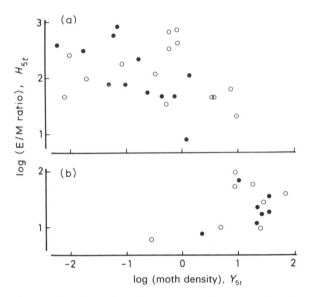

Figure 9.16 Observed relationships between log E/M ratio (H_{5t}) and log moth density (Y_{5t}) at four plots over several years during the Green River study. Graph a: plot G4 ($t = 1945$ to 1958; open circles) and plot G5 ($t = 1946$ to 1957; solid circles). Graph b: plot K1 ($t = 1951$ to 1958; open circles) and plot K2 ($t = 1951$ to 1957; solid circles). After Royama (1984).

In the K plots, however, foliage was still in a poor condition even after the population declined, and so E/M ratios were on average low. This explains a poor correlation between E/M ratio and population density in Fig. 9.16b. In the G plots, density never reached high enough to heavily damage foliage. So the average E/M ratio remained high. However, the ratio still fluctuates from year to year by hypothesis (2). Such a fluctuation in E/M ratio about an average should not depend on local density but on the climate. This explains a poor correlation in Fig. 9.16a.

Finally, if we inspect Fig. 9.16a closely, we see a sign of negative correlation. This does not imply that the year-to-year fluctuation in E/M ratio depended on population density during that interval as in the spatial relationship of Fig. 9.15. This negative correlation arises from an entirely different reason as I now explain.

In Fig. 9.17, I plotted the time series of log E/M ratios $\{H_{5t}\}$ and log egg densities $\{Y_{1t}\}$ from plot G4. Now look at the interval between 1949 and 1956 generations (marked with solid circles) during which the egg-density series stayed more or less at its plateau. There we see a curious negative correlation between the two series (see the solid circles in graph c). If we include the data points outside the above interval (open circles), the correlation deteriorates. As I demonstrated by the simu-

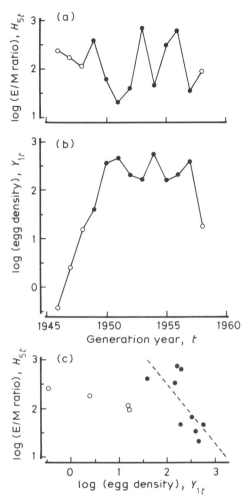

Figure 9.17   A curious negative correlation between log E/M ratio (H_{5t}) and log egg density (Y_{1t}) at plot G4 in the Green River area between 1949 and 1957 generation years. After Royama (1981a).

lation in Fig. 3.19, section 3.3.3, this situation occurs even when the year-to-year fluctuation in E/M ratio is completely independent of population density.

In the simulation, I used the linear second-order density-dependent population process (3.45) in which X is log population density and u is an uncorrelated perturbation effect. Figure 3.19 illustrates that if the series $\{X_t\}$ stays at its plateau for a while, then, during that time interval, the perturbation effect u_t will be negatively correlated with X_t, even though u_t is generated completely independent of X_t! If we compare log egg density (Y_{1t}) with X_t in the simulation and log E/M ratio (H_{5t}) with perturbation effect u_t, then, the observed situation in Fig. 9.17 is practically identical with the simulation in Fig. 3.19.

E/M ratio during the endemic period
The Green River study did not provide information about E/M ratios during the endemic period after the 1958 generation year. However, the log rate of change in density (R_t) continued to fluctuate about the principal oscillation (Fig. 9.5). This must be due to the fluctuation in the E/M ratio (H_{5t}) for the reason already discussed in section 9.6 (Fig. 9.6). Thus, we can conclude that the E/M ratio fluctuated just as much and as frequently during the 10 years of endemic period after 1958 as during the preceding epidemic period. This makes sense if moth dispersal occurred regardless of density and if local population cycles were in synchrony over a wide area which includes the locality where the immigrants originated. The E/M ratio is a net effect of emigration and immigration regardless of the populations being at their epidemic or endemic phase as long as they are in the same phase.

Conclusion
The E/M ratio (recruitment rate) has two components: the mean and deviations from it. The mean depends on the level of defoliation which, in turn, depends on population density nonlinearly and can lag in time. The mean E/M ratio depends only weakly on density when it is low and foliage is little damaged. It depends more strongly as density rises and foliage is damaged. If the forest is severely damaged, the mean ratio would remain low, even after the population declines, until the foliage recovers from the damage. In other words, the mean recruitment rate is a second-order density-dependent process.

The second component, the year-to-year fluctuation of E/M ratio about the mean, is density-independent (because it is caused by a climatic fluctuation). This component acts as vertical perturbation on the annual rate of change in population density (section 1.7.1). Thus it is important to analyse these two components separately, or one may get a nonsensical result.

9.10 THEORY OF OUTBREAK PROCESSES

Now, I shall try to synthesize the results of the foregoing analysis to conclude my investigation. But, first, let us critically look at some ideas that earlier investigators advocated.

9.10.1 Early theories

The theory of spruce budworm outbreak processes most popular up to the early 1980s is based on the hypotheses of climatic release, epicentres and multiple equilibrium structure. The essence of these hypotheses is the following.

A local budworm population is usually maintained at an endemic level by the action of a natural enemy complex. However, a series of years with favourable weather in a healthy forest stand would promote high larval survival resulting in a rapid population increase. It was also hypothesized that larval survival increases as the forest matures. Thus, in some scattered forest stands, the budworm populations would eventually escape the action of natural enemies and head towards extremely high density levels. These populations would act as epicentres from which a large number of moths disperse. These moths infest surrounding areas with eggs, assisting the populations there to also escape from their endemic state, resulting in widespread outbreaks. After several years of severe defoliation, however, the quality and quantity of food supply deteriorate, and larval survival decreases. In the meantime, a spell of unfavourable weather would arrive, and the outbreaks would finally collapse. The populations now return to endemic levels and are maintained there by the action of natural enemies. The populations would stay endemic until the damaged forest fully recovers and a spell of favourable weather returns.

Morris (1963b) idealized the above theory as a multiple-equilibrium scheme (his fig. 40.1 – see my Fig. 2.4, section 2.2.3) originally proposed by Ricker (1954). The theory was further elaborated by a group from the University of British Columbia (Ludwig *et al.*, 1978).

9.10.2 Failure of the early theory

My foregoing analyses of the same field data do not support five major assumptions in the early theory.

First, budworm larval survival was not as sensitive to changes in weather conditions (section 9.8.5) as the theory supposed. Moreover, the supposed weather cycle does not exist. It was artificially created by the 5-year moving-average transformation of a non-cyclic weather series (section 3.4). Also, larval survival does not depend on the maturity of the stand in which they grow. Survival was high in both K1 (mature) and K2 (immature) plots but it was comparatively low in both G4 (mature) and G5 (immature) plots.

Second, food shortage was not a primary universal cause of population decline (section 9.8.4). When budworm populations began to decline, they tended to decline everywhere regardless of the foliage condition. The decline of the budworm population in Cape Breton, Nova Scotia, preceded the occurrence of major tree mortality.

Third, moth invasions enhanced a population increase only when the population was in the upward phase of the cycle. Once the population had begun to decline, even if the foliage was undamaged, moth invasions did not reverse the trend. An example is the population in plot G4. It increased each summer following a big moth invasion in the previous autumn (1954, 1957 and 1961 in Fig. 9.2a). However, the declining trend continued even though the foliage was little damaged.

Also, moth invasions in this plot occurred just as frequently during the endemic period of the 1960s as during the epidemic period in the 1950s (section 9.9.3). Nonetheless, the population stayed low until the late 1960s (Fig. 9.2a). As Fig. 9.5 clearly shows, perturbation caused by the year-to-year fluctuation in the recruitment rate did not alter the principal oscillation of the reproductive rate dictated by survival of feeding larvae and pupae.

Fourth, the idea of 'epicentre' is unfounded. Probably, it was originally conceived by budworm infestation maps (see updated ones by Kettela, 1983; Hardy *et al.*, 1986) showing annual changes in areas (rather than population sizes) of different infestation levels, e.g. severe, moderate or light. It looks as though several scattered 'hot spots' appear first, then grow outward like the growth of fungus colonies.

Stehr (1968), however, suggested an alternative interpretation: an epicentre might be 'merely the spot at which a general and already widespread population surfaces first'. A close inspection of the annual variations in egg-mass density (Fig. 9.3) supports Stehr's alternative. It reveals that budworm populations were in their troughs by the early 1960s, and started increasing again thereafter just about everywhere in New Brunswick. However, in the central region (Blocks B3, B4, C3, C4 and C5 in Fig. 9.3), the trough populations were somehow maintained at much higher levels than in any other areas of the province. Consequently, when all populations in the province increased again in the early 1970s, a level of severe infestation was reached in the central region sooner than in surrounding regions.

Moth dispersal is unlikely to trigger outbreaks in surrounding areas in the manner the early theory hypothesized. Only when the local populations were already in an increasing phase did moth invasions accelerate the rate of increase in these populations to reach epidemic levels earlier or more readily. In other words, moth invasions acted like fertilizers stimulating the 'seeds' of epidemics already growing everywhere. This analogy is consistent with the fact that once populations had begun to decrease (i.e. once the 'plants' had begun to die), further moth invasions (fertilization) did not reverse the trend even if food was still plentiful.

Fifth, the existence of a multiple equilibrium structure is not substantiated by data. Morris (1963b) conceived this structure from Watt's (1963) analysis. Watt regressed survival of feeding larvae on their density and found a tendency for their survival to increase with density up to a point and, then, to fall (Fig. 9.18). Morris envisaged that this graph was like his sinuous multiple-equilibrium graph (his fig. 40.1b – see my fig. 2.4, section 2.2.3) but only truncated in a low-density range for lack of data; so the rising 'tail' section of the sinuous curve could not be seen.

This is not convincing because Watt's graph does not represent a density-dependent structure. He indiscriminately pooled all data sets from as many plots over as many years as the study provided. Then, he divided them into six arbitrary intervals of density spectrum and averaged the data in each section to obtain the six points in Fig. 9.18. One should not do this, however.

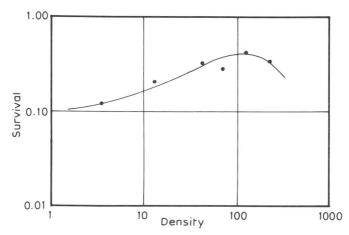

Figure 9.18 'The survival rate of the late-instar larvae (after the removal of the effect of parasitism) as a function of their density' depicted by Watt (1963).

During the interval between 1945 and 1958, all populations in the Green River area increased and then decreased (section 9.5). I deduced that such a budworm population cycle was due to the action of the natural enemy complex operating on feeding larvae and pupae (section 9.8.6). Thus, the budworm survival depends not only on its own density but also on the natural enemy complex, i.e. the budworm cycle is a second-order density-dependent process (section 2.3.2). So the relationship should be represented by a three-dimensional graph (budworm survival against its own density and against density of the natural-enemy complex), although we do not have a good data set to construct such a graph.

If, as in Watt's graph, survival of larvae was plotted only against their density which was cyclic, the series of data points from each plot should exhibit an elliptical trajectory (Fig. 9.19). We only see a fraction of each

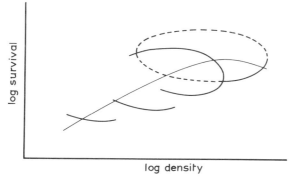

Figure 9.19 A schematic illustration of the relationship between the survival of larvae and their initial density in Watt (1963; fig. 10.5). After Royama (1984).

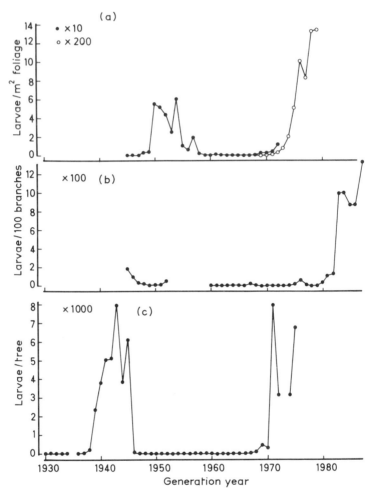

Figure 9.20 Annual changes in spruce budworm density; the same as in Fig. 9.2 but the density is plotted in linear (rather than log) scale. (a): the Green River area; (b): Black Sturgeon area; (c): near Chapleau.

trajectory because the observation did not cover the whole cycle in any one plot. Moreover, in some plots (e.g. the K plots), survival was somehow higher and, accordingly, peak densities grew higher than in other plots (e.g. G plots). [Peak budworm density in each plot depends on the natural enemy complex which is not identical in all plots.] Thus, the trajectories of data sets from different plots have different positions in the graph: those for high-density plots extended towards an upper right area and those for low-density plots stayed around a lower left area. I idealized these situations in Fig. 9.19.

Watt's graph was a collection of such fragmented trajectories from many plots. We do not see these trajectories in his graph because all individual data points were averaged in arbitrary sections of the density spectrum. His regression curve compares with the humped curve in Fig. 9.19. Such a curve does not represent a density-dependent structure by any means, much less a multiple-equilibrium structure.

Perhaps, the idea of multiple-equilibrium structure stemmed from a graph of budworm population changes plotted in a linear scale (Fig. 9.20). It looks as though the populations stayed at an endemic equilibrium state for a prolonged period of time. But it appears so merely because of poor data resolution at a low density level. The same data sets, when plotted in a logarithmic scale (Fig. 9.2), reveal little sign of a prolonged endemic equilibrium state. The rate of change in density between two successive years (the difference between two successive data points in Fig. 9.2) was changing only gradually over time, rather than revealing an abrupt shift from one equilibrium state to another.

Thus, all crucial aspects of the early theory turned out to be unfounded. I now propose my alternative.

9.10.3 Alternative theory

Before proposing an alternative theory, let me summarize the results of foregoing analyses in terms of the process structure as well as a food-web structure centred around the budworm.

The process structure
The net reproductive rate of the budworm (year-to-year rate of change in population density: r) is comprised of two major components: generation survival (h_g) and recruitment rate (E/M ratio: h_5) (section 9.6). After log transformation:

$$R = H_g + H_5$$

in which a capital letter is the log of its lower-case letter as before.

The generation survival is further partitioned into the survival of the early, non-feeding stages (h_{1+2}) and that of the feeding and pupal stages (h_{3+4}), so that, after log transformation:

$$H_g = H_{1+2} + H_{3+4}$$

(section 9.7).

I argued that the key components of the budworm population process were H_{3+4} and H_5. H_{3+4} generates the principal oscillation of a budworm population. The mean of H_5 is defoliation-dependent and a deviation from the mean acts as vertical perturbation on the reproductive rate.

Food-web structure
The major source of mortality during the late stages (feeding larvae and pupae) is a complex of natural enemies (section 9.8). These are, in turn, attacked by their own natural enemies. Thus, a food web formed around the budworm upon the infrastructure of forest can be stratified as follows:

I. Fir and spruce foliages
II. Defoliators:
  a. Spruce budworm
  b. Other
III. Primary natural enemies of budworm:
  a. Univoltine parasitoids and *Nosema* (microsporidia)
  b. Multivoltine parasitoids and omnivorous predators
  c. Pathogenic microbes (viruses, fungi, bacteria and yeasts)
IV. Secondary natural enemies:
  Hyperparasitoids and predators of IIIa and b.

The primary natural enemies of category IIIa can complete their life cycle entirely on budworm, although they usually have many alternative hosts from category IIb. Those of category IIIb must find alternate hosts or prey from category IIb to complete their annual life cycle. The ecology of the microbes in category IIIc is least known. Also, little is known about the life cycles of the hyperparasitoids in category IV; some are probably multivoltine and require alternate hosts from outside category III. [Even the taxonomy of the hyperparasitoids is incomplete.]

Synthesis
Let us first consider an idealized budworm–forest system without natural enemies. The persistence of this basic system must depend on the budworm's self-imposed (intrinsic) population regulation mechanism. If competition among the larvae is strong enough, the population density would be kept sufficiently low so that no defoliation occurs. A first-order logistic process (section 4.3.1) should be realized. The system will, sooner or later, reach an equilibrium state similar to the weevil–bean system of Chapter 7. If the self-imposed regulation is only moderately strong, defoliation may occur. The process is, then, second-order logistic and may oscillate depending on how long it takes for the foliage to recover from the damage (section 4.3.4).

Actually, however, budworm larvae can tolerate extreme crowding. This results in 100% defoliation of new foliage every year. The only self-imposed regulation of a budworm population is its defoliation-dependent recruitment rate. But the rate does not drop sufficiently low to reduce population density so that the trees can recover from the damage quickly. Host trees (particularly, mature firs) do not tolerate severe defoliation for five or more years and will, sooner or later, die. Under these circumstances, there should be frequent extinction of local systems which may be synchronized over a wide area. Then the whole system may not be sustainable. In other words, an equilibrium state is unlikely to exist, let alone the upper equilibrium assumed in the early theory.

Now consider that we introduce some natural enemies from category IIIa. Typical examples are: the parasitoids, *A. fumiferanae*, *G. fumiferanae*, and *W. fumiferanae* (Table 9.2, section 9.8.1); or the microsporidia, *Nosema fumiferanae* (section 9.8.3). These can increase reproductively in response to an increase in budworm density. So, acting as second-order density-dependent factors, these can potentially control the budworm (section 9.8.6). An efficient hunter can control a host population so that no severe defoliation occurs. Such a host–parasitoid system is likely to oscillate like the weevil–wasp interaction of Chapter 8.

The amplitude and length of an oscillation depend on the efficacy of these natural enemies. Because many (probably all) primary parasitoids in the field are attacked by their own enemies in category IV, their efficacy to control budworm is greatly reduced. Also, *Nosema fumiferanae* is known to be of low virulence. So it takes many years before they build up to effectively influence budworm survival. Thus, the system of forest, budworm and natural enemy complex oscillates slowly with high amplitude. When local systems are synchronized due to the Moran effect, a widespread outbreak may occur at more or less regular intervals.

When defoliation becomes severe, the budworm recruitment rate (E/M ratio) drops. This prevents the budworm population from increasing further. However, the recruitment rate does not usually drop to reduce the number of progeny substantially. Consequently, the budworm population stays at a plateau for several years. It fluctuates about the plateau as the E/M ratio (as vertical perturbation) fluctuates. If, in the meantime, the natural enemy complex catches up, the budworm population begins to decline. If not, the forest may be destroyed.

The system behaviour further depends on the presence of the natural enemies in category IIIb, those that require alternate hosts or prey other than the budworm. Typical examples are *M. trachynotus* (Table 9.2, section 9.8.1) and insectivorous birds (section 9.8.2). The ability of these species to respond reproductively to changes in budworm density depends on how abundant these alternate hosts and prey are relative to the budworm. When the alternate species are much less abundant than the budworm, these natural enemies cannot respond reproductively to changes in budworm density but only functionally. In other words, the population densities of these alternate hosts and prey, independent of budworm (although they may be correlated with budworm), dictate the efficacy of the predators and parasitoids of category IIIb and, hence, act as nonlinear perturbation on the budworm dynamics (section 1.7.3).

A spider complex may act somewhat differently. These ambush predators may catch budworm larvae, particularly when L_1 and L_2 disperse, but only in proportion to the larval density. Thus, the spider complex, largely independent of budworm density (section 9.8.2), acts as lateral perturbation on budworm dynamics (section 1.7.2).

Thus, the system's equilibrium state depends on the relative abundances of the alternate hosts and prey: if their densities are low, the budworm should have a higher equilibrium density (and, probably, a higher peak density as well) and vice versa. [The effects of different types of perturbation on population equilibria are illustrated in Figs 1.12 and 1.13, section 1.7, using a simple first-order process.] The involvement of nonlinear and lateral perturbations may further complicate density-dependent aspects of budworm dynamics. This is because the timing of decline from a plateau or that of increase from an endemic state depends **probabilistically** on the status of non-budworm populations in the food web. Then, budworm density (equilibrium, peak or trough) can become higher in certain forest stands than in others. What happens to the system at a given moment will be difficult to predict unless we know the composition and status of the food web.

The system I described above has a unique conditional equilibrium state. For instance, the system of budworm and its natural enemy complex has a unique equilibrium state when the components of the complex are specified and all other elements of the food web are fixed. Let us relax some of such conditions: let, for example, the hyperparasitoid complex respond to the system while the remaining elements are still fixed. The unique equilibrium state of the system will move to its new position. Each time the condition (status of the fixed elements) is changed, the equilibrium state will change. So, if we let every element change with time, so does the equilibrium state. [This is not a multiple equilibrium process as described by the sinuous reproduction curves of Fig. 2.4.] An equilibrium state of every stochastic population process is conditional and, therefore, changes with time as the conditions change. In a stationary process, the equilibrium state itself is a stationary stochastic process. The state may be nonstationary in a nonstationary process. A budworm outbreak occurs as a result of the combined effect of the food-web elements taking certain probabilistic states.

The above outlines the principles of budworm outbreak processes that I envisage. The basic principles in the budworm population dynamics are no more different from those underlying the lynx, snowshoe hare, or Utida's experimental host–parasitoid system. The budworm system differs from those in the degree of complexity in the food-web structure to which the budworm belongs.

Epilogue

'..., we all know that 50 years from now, most of the things we learned here will turn out not to have been quite right.'

Jacob Bronowsky (*The Origins of Knowledge and Imagination*, 1978)

Two things are crucial to the study of animal population dynamics: to carry out an uninterrupted long-term observation of a natural population and to apply theoretical knowledge to the analysis, interpretation and synthesis of the results.

Students of population ecology must acquire sufficient knowledge of elementary mathematics and an ability to apply it to their analytical work. Gathering data and leaving the analysis to a statistician will not work well. Cooperation with a theorist would help or often would be necessary. But one must have, at least, a working knowledge to thoroughly understand what the theorist suggests and to make one's own judgment, rather than swallowing the whole suggestion. Acquiring such working knowledge depends largely on the individual's motivation and commitment.

To carry out a long-term study, on the other hand, is often beyond the capability of the individual. It may take a few decades to obtain meaningful results. This can be done only when the research organization also commits itself. As far as I am aware, only a few research institutes in the world, e.g. the Edward Grey Institute of Field Ornithology of Oxford, the Institute for Ecological Research of the Netherlands, the Smithsonian Tropical Research Institute in Panama, the Department of Wildlife Ecology of Wisconsin, and Canadian Forest Service, have maintained such a long-term research project. In recent years, however, funding for such projects seems to be diminishing.

An increasingly urgent demand to develop better environmental and resource management everywhere in the world, necessitates a firm commitment to promoting and supporting a long-term project by a funding institution and by a local or national authority.

If this book has sufficiently motivated many people I would be satisfied, for a major purpose of the book will have been achieved. I suppose there are many unsatisfactory aspects to this book, but I hope that it will improve with the help of enthusiastic readers.

References

Adamcik, R. S and Keith, L. B. (1978) Regional movements and mortality of great horned owls in relation to snowshoe hare fluctuations. *Canadian Field-Naturalist*, **92**, 228–34.

Adamcik, R. S., Todd, A. W. and Keith, L. B. (1978) Demographic and dietary responses of great horned owls during a snowshoe hare cycle. *Canadian Field-Naturalist*, **92**, 156–66.

Allee, W. C., Emerson, A. E., Park, O. *et al.* (1949). *Principles of Animal Ecology*. Saunders, London.

Andrewartha, H. G. and Birch, L. C. (1954) *The Distribution and Abundance of Animals*. University of Chicago Press, Chicago.

Archibald, H. L. (1977) Is the 10-year wildlife cycle induced by a lunar cycle? *Wildlife Society Bulletin*, **5**, 126–9.

Arditi, R. (1979) Relation of the Canadian lynx, *Lynx canadensis*, cycle to a combination of weather variables. A stepwise multiple regression analysis. *Oecologia*, **41**, 219–34.

Arthur, A. P. and Wylie, H. G. (1959) Effects of host size on sex ratio, development time and size of *Pimpla turionellae* (L.) (Hymenoptera: Ichneumonidae). *Entomophaga*, **4**, 297–301.

Assem, J. van den (1971) Some experiments on sex ratio and sex regulation in the pteromalid *Lariophagus distinguendus*. *Netherlands Journal of Zoology*, **21**, 373–402.

Aubert, Jacques-F. (1961) L'expérience de la baurre de cotton demontre que la volume de l'hôte intervient en tant que facteur essential dans la détérmination du sexe chez les Ichneumonides Pimplinae. *Bulletin de la Société Entomologique de France*, **66**, 89.

Balch, R. E. and Bird, F. T. (1944) A disease of the European spruce sawfly, *Gilpinia hercyniae* (Htg.), and its place in natural control. *Science in Agriculture*, **25**, 65–80.

Balen, J. H. van (1980) Population fluctuations of the great tit and feeding conditions in winter. *Ardea*, **68**, 143–64.

Bartlett, M. S. (1946) On the theoretical specification of sampling properties of autocorrelated time-series. *Journal of the Royal Statistical Society*, **B**, 8, 27–41.

Bartlett, M. S. (1966) *An Introduction to Stochastic Processes with Special Reference to Methods and Applications* (2nd edn). Cambridge University Press, Cambridge.

Beddington, J. R., Free, C. A. and Lawton, J. H. (1975) Dynamic complexity in predator-prey models framed in difference equations. *Nature*, **255**, 58–60.

Blais, J. R. (1952) The relationship of the spruce budworm (*Choristoneura fumiferana*, Clem.) to the flowering condition of balsam fir (*Abies balsamea* (L.) Mill.) *Canadian Journal of Zoology*, **30**, 1–29.

Blais, J. R. (1958) Effects of defoliation by spruce budworm on radial growth at breast height of balsam fir and white spruce. *Forestry Chronicle*, **34**, 39–47.

Blais, J. R. (1962) Collection and analysis of radial growth data from trees of evidence of past spruce budworm outbreaks. *Forestry Chronicle*, **38**, 474–84.

Blais, J. R. (1965) Spruce budworm outbreaks in the past three centuries in the Laurentide Park, Quebec. *Forest Science*, **11**, 130–8.

Blais, J. R. (1981) Mortality of balsam fir and white spruce following a spruce budworm outbreak in the Ottawa River watershed in Quebec. *Canadian Journal of Forest Research*, **11**, 620–9.

Bodenheimer, F. S. (1938) *Problems of Animal Ecology*. Oxford University Press, Oxford.

Bodenheimer, F. S. (1958) *Animal Ecology To-day*. Junk, Den Haag.

Bookhout, T. A. (1965) The snowshoe hare in upper Michigan—its biology and feeding coactions with white-tailed deer. Michigan State Department of Conservation, Research Division Report, 38.

Boutin, S., Krebs, C. J., Sinclair, A. R. E. and Smith, J. N. M. (1986) Proximate causes of losses in a snowshoe hare population. *Canadian Journal of Zoology*, **64**, 606–10.

Box, G. E. P. and Jenkins, G. M. (1970) *Time Series Analysis: Forecasting and control*. Holen-Day, San Francisco.

Brand, C. J. and Keith, L. B. (1979) Lynx demography during a snowshoe hare decline in Alberta. *Journal of Wildlife Management*, **43**, 827–49.

Brand, C. J., Vowles, R. H. and Keith, L. B. (1975) Snowshoe hare mortality monitored by telemetry. *Journal of Wildlife Management*, **39**, 741–7.

Brand, C. J., Keith, L. B. and Fischer, C. A. (1976) Lynx responses to changing snowshoe hare densities in central Alberta. *Journal of Wildlife Management*, **40**, 416–28.

Brockwell, P. J. and Davis, R. A. (1987) *Time Series: theory and methods*. Springer-Verlag, New York.

Brown, C. E. (1970) A cartographic representation of spruce budworm *Choristoneura fumiferana* (Clem.) infestation in eastern Canada, 1909–1966. Canadian Forestry Service (Forestry Canada) Publication No. 1263, 1–4.

Bryant, J. P. and Kuropat, P. J. (1980) Selection of winter forage by sub-arctic browsing vertebrates: the role of plant chemistry. *Annual Review of Ecology and Systematics*, **11**, 261–85.

Bryant, J. P., Wieland, G. D., Clausen, T. and Kuropat, P. (1985) Interactions of snowshoe hare and feltleaf willow in Alaska. *Ecology*, **66**, 1564–73.

Bump, G., Darrow, R. W., Edminister, F. C. and Crissey, W. F. (1947) *The Ruffed Grouse*. New York State Department of Conservation.

Bulmer, M. G. (1974) A statistical analysis of the 10-year cycle in Canada. *Journal of Animal Ecology*, **43**, 701–18.

Bulmer, M. G. (1975) The statistical analysis of density-dependence. *Biometrics*, **31**, 901–11.

Bulmer, M. G. (1976) The theory of predator–prey oscillations. *Theoretical Population Biology*, **9**, 137–50.

Campbell, M. J. and Walker, A. M. (1977) A survey of statistical work on the Mackenzie River series of annual Canadian lynx trappings for the year 1821–1934 and a new analysis. *Journal of Royal Statistical Society*, Series A, **140** (Part 4), 411–31.

Cary, J. R. and Keith, L. B. (1979) Reproductive change in the 10-year cycle of snowshoe hares. *Canadian Journal of Zoology*, **57**, 375–90.

Chapman, R. N. (1928) The quantitative analysis of environmental factors. *Ecology*, **9**, 111–22.

Chatfield, C. (1984) *The Analysis of Time Series: An Introduction* (3rd edn). Chapman and Hall, London.

Chewyreuv, I. (1913). Le rôle des femelles dans la détermination du sexe de leur descendance dans le groupe des Ichneumonides. *Comptes Rendus des séances de la Société de Biologie (Paris)*, **74**, 695–9.

Chitty, D. and Nicholson, M. (1943) The snowshoe rabbit enquiry, 1940–41. *Canadian Field-Naturalist*, **57**, 64–8.

Christian, J. J. (1950) The adreno-pituitary system and population cycle in mammals. *Journal of Mammalogy*, **31**, 247–59.

Crawford, H. S. and Jennings, D. T. (1989) Predation by birds on spruce budworm *Choristoneura fumiferana*: functional, numerical, and total responses. *Ecology*, **70**, 152–63.

Criddle, N. (1930) Some natural factors governing the fluctuations of grouse in Manitoba. *Canadian Field-Naturalist*, **44**, 77–80.

Cunningham, W. J. (1954) A nonlinear differential–difference equation of growth. *Proceedings of National Academy of Sciences*, **40**, 708–13.

Den Boer, P. J. and Reddingius, J. (1989) On the stabilization of animal numbers. Problems of testing. 2. Confrontation with data from the field. *Oecologia*, **79**, 143–9.

Devaney, R. L. (1990) *Chaos, Fractols, and Dynamics*. Addison-Wesley, New York.

Dixon, W. J. (Ed.) (1981) *BMDP Statistical Software*. University of California Press, Berkeley.

Dominion Bureau of Statistics (1965) Catalogue 23-207, fur production. Ottawa.

Dowden, P. B. and Carolin, V. M. (1950) Natural control of factors affecting the spruce budworm in the Adirondacks during 1946–1948. *Journal of Economic Entomology*, **43**, 774–83.

Dowden, P. B., Jaynes, H. A. and Carolin, V. M. (1953) The role of birds in a spruce budworm outbreak in Maine. *Journal of Economic Entomology*, **46**, 307–12.

Elton, C. (1924) Fluctuations in the numbers of animals: their causes and effects. *British Journal of Experimental Biology*, **2**, 119–63.

Elton, C. and Nicholson, M. (1942) The ten-year cycle in numbers of the Lynx in Canada. *Journal of Animal Ecology*, **11**, 215–44.

Eveleigh, E., Strongman, D. and Royama, T. (1989) B.t. in spruce budworm populations in New Brunswick: where did it originate? 13th Annual Eastern Spruce Budworm Research Work Conference, Fredericton, New Brunswick, 24–26 January 1989.

Eveleigh, E., McCarthy, P., McDougall, G. and Royama, T. (1990) Mortality factors during the feeding stage of budworm larvae in outbreak and declining populations. 14th Annual Eastern Spruce Budworm Research Work Conference, Fredericton, New Brunswick, 23–25 January 1990.

Feller, W. (1952) *An Introduction to Probability Theory and Its Applications*. Chapman & Hall, London.

Finerty, J. P. (1979) Cycles in Canadian lynx. *American Naturalist*, **114**, 453–5.

Fox, J. F. (1978) Forest fires and the snowshoe hare–Canada lynx cycle. *Oecologia*, **31**, 349–74.

Fujii, K., Gatehouse, A. M. R., Johnson, C.D. *et al.* (Eds) (1990) *Bruchids and Legumes: Economics, Ecology, and Coevolution*. Proceedings of the Second International Symposium on Bruchids and Legumes, Okayama, 1989. Kluwer Academic Publishers, London.

Fujita, H. and Utida, S. (1952) The effect of population density in the growth of an animal population. *Researches on Population Ecology*, **1**, 1–14 (in Japanese with English summary).

Fujita, H. and Utida, S. (1953) The effect of population density on the growth of an animal population. *Ecology*, **34**, 488–98. (English translation of Fujita and Utida, 1952).

Gage, S. H. and Miller, C. A. (1978) A long-term bird census in spruce budworm-prone balsam fir habitats in northwestern New Brunswick. Information Report M-X-84, Forestry Canada-Maritimes Region, Fredericton, New Brunswick, Canada.

Gause, G. F. (1934) *The Struggle for Existence*. Williams and Wilkins, Baltimore.

Gilpin, M. E. (1973) Do hares eat lynx? *American Naturalist*, **107**, 727–30.

Gottman, J. M. (1981) *Time-series Analysis: A Comprehensive Introduction for Social Scientists*. Cambridge University Press, Cambridge.

Grange, W. B. (1949) *The Way to Game Abundance*. Scribner's, New York.

Grange, W. B. (1965) Fire and Tree Growth Relationships to Snowshoe Rabbits. *Proceedings of Annual Tall Timbers Fire Ecology Conference*, **4**, 110–25.

Green, R. G. and Evans, C. A. (1940) Studies on a population cycle of snowshoe hares on the Lake Alexander area. *Journal of Wildlife Management*, **4**, 220–38; **4**, 247–58; **4**, 267–78.

Green, R. G., Larson, C. L. and Bell, J. F. (1939) Shock disease as the cause of the periodic decimation of the snowshoe hare. *American Journal of Hygiene*, **30(B)**, 83–102.

Greenbank, D. O. (1963a) The development of the outbreak, in R. F. Morris (Ed.) The dynamics of epidemic spruce budworm populations. *Entomological Society of Canada Memoir*, **31**, pp. 19–23.

Greenbank, D. O. (1963b) The analysis of moth survival and dispersal in the unsprayed area, in R. F. Morris (Ed.) The dynamics of epidemic spruce budworm populations. *Entomological Society of Canada Memoir*, **31**, pp. 87–99.

Greenbank, D. O., Schaefer, G. W. and Rainey, R. C. (1980) Spruce budworm (Lepidoptera: Tortricidae) moth flight and dispersal: new understanding from canopy observations, radar, and aircraft. *Entomological Society of Canada Memoir*, **110**.

Haggan, V. and Ozaki, T. (1981) Modeling nonlinear random variations using an amplitude-dependent autoregressive time series model. *Biometrika*, **68**, 189–196.

Hardy, Y., Mainville, M. and Schmitt, D. M. (1986) *An Atlas of Spruce Budworm Defoliation in Eastern North America, 1938–80*. United States Department of Agriculture, Forest Service, Co-operative State Research Service, Miscellaneous Publication No. 1449.

Hassell, M. P. (1971) Mutual interference between searching insect parasites. *Journal of Animal Ecology*, **40**, 473–86.

Hassell, M. P. (1975) Density-dependence in single-species populations. *Journal of Animal Ecology*, **44**, 283–95.

Hassell, M. P. and Varley, G. C. (1969) New inductive population model for insect parasites and its bearing on biological control. *Nature*, **223**, 1133–7.

Hassell, M. P., Latto, J. and May, R. M. (1989) Seeing the wood for the trees: detecting density dependence from existing life-table studies. *Journal of Animal Ecology*, **58**, 883–92.

Hess, Q. F. (1946) A trapper's record of animal abundance in the Oba-Hearst area of Ontario for the years 1931–1944. *Canadian Field-Naturalist*, **60**, 31–3.

Hoel, P. G. (1947) *Introduction to Mathematical Statistics*. Chapman and Hall, London.

Holling, S. C. (1959) Some characteristics of simple types of predation and parasitism. *Canadian Entomologist*, **91**, 385–98.

Huntington, E. (1945) *Mainspring of Civilization*. John Wiley & Sons, New York.

Hutchinson, G. E. (1948) Circular causal systems in ecology. *Annals of the New York Academy of Sciences*, **50**, 221–46.

Ishikura, H. (1941) Influence of temperature and humidity on the total developmental time of the azuki bean weevil. *Oyo Dobutsugaku Zasshi* (Applied Zoological Magazine), **13**, 118–31 (in Japanese with English title).

Itô, Y. (1978) *Comparative Ecology*. Iwanami Shoten, Tokyo. English edition (1980), Cambridge University Press, Cambridge.

Ives, K. H. and Gibbons, J. D. (1967) A correlation measure for nominal data. *American Statistician*, **21**(5), 16–17.

Ivlev, V. S. (1955) *Experimental Ecology of the Feeding of Fishes* (English translation, 1961). Yale University Press, New Haven.

Jenkins, G. M. and Watts, D. G. (1968) *Spectral Analysis and its Applications*. Holden-Day, San Francisco.

Jensen, R. V. (1987) Classical chaos. *American Scientist*, **75**, 168–81.

Jones, J. W. (1914) Fur-farming in Canada (2nd edn). Committee on Fisheries, Game and Fur-bearing Animals, Commission of Conservation Canada. Mortimer, Ottawa.

Jones, W. T. (1982) Sex ratio and host size in a parasitoid wasp. *Behavioural Ecology and Sociobiology*, **10**, 207–10.

Keith, L. B. (1963) *Wildlife's Ten-Year Cycle*. University of Wisconsin Press, Wisconsin.

Keith, L. B. (1974) Some features of population dynamics of mammals. *Proceedings of the International Congress of Game Biologists*, **11**, 17–58.

Keith, L. B. (1983) Role of food in hare population cycles. *Oikos*, **40**, 385–95.

Keith, L. B. and Windberg, L. A. (1978) A demographic analysis of the snowshoe hare cycle. *Wildlife Monographs*, **58**.

Keith, L. B., Todd, A. W., Brand, C. J. *et al.* (1977) An analysis of predation during a cyclic fluctuation of snowshoe hares. *Proceedings of the International Congress of Game Biologists*, **13**, 151–75.

Keith, L. B., Cary, J. R., Rongstad, O. J. and Brittingham, M. C. (1984) Demography and ecology of a declining snowshoe hare population. *Wildlife Monographs*, **90**.

Kendall, M. G. (1954) Notes on bias in the estimation of auto-correlation. *Biometrika*, **41**, 403–4.

Kendall, M. G. and Stuart, A. (1968) *The Advanced Theory of Statistics* Vol. III (2nd edn). Hafner, New York.

Kendeigh, S. C. (1947) Bird population studies in the coniferous forest biome during a spruce budworm outbreak. Ontario Department of Lands and Forests, Division of Research, Biological Bulletin No. 1.

Kennedy, P. (1984). *A Guide to Econometrics* (5th edn). MIT Press, Cambridge, Massachusetts.

Kettela, E. G. A.. (1983) A cartographic history of spruce budworm defoliation from 1967 to 1981 in eastern North America. Information Report DCP-X-14, Forestry Canada, Ottawa, Ontario, Canada.

Kingsland, S. E. (1982) The refractory model: the logistic curve and the history of population ecology. *Quarterly Review of Biology*, **57**, 29–52.

Kingsland, S. E. (1985) *Modeling Nature*. The University of Chicago Press, Chicago.

Krebs, C. J. (1972) *Ecology*. Harper and Row, London.

Krebs, C. J., Cilbert, B. S., Boutin, S. *et al.* (1986) Population biology of snowshoe hares. I. Demography of food-supplemented populations in the southern Yukon, 1976–84. *Journal of Animal Ecology*, **55**, 963–82.

Kuno, E. (1971) Sampling error as a misleading artifact in key factor analysis. *Researches on Population Ecology*, **13**, 28–45.

Kuno, E. (1973) Statistical characteristics of the density-dependent population fluctuation and the evaluation of density-dependence and regulation in animal populations. *Researches on Population Ecology*, **15**, 99–120.

Kuno, E. (1991) Some strange properties of the logistic equation defined with r and K: Inherent defects or artifacts? *Researches on Population Ecology*, **33**, 33–9.

Lack, D. (1954) Cyclic mortality. *Journal of Wildlife Management*, **18**, 25–37.

Lauckhart, J. B. (1957) Animal cycles and food. *Journal of Wildlife Management*, **21**, 230–4.

Leigh, E. R. (1968) The ecological role of Volterra's equations, in M. Gerstenhaber (Ed.) *Some Mathematical Problems in Biology*. American Mathematical Society, Providence, Rhode Island.

Leopold, A. (1933) *Game Management*. C. Scribner's Sons, New York.

Loughton, B. G., Derry, C. and West, A. S. (1963) Spiders and the spruce budworm, in R. F. Morris (Ed.) The dynamics of epidemic spruce budworm populations. *Canadian Entomological Society Memoir*, **31**, pp. 249–68.

Ludwig, D., Jones, D. D. and Holling, S. C. (1978) Qualitative analysis of insect outbreak systems: the spruce budworm and forest. *Journal of Animal Ecology*, **47**, 315–32.

MacFadyen, A. (1957) *Animal Ecology: aims and methods*. Pitman and Sons, London.

MacLean, D. A. (1980) Vulnerability of fir-spruce stands during uncontrolled spruce budworm outbreaks: A review and discussion. *Forestry Chronicle*, **56**, 213–21.

MacLulich, D. A. (1937o) Fluctuations in the numbers of the varying hare (*Lepus americanus*). University of Toronto Studies, Series No. 43, University of Toronto Press, Toronto.

MacLulich, D. A. (1957) The place of chance in population processes. *Journal of Wildlife Management*, **21**, 293–9.

Major, J. T. and Sherburne, J. A. (1987) Interspecific relationships of coyotes, bobcats, and red foxes in western Maine. *Journal of Wildlife Management*, **51**, 606–16.

Maltais, J., Régnière, J., Cloutier, C. *et al.* (1989) Seasonal biology of *Meteorus trachynotus* Vier. (Hymenoptera: Braconidae) and of its over-wintering host *Choristoneura rosaceana* (Harr.) (Lepidoptera: Tortricidae). *Canadian Entomologist*, **121**, 745–56.

Marriott, F. H. C. and Pope, J. A. (1954) Bias in the estimation of auto-correlations. *Biometrika*, **41**, 390–402.

Marshall, W. H. (1954) Ruffed grouse and snowshoe hare populations on the Cloquet Experimental Forest, Minnesota. *Journal of Wildlife Management*, **18**, 109–12.

May, R. M. (1976) Simple mathematical models with very complicated dynamics. *Nature*, **261**, 459–67.

May, R. M. (1981) The dynamics of natural and managed populations, in R. W. Hirons and D. Cooke (Eds) *The Mathematical Theory of the Dynamics of Biological Populations*. II. Academic Press, London.

Maynard Smith, J. (1968) *Mathematical Ideas in Biology*. Cambridge University Press, Cambridge.

Maynard Smith, J. and Slatkin, M. (1973) The stability of predator–prey systems. *Ecology*, **54**, 384–91.

Mealzer, D. A. (1970) The regression of log N_{n+1} on log N_n as a test of density dependence: an exercise with computer-constructed density-dependent populations. *Ecology*, **51**, 810–22.

Meslow, E. C. and Keith, L. B. (1968) Demographic parameters of a snow-shoe hare population. *Journal of Wildlife Management*, **32**, 812–34.

Miller, C. A. (1958) The measurement of spruce budworm populations and mortality during the first and second larval instars. *Canadian Journal of Zoology*, **36**, 409–22.

Miller, C. A. (1963a) The analysis of fecundity proportion in the unsprayed area, in R. F. Morris (Ed.) The dynamics of epidemic spruce budworm populations. *Entomological Society of Canada Memoir*, **31**, pp. 75–87.

Miller, C. A. (1963b). Parasites and the spruce budworm, in R. F. Morris (Ed.) The dynamics of epidemic spruce budworm populations. *Entomological Society of Canada Memoir*, **31**, pp. 228–44.

Miller, C. A. (1977) The feeding impact of spruce budworm on balsam fir. *Canadian Journal of Forest Research*, **7**, 76–84.

Miller, C. A. and Renault, T. R. (1981) The use of experimental populations to assess budworm larval mortality at low densities. Information Report M-X-115, Forestry Canada-Maritime Region, Fredericton, New Brunswick, Canada.

Mitchell, R. T. (1952) Consumption of spruce budworms by birds in a Maine spruce-fir forest. *Journal of Forestry*, **50**, 387–9.

Mook L. J. (1963) Birds and the spruce budworm, in R. F. Morris (Ed.) The dynamics of epidemic spruce budworm populations. *Entomological Society of Canada Memoir*, **31**, pp. 268–71.

Moran, P. A. P. (1948) Some theorems in time series. II. The significance of the serial correlation coefficient. *Biometrika*, **35**, 255–60.

Moran, P. A. P. (1953a) The statistical analysis of the Canada lynx cycle. I. Structure and prediction. *Australian Journal of Zoology*, **1**, 163–73.

Moran P. A. P. (1953b) The statistical analysis of the Canadian lynx cycle. II. Synchronization and meteorology. *Australian Journal of Zoology*, **1**, 291–98.

Morisita, M. (1954) Estimation of population density by spacing method. Memoirs of Faculty of Science, Kyushu University, Series E, **1**, 187–97.

Morisita, M. (1965) The fitting of the logistic equation to the rate of increase of population density. *Researches on Population Ecology*, **7**, 52–5.

Morris, R. F. (1959) Single-factor analysis in population dynamics. *Ecology*, **40**, 580–8.

Morris, R. F. (Ed.) (1963) The dynamics of epidemic spruce budworm populations. *Entomological Society of Canada Memoir*, **31**.

Morris, R. F. (1963a) The development of predictive equations for the spruce budworm based on key-factor analysis, in R. F. Morris (Ed.) The dynamics of epidemic spruce budworm populations. *Entomological Society of Canada Memoir*, **31**, pp. 116–129.

Morris, R. F. (1963b) Résumé, in R. F. Morris (Ed.) The dynamics of epidemic spruce budworm populations. *Entomological Society of Canada Memoir*, **31**, pp. 311–20.

Morris, R. F., Cheshire, W. F., Miller, C. A. and Mott, D. G. (1958) The numerical response of avian and mammalian predators during the gradation of the spruce budworm. *Ecology*, **39**, 487–94.

Morrison, M. B. (1938) *The Forest of New Brunswick*. Dominion Forest Service Bulletin 91, Canada Department of Mines and Resources (Forestry Canada), Ottawa, Ontario, Canada.

Murdoch, W. W. and Oaten, A. (1975) Predation and population stability. *Advances in Ecological Research*, **9**, 1–131.

Neave, F. (1953) Principles affecting the size of pink and chum salmon populations in British Columbia. *Journal of Fisheries Research Board of Canada*, **9**, 450–91.

Nellis, C. H., Wetmore, S. P. and Keith, L. B. (1972) Lynx–prey interactions in central Alberta. *Journal of Wildlife Management*, **36**, 320–9.

Newman, P. C. (1985) *Company of Adventurers* (Vol. 1). Viking, Markham.

Nicholson, A. J. and Bailey, V. A. (1935) The balance of animal populations. Part I. *Proceedings of the Zoological Society of London* (1935), Part 3, 551–98.

Orcutt, G. H. (1948) A study of the autoregressive nature of the time series used for Tinbergen's model of the economic system of the United States, 1919–1932. *Journal of the Royal Statistical Society*, B, **10**, 1–35.

Oshima, K., Honda, H. and Yamamoto, I. (1973) Isolation of an oviposition marker from azuki bean weevil, *Callosobruchus chinensis* (L.). *Agricultural and Biological Chemistry*, **37**, 2679–80.

Ostaff, D. P. and MacLean, D. A. (1989) Spruce budworm populations, defoliation, and changes in stand condition during an uncontrolled spruce budworm outbreak on Cape Breton Island, Nova Scotia. *Canadian Journal of Forest Research*, **19**, 1077–86.

Park, T. (1932) Studies in population physiology: I. The relation of numbers to initial population growth in the flour beetle *Tribolium confusum* Duval. *Ecology*, **13**, 172–81.

Pearl, R. (1927) The growth of populations. *Quarterly Review of Biology*, **2**, 532–48.

Pearl, R. and Reed, L. J. (1920) On the rate of growth of the population of the United States since 1790 and its mathematical representation. *Proceedings of the Academy of Sciences*, **6**, 275–88.

Pease, J. L., Vowles, R. H. and Keith, L. B. (1979) Interaction of snowshoe hares and woody vegetation. *Journal of Wildlife Management*, **43**, 43–60.

Perrins, C. M. (1979) *British Tits*. William Collins Sons, London.

Perry, D. F. and Régnière, J. (1986) The role of fungal pathogens in spruce budworm population dynamics: frequency and temporal relationships, in R. A. Samson, J. M. Vlak, and D. Peters (Eds) *Fundamental and Applied Aspects of Invertebrate Pathology*. Wageningen, pp. 167–70.

Perry, D. F., Nadeau, M. and Arella, M. (1989a) *Choristoneura fumiferana* nuclear polyhydrosis virus, relationship with the important braconid parasitoid, *Meteorus trachynotus*. Society for Invertebrate Pathology XXII Annual Meeting, Maryland, 20–24 August 1989.

Perry, D. F., Merzouki, A. and Arella, M. (1989b) Granulosis in declining populations of *Choristoneura fumiferana*. Society for Invertebrate Pathology XXII Annual Meeting, Maryland, 20–24 August 1989.

Peterman, R. M. (1977) A simple mechanism that causes collapsing stability regions in exploited salmonid populations. *Journal of the Fisheries Research Board of Canada*, **34**, 1130–42.

Pielou, E. C. (1977) *Mathematical Ecology*. John Wiley and Sons, London.

Piene, H. (1980) Effects of insect defoliation on growth and foliar nutrients of young balsam fir. *Forest Science*, **26**, 665–73.

Piene, H. (1989) Spruce budworm defoliation and growth loss in young balsam fir: defoliation in spaced and unspaced stands and individual tree survival. *Canadian Journal of Forest Research*, **19**, 1211–17.

Pimm, S. L. (1982) *Food Webs*. Chapman and Hall, London.

Poland, H. (1892) *Fur-bearing Animals in Nature and in Commerce*. Gurney and Jackson, London.

Pollard, E., Lakhani, K. H. and Rothery, P. (1987) The detection of density-dependence from a series of annual censuses. *Ecology*, **68**, 2046–55.

Poole, R. W. (1974) *An Introduction to Quantitative Ecology*. McGraw-Hill, New York.

Price, P. W. (1984) *Insect Ecology* (2nd edn). John Wiley and Sons, New York.

Priestley, M. B. (1981) *Spectral Analysis and Time Series*. Vol. 1. *Univariate Series*. Academic Press, London.

Quenouille, M. H. (1956) Notes on bias in estimation. *Biometrika*, **43**, 353–60.

Reddingius, J. and Den Boer, P. J. (1989) On the stabilization of animal numbers. Problems of testing. 1. Power estimates and estimation errors. *Oecologia*, **78**, 1–8.

Régnière, J. (1984) Vertical transmission of diseases and population dynamics of insects with discrete generations: a model. *Journal of Theoretical Biology*, **107**, 287–301.

Reichardt, P. B., Bryant, J. P., Clausen, T. P. and Wieland G. D. (1984) Defense of winter-dormant Alaska paper birch against snowshore hares. *Oecologia (Berlin)*, **65**, 58–69.

Renault, T. R. and Miller, A. C. (1972) Spiders in a fir-spruce biotype: abundance, diversity, and influence on spruce budworm densities. *Canadian Journal of Zoology*, **50**, 1039–46.

Ricker, W. E. (1954) Stock and recruitment. *Journal of Fisheries Research Board of Canada*, **II(5)**, 559–623.

Rosenzweig, M. L. (1977) Aspects of biological exploitation. *Quarterly Review of Biology*, **52**, 371–80.

Rosenzweig, M. L. and MacArthur, R. H. (1963) Graphical representation and stability conditions of predator–prey interactions. *American Naturalist*, **97**, 209–23.

Rowan, W. (1950) Canada's premier problem of animal conservation. *New Biology*, **9**, 38–57.

Rowan, W. (1954) Reflections on the biology of animal cycles. *Journal of Wildlife Management*, **18**, 52–60.

Royama, T. (1970) Factors governing the hunting behaviour and selection of food by the great tit (*Parus major* L.). *Journal of Animal Ecology*, **39**, 619–68.

Royama, T. (1971) Comparative study of models for predation and parasitism. *Researches on Population Ecology, Supplement 1*.

Royama, T. (1981a) Fundamental concepts and methodology for the analysis of animal population dynamics, with particular reference to univoltine species. *Ecological Monographs*, **51**, 473–93.

Royama, T. (1981b) Evaluation of mortality factors in insect life table analysis. *Ecological Monographs*, **51**, 495–505.

Royama, T. (1984) Population dynamics of the spruce budworm *Choristoneura fumiferana*. *Ecological Monographs*, **54**, 429–62.

Rusch, D. H. (1976) Wildlife cycle in Manitoba. Manitoba Department of Mines, Resources and Environmental Management Information Series No. 11.

Rusch, D. H. and Keith, L. B. (1971) Seasonal and annual trends in numbers of Alberta ruffed grouse. *Journal of Wildlife Management*, **35**, 803–22.

Rusch, D. H., Meslow, E. C., Doerr, P. D. and Keith, L. B. (1972) Response of great horned owl populations to changing prey densities. *Journal of Wildlife Management*, **36**, 282–96.

Rusch, D. H., Gillespie, M. M. and McKay, D. I. (1978) Decline of a ruffed grouse population in Manitoba. *Canadian Field-Naturalist*, **92**, 123–7.

St. Amant, J. L. S. (1970) The detection of regulation in animal populations. *Ecology*, **51**, 823–8.

Sanders, C. J. (1988) Monitoring spruce budworm population density with sex pheromone traps. *Canadian Entomologist*, **120**, 175–83.

Sandlan, K. (1979) Sex ratio regulation in *Coccygomimus turionella* Linnaeus (Hymenoptera: Ichneumonidae) and its ecological implications. *Eco--logical Entomology*, **4**, 365–78.

Seton, E. T. (1909) *Life Histories of Northern Animals*, Vols 1 and 2. C. Scribner's Sons, New York.

Seton, E. T. (1928) *Lives of Game Animals*, Vol. 4. Doubleday, New York.

Siivonen, L. and Koskimies, J. (1955) Population fluctuations and the lunar cycle. Finnish Game Foundation, Papers on Game Research, No. 14.

Sinclair, A. R. E. and Smith, J. N. M. (1984) Do plant secondary compounds determine feeding preferences of snowshoe hares? *Oecologia (Berlin)*, **61**, 403–10.

Sinclair, A. R. E., Krebs, C. J. and Smith, J. N. M. (1982) Diet quality and food limitation in herbivores: the case of the snowshoe hare. *Canadian Journal of Zoology*, **60**, 889–97.

Sinclair, A. R. E., Krebs, C. J. Smith, J. N. M. and Boutin, S. (1988) Population biology of snowshoe hares. III. Nutrition, plant secondary compounds and food limitation. *Journal of Animal Ecology*, **57**, 787–806.

Skellam, J. G. (1952) Studies in statistical ecology: I. Spatial pattern. *Biometrika*, **39**, 346–62.

Slobodkin, L. B. (1961) *Growth and Regulation of Animal Populations*. Holt, Rinehart and Winston, New York (Dover edition, 1980).

Slutzky, E. (1927) The summation of random causes as the source of cyclic processes. English version in *Econometrica* (1937), **5**, 105–46.

Smith, J. N. M., Krebs, C. J., Sinclair, A. R. E. and Boonstra, R. (1988) Population biology of snowshoe hares. II. Interactions with winter food plants. *Journal of Animal Ecology*, **57**, 269–86.

Smith, R. L. (1980) *Ecology and Field Biology* (3rd edn). Harper and Row, New York.

Solomon, M. E. (1969) *Population Dynamics*. Edward Arnold, London.

Southwood, T. R. E. (1978) *Ecological Methods* (2nd edn). Chapman and Hall, London.

Statistics Canada (1983) Catalogue 23-207, fur production. Ottawa.

Stehr, G. (1968) On some concepts in the population biology of the spruce budworm. *Proceedings of the Entomological Society of Ontario*, **99**, 54–6.

Stiling, P. (1988) Density-dependent processes and key factors in insect populations. *Journal of Animal Ecology*, **57**, 581–93.

Swaine, J. M. and Craighead, F. C. (1924) Studies on the spruce budworm (*Cacoecia fumiferana* Clem.). Agriculture Canada Technical Bulletin, Ottawa, Ontario, Canada, 37.

Tanner, J. T. (1978) *Guide to the Study of Animal Populations.* University of Tennessee Press, Knoxville.

Thomas, A. W., Borland, S. A. and Greenbank, D. O. (1980) Field fecundity of the spruce budworm (Lepidoptera: Tortricidae) as determined from regression relationships between egg complement, fore wing length, and body weight. *Canadian Journal of Zoology,* **58**, 1608–11.

Thomson, H. M. (1958) Some aspects of the epidemiology of a microsporidian parasite of the spruce budworm, *Choristoneura fumiferana. Canadian Journal of Zoology,* **36**, 309–16.

Thomson, H. M. (1960) The possible control of a budworm infestation by a microsporidian disease. *Canada Department of Agriculture Biomonthly Progress Report* **16**, 1.

Todd, A. W., Keith, L. B. and Fischer, C. A. (1981) Population ecology of coyotes during a fluctuation of snowshoe hares. *Journal of Wildlife Management,* **45**, 629–40.

Todd, A. W., and Keith, L. B. (1983) Coyote demography during a snowshoe hare decline in Alberta. *Journal of Wildlife Management,* **47**, 394–404.

Tong, H. (1977) Some comments on the Canadian lynx data. *Journal of Royal Statistical Society,* Series A, **140**, 432–6.

Tothill, J. D. (1922) Notes on the outbreaks of spruce budworm, forest tent caterpillar, and larch sawfly in New Brunswick. *Proceedings of the Acadian Entomological Society,* **8**, 172–82.

Turchin, P. (1990) Rarity of density dependence or population regulation with lags. *Nature,* **344**, 660–3.

Umeya, K. (1966) Studies on the comparative ecology of bean weevils. I. On the egg distribution and the oviposition behaviours of three species of bean weevils infesting azuki bean. *Research Bulletin of the Plant Protection Service Japan,* **3**, 1–11 (in Japanese with English summary.)

Utida, S. (1941a) Studies on experimental population of the azuki bean weevil, *Callosobruchus chinensis* (L.). I. The effect of population density on the progeny populations. *Memoirs of the College of Agriculture, Kyoto Imperial University,* **48** (Entomological Series No. 6), 1–30.

Utida, S. (1941b) Studies on experimental population of the azuki bean weevil, *Callosobruchus chinensis* (L.). II. The effect of population density on progeny populations under different conditions of atmospheric moisture. *Memoirs of the College of Agriculture, Kyoto Imperial University,* **49** (Entomological Series No. 7), 1–20.

Utida, S. (1941c) Studies on experimental population of the azuki bean weevil, *Callosobruchus chinensis* (L.). III. The effect of population density upon the mortalities of different stages of life cycle. *Memoirs of the College of Agriculture, Kyoto Imperial University,* **49** (Entomological Series No. 7), 21–42.

Utida, S. (1941d) Studies on experimental population of the azuki bean weevil, *Callosobruchus chinensis* (L.). IV. Analysis of density effect with respect to fecundity and fertility of eggs. *Memoirs of the College of*

Agriculture, Kyoto Imperial University, **51** (Entomological Series No. 8), 1–26.

Utida, S. (1941e) Studies on experimental population of the azuki bean weevil, *Callosobruchus chinensis* (L.). V. Trend of population density at the equilibrium position. *Memoirs of the College of Agriculture, Kyoto Imperial University*, **51** (Entomological Series No. 8), 27–34.

Utida, S. (1942a) Studies on experimental population of the azuki bean weevil, *Callosobruchus chinensis* (L.). VI. The relations between the size of environment and the rate of population growth. *Memoirs of the College of Agriculture, Kyoto Imperial University*, **53** (Entomological Series No. 9), 1–18.

Utida, S. (1942b) Studies on experimental population of the azuki bean weevil, *Callosobruchus chinensis* (L.). VII. Analysis of the density effect in the preimaginal stage. *Memoirs of the College of Agriculture, Kyoto Imperial University*, **53** (Entomological Series No. 9), 21–31.

Utida, S. (1943a) Studies on experimental population of the azuki bean weevil, *Callosobruchus chinensis* (L.). VIII. Statistical analysis of the emerging weevils on beans. *Memoirs of the College of Agriculture, Kyoto Imperial University*, **54** (Entomological Series No. 10), 1–22.

Utida, S. (1943b) Studies on experimental population of the azuki bean weevil, *Callosobruchus chinensis* (L.). IX. General consideration and summary of the series reports from I to VIII. *Memoirs of the College of Agriculture, Kyoto Imperial University*, **54** (Entomological Series No. 10), 23–40.

Utida, S. (1944) Host–parasite interaction in the experimental population of the azuki bean weevil, *Callosobruchus chinensis* (L.). II. The effect of parasite density on the growth of host and parasite populations. *Oyo Dobutsugaku Zasshi* (Applied Zoological Magazine), **15**, 1–18 (in Japanese only).

Utida, S. (1947) Studies on experimental population of the azuki bean weevil, *Callosobruchus chinensis* (L.). X. The effect of different sex ratios upon the growth of population. *Seiri Seitai* (Physiology and Ecology), **1**, 67–78 (in Japanese with English summary).

Utida, S. (1948) Host-parasite interaction in the experimental population of the azuki bean weevil, *Callosobruchus chinensis* (L.). IV. The effect of host density on the growth of host and parasite populations. *Oyo Kontyu* (Applied Entomology), **4**, 164–74 (in Japanese).

Utida, S. (1956) Long-term fluctuation of population in the system of host-parasite interaction. *Researches on Population Ecology*, **3**, 52–9 (in Japanese with English summary).

Utida, S. (1957a) Cyclic fluctuation of population density intrinsic to the host-parasite system. *Ecology*, **38**, 442–9.

Utida, S. (1957b) Population fluctuation, an experimental and theoretical approach. *Cold Spring Harbor Symposia on Quantitative Biology*, **22**, 139–51.

Utida, S. (1959) Sequential frequency of the emergence of adult insect in relation to the change of environmental conditions. *Japanese Journal of Ecology*, **9**, 139–43 (in Japanese with English summary).

Utida, S. (1967) Damped oscillation of population density at equilibrium. *Researches on Population Ecology*, **9**, 1–9.

Utida, S. (1972) *Animal Populations: Ecology of over- and undercrowding.* NHK Books, Tokyo (in Japanese only).

Utida, S. and Nagasawa, S. (1949) On the developmental period and that of adult life to *Neocatolaccus mamezophagus*, a pteromalid parasite of the azuki bean weevil. *Kontyu* (Insects), **17**, 1–15 (in Japanese with English title).

Van Driesche, R. G. (1983) Meaning of 'percent parasitism' in studies of insect parasitoids. *Environmental Entomology*, **12**, 1611–22.

Varley, G. C. and Gradwell, G. R. (1960) Key factors in population studies. *Journal of Animal Ecology*, **29**, 399–401.

Varley, G. C. and Gradwell, G. R. (1963) The interpretation of insect population changes. *Proceedings of the Ceylon Association of Advancement of Sciences*, **18**, 142–56.

Varley, G. C., Gradwell, G. R. and Hassell, M. P. (1973) *Insect Population Ecology.* University of California Press, California.

Vaughan, M. R. and Keith, L. B. (1981) Demographic response of experimental snowshoe hare populations to overwinter food shortage. *Journal of Wildlife Management*, **45**, 354–80.

Verhulst, P.-F. (1838) Notice sur la loi que la population suit dans son accroissement. *Correspondence Mathématique et Physique*, **10**, 113–21.

Verhulst, P.-F. (1845). Recherches mathématiques sur la loi d'accroissement de la population. *Mémoires de l'Académie Royal de Bruxelles*, **18**, 1–38.

Wangersky, P. J. and Cunningham, W. J. (1956) On time lags in equation of growth. *Proceedings of National Academy of Sciences*, **42**, 699–702.

Wangersky, P. J. and Cunningham, W. J. (1957) Time lag in population models. *Cold Spring Harbor Symposia on Quantitative Biology*, **22**, 329–38.

Ward, R. M. P. and Krebs, C. J. (1985) Behavioural responses of lynx to declining snowshoe hare abundance. *Canadian Journal of Zoology*, **63**, 2817–24.

Watt, K. E. F. (1963) The analysis of the survival of large larvae in the unsprayed area, in R. F. Morris (Ed.) *The Dynamics of Epidemic Spruce Budworm Populations.* Entomological Society of Canada Memoir, **31**, pp. 52–63.

Weinstein, M. S. (1977) Hares, lynx, and trappers. *American Naturalist*, **111**, 806–8.

Wellington, W. G., Fettes, J. J., Turner, K. B. and Belyea, R. M. (1950) Physical and biological indicators of the development of outbreaks of the spruce budworm. *Canadian Journal of Research*, D, **28**, 308–31.

Williams, G. R. (1954) Population fluctuations in some northern hemisphere game birds (Tetraonidae). *Journal of Animal Ecology*, **23**, 1–34.

Williamson, M. (1972) *The Analysis of Biological Populations*. Edward Arnold, London.

Williamson, M. (1975) The biological interpretation of time series analysis. *Bulletin of Institute of Mathematics and its Applications*, **11**, 67–9.

Wilson, E. O. and Bossert, W. H. (1971) *A Primer of Population Biology*. Sinauer, Stamford.

Wilson, G. G. (1973) Incidence of microsporidia in a field population of spruce budworm. *Forestry Canada Bimonthly Research Notes*, **29**, 35–6.

Wilson, G. G. (1977) Observations on the incidence rates of *Nosema fumiferanae* (Microsporidia) in a spruce budworm (*Choristoneura fumiferana*) (Lepidoptera: Tortricidae) population. *Proceedings of the Entomological Society of Ontario*, **108**, 144–5.

Winterhalder, B. P. (1980) Canadian fur bearer cycles and Cree-Objibwa hunting and trapping practices. *American Naturalist*, **115**, 870–9.

Yoshida, T. (1961) Oviposition behaviour of the azuki bean weevil and southern cowpea weevil and their specific relationship. Bulletin of the Faculty of Liberal Arts, Miyazaki University, Natural Science Series, **11**, 41–65 (in Japanese with English title).

Yule, G. U. (1926) Why do we sometimes get nonsense-correlation between time-series?—a study in sampling and the nature of time-series. *Journal of the Royal Statistical Society*, **89**, 1–64.

Yule, G. U. (1927) On a method of investigating periodicities in disturbed series, with special reference to Wolfer's sunspot numbers. *Philosophical Transactions of Royal Society of London*, Series A, **226**, 267–98.

Zar, J. H. (1984) *Biostatistical Analysis* (2nd edn). Prentice-Hall, Englewood Cliffs.

Zwölfer, W. (1931) Studien zur Ökologie und Epidemiologie der Insecten. I. Die Kieferneule, *Panolis flammea* Schiff. *Zeitschrift für Angewandte Entomologie*, **17**, 475–562.

Zwölfer, W. (1932) Methoden zur Regulierung von Temperatur und Luftfeuchtigkeit. *Zeitschrift für Angewandte Entomologie*, **19**, 497–513.

Author index

Subject index